P9-DOG-578

# THE HANDY OCEAN ANSWER BOOK

# THE HANDY OCEAN ANSWER BOOK™

Thomas E. Svarney • Patricia Barnes-Svarney

# THE HANDY OCEAN ANSWER BOOK™

Copyright 2000, 2003 by Visible Ink Press™

This publication is a creative work fully protected by all applicable copyright laws, as well as by misappropriation, trade secret, unfair competition, and other applicable laws.

No part of this book may be reproduced in any form without permission in writing from the publisher, except by a reviewer who wishes to quote brief passages in connection with a review written for inclusion in a magazine or newspaper.

All rights to this publication will be vigorously defended.

Visible Ink Press™
43311 Joy Road #414
Canton, MI 48187-2075

Visible Ink Press is a trademark of Visible Ink Press LLC.

Most Visible Ink Press books are available at special quantity discounts when purchased in bulk by corporations, organizations, or groups. Customized printings, special imprints, messages, and excerpts can be produced to meet your needs. For more information, contact Special Markets Director, Visible Ink Press, at www.visibleink.com.

Art Director: Pamela A. E. Galbreath
Typesetting: The Graphix Group
ISBN 1-57859-063-9

**Library of Congress Cataloging-In-Publication Data**

Barnes-Svarney, Patricia L.
The handy ocean answer book / Patricia Barnes-Svarney & Thomas E. Svarney.
p. cm.
Includes bibliographical references and index.
ISBN 1-57859-063-9 (softcover)
1. Oceanography-Miscellanea. 2. Ocean-Miscellanea. 3. Marine ecology-Miscellanea. I. Svarney, Thomas E. II. Title.
GC28.B37 2000
551.46-dc21

99-048151

Printed in the United States of America
All rights reserved
10 9 8 7 6 5 4

*To Gail Greco and Tom Bagley—*
*good friends, good times, good food, and the C-Bay life....*

## The Handy Answer Book™ Series

The Handy Answer Book for Kids (and Parents)
The Handy Bug Answer Book
The Handy Dinosaur Answer Book
The Handy Geography Answer Book
The Handy History Answer Book
The Handy Ocean Answer Book
The Handy Physics Answer Book
The Handy Politics Answer Book
The Handy Religion Answer Book
The Handy Science Answer Book
The Handy Space Answer Book
The Handy Sports Answer Book
The Handy Weather Answer Book

**Please visit Visible Ink Press at visibleink.com.**

# Contents

# LIFE IN THE OCEANS

# HUMANS AND THE OCEANS

# Introduction

It is difficult for those of us who live inland to understand the immensity and effects of the world's oceans. These vast bodies of water take up 70 percent of the surface area of our globe, with a volume close to 350 million cubic miles. And at present, ours is the only known planet in the solar system to contain such waters.

Even those of us living in the middle of a continent are affected by these immense bodies of water in a direct or indirect way. For example, the Sun heats the surface of the oceans, evaporating waters to form clouds and mixing the air currents to produce our weather systems. The Sun's heat also stirs the surface waters, creating waves and currents that sculpt our shorelines. Seasonal temperature changes of the ocean waters alternately increase or decrease populations of marine organisms, including many that humans harvest for food.

Most importantly, without the oceans, there would be no life. It was in the oceans, billions of years ago, that life began—probably in shallower waters or around deep hydrothermal vents. Either way, life evolved under water, eventually reaching the edges of the oceans and making its way to land. As amazing as it seems, our entire population of terrestrial and marine organisms all had a common beginning—in our oceans.

To most of us, the ocean is beautiful, awesome, sometimes deadly—and at the same time, misunderstood. The reason it is not well understood is, perhaps, obvious: We cannot see below the surface of the ocean without diving in; and because of our physical limitations, we can only dive so far and for so long. We can only touch and examine the fringes of this huge body of water, and, so far, have made relatively few forays into its depths. As many people have pointed out, what lies below the surface of the Earth's oceans is almost an alien planet to us.

It is this alienness that *The Handy Ocean Answer Book* addresses, filling the gaps in the reader's understanding of the most important part of our world. These pages present answers to the most common questions about our oceans—covering features and organisms from the shoreline to the open ocean. Here, we examine the physical attributes of the oceans, marine animals and plants, and finally, the human ties to (and our effect on) the oceans.

Many people call the oceans our most important natural resource. The waters have furnished humans with food for centuries—from a plethora of fish species to certain marine plants. But people have also adversely affected the oceans: Over-fishing, coastal erosion caused by development, and pollution threaten this natural habitat. We need to keep the oceans in balance in the future—especially since we will continue to need the oceans and their bounty if we are to survive as a species.

The world ocean continues to hold many secrets; there are numerous questions still to be answered. For example, what kinds of organisms are found in the ocean's deepest waters? How many fish thought to be extinct are actually still alive? What species are important to coral reef growth? How do microorganisms live in the coldest waters of the Arctic? Other questions have to do with the connections—and interdependencies—between humans and the ocean. For example, are plankton (one of the most important organisms in the marine food chain) able to withstand environmental stresses such as the ozone hole? Can humans continue to harvest the oceans without disrupting the balance between the organisms and their environment? Scientists hope to answer some of these questions in the near future, not only by using better technology that allows humans to dive into and explore the oceans for longer periods, but through new satellites that watch the global waters—tracking changes over time.

We hope these pages describe an underwater world that will inform and inspire you, and, perhaps, engage your interest enough that you'll want to discover even more about the wonders of the ocean. This is the mysterious and largely unexplored territory where life originated—and it will be an integral part of our future.

# Acknowledgments

As with every book, there are many people who contributed stories and information—a list, in this case, that seems to go on forever. In particular, our thanks go out to the many oceanographers and scientists who have charted the oceans, dived into the depths, and studied marine life for the past decades. Without their hard work and dedication, this book could not have been written.

We would also like to thank the following groups and people: Woods Hole Oceanographic Institution, Scripps Institution of Oceanography, Mystic Aquarium, the National Aquarium (Baltimore), the Smithsonian Institution, many of the National Marine Sanctuaries, Lamont-Doherty Geological Observatory, Shoals Marine Lab, the Naval Research Laboratory, U.S. Naval Academy, Virginia Institute of Marine Science, University of South Florida at St. Petersburg, the Monterey Bay Aquarium, the National Oceanic and Atmospheric Administration (including the U.S. National Oceanographic Data Center), the National Aeronautics and Space Administration, Eugenie Clark, Ann McGovern, and Kristen Kusek.

Our gratitude to the people at Visible Ink Press and the Gale Group who assisted us in many ways: Pamela Galbreath for her book design work and Cindy Baldwin for her artistic guidance; Matt Nowinski and Justin Karr for helping us locate images; Edna Hedblad for securing permissions; Pam Reed, Randy Bassett, and Robert Duncan for preparing the images—and Mary Grimes and Leatha Etheridge-Sims for cataloging them; Marco Di Vita of the Graphix Group for typesetting; and David Deis at Dreamline Cartography for the color map of the world appearing in the insert. And a very special thanks to our ever-patient editors Jeff Hermann and Rebecca Ferguson for all their help and great work on this book; and also to Julia Furtaw, Marty Connors, and Christa Brelin, who believed in us and let us "play in the ocean." We could not have written this book without their help.

# Acknowledgments

[illegible]

# THE HANDY OCEAN ANSWER BOOK

# THE PHYSICAL OCEANS

# OCEAN BASICS

## What is the **world ocean**?

The world ocean is the large, interconnected body of saltwater that covers just over 70 percent of our planet. This vast world ocean is composed of five large, localized oceans (the Pacific, Atlantic, Indian, Antarctic, and Arctic oceans), and numerous seas. The more familiar, individual oceans are bodies of water that have well-defined basins, bottom topography (landscape of the seafloor), wind patterns above the ocean surface, and vertical and horizontal water movement in the form of currents and waves. The definitions and boundaries of the global oceans are determined by the International Hydrographic Bureau, located in Monaco.

The depressions on the ocean floor holding most of the planet's water are called ocean basins. Landscapes of these depressions are unique and changing—and are an often-neglected, integral part of the ocean.

## Where did the **word "ocean" originate**?

The word ocean is from the Greek word *okeanos,* meaning "river." The early Greeks believed the ocean to be a great river flowing around the Earth. They also believed the Mediterranean, their local body of water, was a sea within the river.

## What is the **total volume** of the **world ocean**?

Since much of the world ocean has not been mapped yet, the total volume of the oceans and seas cannot be absolutely determined. However, scientists estimate the total volume of the world's oceans and seas is about 350 million cubic miles (1.5 x $10^9$ cubic kilometers)—making our globe a definite water planet. The Earth has more ocean volume than any other planet in the solar system. Jupiter's moons Europa and Callisto may harbor oceans under their icy surfaces—a highly debated theory currently under study—but these oceans would be much smaller than the Earth's.

## Why are **oceans important** to the global **environment**?

The oceans are the cradles and nurturers of life on this planet—without them, you would not be here. The oceans are the main drivers of the Earth's water cycle, supplying most of the moisture that influences the global climate. Without moisture from the oceans, the planet would be parched, similar to oceanless Mars; or it would be a staggeringly hot and carbon dioxide–filled environment similar to Venus.

The differential heating of the oceans by the Sun's rays produces temperature and pressure changes in the atmosphere, which in turn generate the winds and create our weather. The long-term distribution of this energy by the oceans creates the localized climates that support the varied life forms found in our world.

In addition, the ocean currents transport heat, oxygen, nutrients, plants, and animals around the globe, and the ocean itself provides us with food and other necessities. The oceans sustain a large and complex food chain—a major part of the planet's overall food web.

## Why does the **Earth** look **blue from space**?

The Earth looks blue from space because of its oceans. Using a prism, light can be broken up into an array of colors—red, orange, yellow, green, blue, and violet—called a spectrum. The ocean waters act as a prism, but as light travels into the water, the red, orange, and yellow colors are absorbed more quickly than the blue light. Blue light can pene-

**The Black Sea living up to its name: High levels of hydrogen sulfide in its waters make the sea look black—particularly when viewed against the backdrop of Turkey's white beaches. (For color image, *see* insert.)** ***CORBIS/Chris Hellier***

trate below about 100 feet (30 meters) in depth—which is why the clear, mid-ocean water looks so blue on a sunny day. In many coastal and polar water areas—in which animals and plant life are plentiful and sediment arrives from rivers—the blue light is more readily absorbed, causing the water to look more greenish in color.

## Can an **ocean** be a **color** other than blue?

Yes, and many times, depending on the circumstances, an ocean seems to change color. For example, during hurricanes in the Atlantic Ocean, the water often looks green, as it mixes with the yellow pigments of floating plants (the colors yellow and blue mixed together equal green). Other times, parts of the Atlantic surf will look almost a milky brown, as a storm churns up sand and sediment is swept out of rivers by fast-moving currents. Other oceans and seas also seem to vary in color, including the Red Sea in the Middle East, in which seasonal growth of red algae turns the waters red; the Black Sea (north of Turkey), which contains so much hydrogen sulfide that the water truly looks black; and the Yellow

Sea (off the coast of China), which looks yellow because of the nearby mud and sediment-filled rivers that empty into it.

The Yellow Sea does in fact look yellow at times; rivers that dump mud and other sediments into the sea are the cause. Here fishermen ply the Yellow Sea waters while a South Korean naval ship stands guard after a June 1999 conflict with North Korea. *Associated Press/Yonhap*

## What **percentage** of the Earth's surface is **oceans** and what percentage is **land**?

An estimated 70.78 percent of the Earth's surface is covered by oceans, or 139.4 million square miles (360.9 million square kilometers); only 29.22 percent is dry land, or 57.5 million square miles (148.9 million square kilometers). The amount in each hemisphere also differs: In the Southern Hemisphere, the proportion of water to land is approximately 4:1; in the Northern Hemisphere, it is approximately 3:2. Put another way, about 81 percent of the surface of the Southern Hemisphere is covered by oceans; while only about 61 percent of the Northern Hemisphere is covered by oceans.

## What **percentage** of the world's **water** is held in the oceans?

The oceans contain about 97 percent of the total water on Earth. In contrast, the ice caps and glaciers hold about 2 percent, and less than 0.3 percent is carried in the atmosphere in the form of clouds, rain, and snow. Our inland seas, lakes, and channels account for a mere 0.02 percent of the Earth's water.

## What is the **total weight** of all the water in the oceans?

The approximate total weight of all the water in the Earth's oceans is $1.45 \times 10^{18}$ tons. This represents about 0.022 percent of the total weight of the planet.

**The Floor of the Oceans: A view of what the world's oceans would look like if all the water were drained out of them.** *Copyright Maria Tharpe*

## What is the **average depth** of the **world ocean**?

The average worldwide ocean depth is about 12,460 feet (3,798 meters) below sea level, or about five times the average elevation on land. The average land height is about 2,757 feet (840 meters) above sea level.

## What are the **average depths** of the **major individual oceans**?

The average depths of the major oceans vary greatly. The Pacific Ocean has an average depth of 13,740 feet (4,188 meters), the Indian Ocean averages 12,704 feet (3,872 meters), and the Atlantic Ocean has an average depth of 12,254 feet (3,735 meters). The average depth of the Arctic Ocean is only 3,407 feet (1,038 meters).

## Which **ocean** is the **deepest**?

The Pacific Ocean is the deepest, with an average depth of 13,740 feet (4,188 meters). It also contains most of the ocean trenches, which are the deepest places known on the Earth's surface. The greatest depth in this body of water is 36,198 feet (11,033 meters).

## What is the **shallowest ocean**?

The Arctic Ocean—the sea and ice north of the Arctic Circle with the North Pole at its center—is the shallowest, with an average depth of 3,407 feet (1,038 meters). The greatest depth in this body of water is 17,848 feet (5,440 meters).

## What is the **deepest point** on the **ocean floor**?

The deepest known part of the ocean is a place called the Challenger Deep, located at the bottom of the Mariana Trench in the Pacific Ocean. It measures 36,198 feet (11,033 meters) below sea level. If Mount Everest—the tallest mountain on the Earth's surface, measuring 29,022 feet, 7 inches (8,846 meters) in height—were dropped into this trench, it would be covered with more than a mile (or more than one and a half kilometers) of seawater.

# SEAWATER

## What is **seawater**?

Water found in an ocean or sea is called seawater (or sea water). It is composed of water—a molecule formed by the combination of one oxygen atom with two hydrogen atoms—and ions (electrically charged atoms) of salt and other dissolved materials (trace elements).

## Does the **composition** of seawater **vary**?

Yes and no: While there are local differences in temperature and salinity (saltiness), and some differences in concentrations of certain substances

## What is the composition of seawater?

French scientist Antoine-Laurent Lavoisier (1743–94) determined that in addition to water ($H_2O$), seawater is made up of ions (electrically charged atoms) of sodium, calcium, and magnesium chlorides; sodium, magnesium, and calcium sulfates; and calcium carbonate. These ions account for more than 99.95 percent of the total weight of all ions present in seawater; the rest are found in trace amounts (trace amounts are measured in parts per million, abbreviated ppm). These trace materials include almost all of the elements.

from ocean to ocean, in general the composition, or content, of seawater is fairly uniform throughout the oceans of the world. This fact was first confirmed by William Dittmar, an English chemist who sailed on the H.M.S. *Challenger* during its round-the-world expedition from 1872 to 1876.

In addition, studies of marine deposits from all geological time periods show that the composition of seawater has remained fairly consistent over the long history of our planet. This would make sense, since the early ocean currents flowed unimpeded by continental barriers—thus the seawater easily mixed throughout the global oceans.

## What are some of the **unique properties** of **water**?

Water is the only known substance on Earth present in nature as a liquid, gas, and solid. In its liquid state, water is the foremost dissolver of material, out-distancing the ability of any other liquid; water vapor "humidifies" the atmosphere, and helps create the weather we experience on the surface of Earth—part of the planet's complex water cycle. And finally, the properties of water as a solid (or frozen) are critical to life in the oceans: Ice is less dense than its liquid form, so it floats (which, by the way, is a property shared only by three known elements). Because of this, a body of water in the colder regions has a cap of ice, with life-sustaining liquid below it. If ice were not less dense than water, it would sink to the bottom, killing much of the marine animals and plants below—and the water would continue to build up a thick layer of ice from bottom to top, solidifying the entire column of water.

## How much **water evaporates** from the oceans and landmasses **annually**?

If all the evaporated water were collected and condensed at one time, it is estimated that it would form a layer about 40 inches (102 centimeters) deep over the entire planet. Annual water evaporation from the oceans and landmasses greatly affects the Earth's overall climate and weather, but does not significantly change the overall volume of water in the oceans. This is because it rains in one place (such as a hurricane in the east Pacific Ocean) while somewhere else (such as the Atlantic Ocean coast), water is evaporating—keeping a balance.

## What is **salinity**?

The salinity of seawater is the amount of dissolved inorganic minerals (salts), such as sodium and magnesium chlorides, in seawater. Dissolved salt, or sodium chloride (NaCl), is the most prevalent substance in seawater.

## How **saline (salty)** is **seawater**?

The salt content of open ocean water varies from 33 to 37.5 parts per thousand by mass, with the average being 35 parts per thousand by mass (that means there are 35 parts of salt for every thousand parts of water). That may not seem like much, but our taste buds are very sensitive to even a little salt—which is why the ocean tastes so salty to us.

## Which oceans are the **saltiest** and **least-saltiest**?

On the average, the saltiest ocean is thought to be the northern subtropical portions of the Atlantic Ocean, with an average of about 37.5 parts per thousand salinity. The Pacific Ocean is probably the least salty ocean, as many deeper spots measure about 33 parts per thousand salinity.

## What is the **difference** between **saltwater** and **freshwater**?

The scientific "dividing line" between saltwater and freshwater is 1,000 parts of dissolved salts per million parts of water. Water containing more than this amount of dissolved salts is classified as saline (saltwater); water with less than this amount of dissolved salts is considered fresh.

## Where did the **salt** in seawater **originate**?

Salt in the oceans is thought to have originated from the natural erosion of rocks on our planet. Salts were eroded from rock by rains and snows, became dissolved in the water, and eventually reached the oceans by way of the planet's rivers.

## What are the most **common ions** found in **seawater**?

The most common ions (electrically charged atoms) found in seawater are those from salt, and are illustrated in the following table.

| Ion | Percent of Total Salt (by weight) |
|---|---|
| Chlorine (Cl-) | 55.04 |
| Sodium (Na+) | 30.61 |
| Sulfate ($SO_4$-2) | 7.68 |
| Magnesium (Mg+2) | 3.69 |
| Calcium (Ca+2) | 1.16 |
| Potassium (K+) | 1.10 |
| Bicarbonate ($HCO_3$-) | 0.41 |
| Bromine (Br-) | 0.19 |

## What **trace elements** are found in seawater?

Almost every element known is found in seawater, but most in very small concentrations. Some of the most common trace elements are strontium (an average of 8 parts per million), silicon (an average of 3 parts per million), and fluorine (1.3 parts per million).

### Is there gold in seawater?

Yes, the mineral gold is suspended in seawater, but don't rush out to go prospecting! There are only very minute amounts of gold present in seawater. If you had the time and patience, the total amount of gold you could find in all of the planet's seawater would be enough to provide 9 pounds (4 kilograms) of gold for every person on Earth.

## What **gases** are present in **seawater**?

The same gases present in the Earth's atmosphere are also found in the waters of the oceans and seas, but in different proportions. To compare, the Earth's atmosphere is composed of approximately 76 percent nitrogen, 21 percent oxygen, and 1 percent argon, with all the other gases (such as carbon dioxide, at 0.032 percent) representing approximately 0.25 percent of the total amount by weight. In the ocean, the most common gases present are oxygen (4.6 to 7.5 parts per million, and varying with depth, with the greatest concentration at the surface); nitrogen (0.05 parts per million); argon (0.5 parts per million); and only trace amounts of carbon dioxide.

## What is **density**?

Density is a measurement of the mass of a substance per unit volume. Oceanographers use the metric system to measure the density of water, expressing it in terms of grams per cubic centimeter. The density of pure water at 39° F (4° C) is equal to 1; the average value scientists give to seawater is about 1.026 to 1.028 grams per cubic centimeter. But the density of seawater can differ depending on region.

## What causes the **density** of seawater to **vary**?

The density of seawater is dependent on temperature, pressure, and salinity. A change in any one of these, or a change in some or all of them, will generate a change in density. For example, seawater at a constant temperature and pressure, but with changes in salinity, will vary in density: greater salinity equals a higher density, and lesser salinity equals a lower density.

Many factors cause changes in the density of the seawater at the ocean's surface: The Sun's rays can heat the surface waters; precipitation can add water to the surface; or freshwater can pour into the oceans from land, rivers, and melting ice. The introduction of heat or freshwater decreases seawater density—with this surface water having lower density than the water below.

Conversely, if there are large amounts of evaporation of the surface waters, if ice forms and takes up water, or if the surface waters cool, then the density increases. If this water layer becomes denser than the water below, it will sink until it reaches a layer of water with the same density. At this depth, the dense water will spread out, forming a separate layer; or will increase the thickness of the existing layer. The sinking of this dense layer of seawater creates convective circulation, with lighter water rising to make room for the more dense water, and the surface water moving in to replace the descending water. This convection motion continues until the density becomes uniform—from the surface to the depth at which the more dense seawater layer resides.

## What is a **pycnocline**?

The term pycnocline describes the phenomena of a rapid change in the density of seawater in a vertical direction. In other words, it is a water layer that rapidly increases in density with an increase in depth.

## What happens to the **very dense seawater** that forms at the **polar regions**?

The very dense seawater that forms at the poles sinks all the way to the ocean floor. The great increase in seawater density occurring at the polar regions is due to the presence of very cold air and the formation of large amounts of ice. This cold, dense seawater—after sinking to the bottom—spreads out to other ocean basins around the globe, eventually working its way to the lower latitudes. This is why the ocean floors are all covered by a bottom layer of dense and cold polar seawater.

## Why do we have **water** and **not ice** in our oceans?

The Earth possesses a "geochemical thermostat" called the carbonate-silicate cycle. The very nature of this cycle works to hold the planet's surface temperature in the range that allows water to remain a liquid—in other words, enough greenhouse heating to maintain liquid water.

It works this way: Carbon dioxide in the atmosphere dissolves in rain, which then reacts with and erodes silicate rocks on the Earth's surface. The silicate materials absorb much of the carbon dioxide gas, and as they are eroded from the continental rock, calcium and bicarbonate ions are released into streams and rivers. The ions (electrically charged atoms) are then carried to the oceans where they form solid calcium carbonate or are used by many marine organisms to build their hard calcium carbonate shells or exoskeletons. These animals eventually die and fall to the ocean floor, adding to the carbonates. Over millions of years, these carbonate deposits travel with the moving continental plates, and eventually, these layers are pushed under other plates. The deposits are "cooked" by the high temperatures and pressures created by this subduction process (in which one continental crustal plate moves underneath another), releasing carbon dioxide gas through volcanoes that form near subduction zones. The gases reenter the atmosphere to start the cycle again. This balance of the carbon dioxide helps maintain the temperature of the Earth's atmosphere—and maintains water as a liquid.

## Do any other planets or satellites in our **solar system** have **water**?

Scientists believe there may be water on several planets and satellites in our solar system—but not in the same way as water is present on Earth. For example, two spacecraft sent to the Moon, *Clementine* and the *Lunar Explorer,* recently discovered that one of the Moon's poles (and perhaps both poles) may have ice present deep inside craters.

Past and recent images from spacecraft orbiting or landing on Mars (including the orbiting *Mars Global Surveyor* and the *Pathfinder* lander) reveal that older parts of the Martian surface are laced with long channels carved by liquid water. In the northern plains of the planet there are also exceptionally flat areas suggestive of the smooth surfaces of the Earth's deep-ocean floor. Thus, some scientists believe that Mars once had an ocean. But just where the ocean went—or, for that matter, where any water on the red planet went—is still a mystery.



Farther out in the solar system, there is less of a chance to find large amounts of water. At least scientists once thought this was true. Recent images from the *Galileo* spacecraft revealed that two of Jupiter's moons, Europa and Callisto, could have a great deal of liquid water.

## Does the presence of **water** mean there is **life on Europa**?

Europa is a snowy white sphere with long webs of ridges cutting across it. It is just smaller than the Earth's Moon, and the surface ice seems relatively smooth. Scientists believe that when cracks appear at the surface, something fills in the cracks—and that something is thought to be water. If Europa truly has a sea under the ice, there may be a chance of life on the moon. After all, on our own planet, a great deal of life evolved and lives in the deep oceans near volcanic vents; Europa may be volcanically active as it is gravitationally tugged by the gas giant Jupiter and its larger moons. Not only could such vents on Europa host life, but internal heat and organic compounds from comets striking the moon could possibly help create life as well.

## What does it mean if there is **water** on other **planets** and **satellites**?

If there is ice on the Moon and Mars, there may be a way to extract water and use it to help colonize our solar system. After all, water, for humans, is the most precious commodity in space; it would take a great deal of time and fuel energy to launch water-containing rockets for use by colonies. If there is water already on the planetary body, it would make life that much easier for colonists.

The presence of water on Jupiter's moons would have a more fundamental result: It would perhaps answer the question of how easy it is for life to form under conditions different than on Earth. Many scientists suggest that life could grow in such exotic locales, similar to how life grows in deep-ocean vents on the Earth's ocean floors—far from any light or warmth from the Sun.

### How long does it take the water in the **world ocean** to thoroughly **mix**?

The world ocean—in other words, all the oceans around the world—thoroughly mixes itself (by circulation) about every 2,000 years.

## SEA LEVEL

### Why does the **ocean** seem **level**?

Aside from waves and other disturbances of nature, the ocean's surface, indeed any body of water, appears level. There are no high or low points in the oceans and seas. This is due to the action of gravity, which exerts a force on everything and everyone, pulling down toward the center of the Earth. Water, being a structureless liquid, flows down into the lowest areas and spreads out flat due to the action of this force. Water, in other words, finds its own level—and this is commonly known as sea level.

Sea level is used as a reference point, to measure how high or deep something is—whether it is on the ocean, or far away from water. Many heights are referenced to mean sea level (MSL). For example, a mountain may be 10,000 feet (3,048 meters) above mean sea level.

### What is **mean sea level (MSL)**?

The average water level at a certain place, taking into consideration all tidal and wave conditions, is called the mean sea level (MSL). Because the ocean is one continuous body of water, the ocean's surface attempts to seek the same level throughout the entire world. But there are many factors—such as tides, landmasses, river discharges, waves, and even variations in gravity—that prevent the surface from being level. But in order to determine variations in sea level, a baseline measurement had to be made—and thus, the concept of local mean sea level was developed.

## What is sea level?

Sea level is the height of the ocean's surface at a certain spot, and is dependent on changing conditions. In the United States, sea level was determined by taking hourly measurements of sea levels in various coastal locations over a period of 19 years. The readings were then averaged for all the measurements, and sea levels were determined. The 19-year period was used because of the Metonic cycle, a natural lunar cycle mostly based on the Moon's declination (the height of the Moon in the sky).

## What are some of the **highest and lowest mean sea level records** around the world?

Heights above and below mean sea level vary greatly from place to place. For example, between Nova Scotia and Florida, there is a difference in sea level of 16 inches (41 centimeters); between southeast Japan and the Aleutian Islands, the difference is about 3 feet (1 meter). The highest point on land is Mount Everest (Nepal-Tibet), which measures 29,022 feet, 7 inches (8,846 meters) above sea level; this measurement was established in 1993. The lowest surface level is the Dead Sea (Israel-Jordan), which is really a salt lake and measures 1,299 feet (396 meters) below sea level. The lowest point on land in the United States is Death Valley, California, which measures 282 feet (86 meters) below sea level. The greatest depth below mean sea level in the oceans is the Mariana Trench (in the Pacific Ocean), measuring 36,198 feet (11,033 meters) below sea level.

## Have **sea levels changed** over the past thousands of years?

Yes, sea levels have changed over time. During the last Ice Age, about 18,000 years ago—when much of the planet's water was locked up in thick polar ice sheets—the oceans were some 330 feet (100 meters) lower than they are today. Sea levels also fluctuated during that time. If the seas continue to rise over the next few thousand years as they have over the past century, sea level measurements will also continue to change.

## What are **eustatic changes**?

Eustatic changes are those in the worldwide sea level, as opposed to localized changes. Eustatic changes occur slowly over many, many years. At this time, the worldwide sea level is rising due to a variety of factors, which will result in a eustatic change.

## What caused **worldwide sea levels to fluctuate** in the past century?

During the past century, the global sea level has apparently risen as much as 8 inches (20 centimeters), although this number has been debated by some scientists who believe the number is lower. No one knows what caused, and continues to cause, this rise in sea level. Many scientists believe that global warming, caused by human pollutants and particulates released into the atmosphere, is responsible for the sea level changes. If this is true, global warming will continue to raise global temperatures, melting the ice in the polar regions; the warmer temperatures will also expand the sea water—thus, causing the global sea level to rise.

Other scientists believe that the changes in sea levels may be from the rise in landmasses. As the areas rise—such as parts of Scandinavia and northern North America that are rebounding from the pressures of the past Ice Age ice sheets—it could cause sea level changes.

## What are some **examples of elevations and depths** around the world in relationship to **sea level**?

The following table lists the highest elevations above sea level and the lowest points below sea level on the world's continents.

| Location | Elevation above Sea Level (feet / meters) |
|---|---|
| Africa, Mount Kilimanjaro (Kibo), Tanzania | 19,340 / 5,895 |
| Antarctica, Vinson Massif | 16,067 / 4,897 |

| Location | Elevation above Sea Level (feet / meters) |
|---|---|
| Asia, Mount Everest | 29,022 ft., 7 in. / 8,846 |
| Australia, Mount Kosciusko | 7,310 / 2,228 |
| Europe, Mount El'brus, Russia | 18,510 / 5,642 |
| North America, Mount McKinley, USA | 20,320 / 6,194 |
| South America, Mount Aconcagua, Argentina | 22,834 / 6,960 |

| Location | Depth below Sea Level (feet / meters) |
|---|---|
| Asia, Dead Sea, Israel-Jordan | 1,299 / 396 |
| Australia, Lake Eyre | 52 / 16 |
| Europe, Caspian Sea | 92 / 28 |
| North America, Death Valley, USA | 282 / 86 |
| South America, Salina Grandes, Argentina | 131 / 40 |

## What is the **Global Sea Level Observing System** and how does it work?

The Global Sea Level Observing System (GLOSS) was established to obtain data on changes in sea level from around the planet. This international network systematically measures the sea level in many parts of the world, generally using standard techniques such as tide gauges. The collected data will be pooled for use in computer models, making analyses widely available.

GLOSS was created in 1986, as part of UNESCO's (United Nations Educational, Scientific, and Cultural Organization's) Intergovernmental Oceanographic Commission (IOC), and now has 85 countries taking part in the program. There are approximately 287 operational tide gauges worldwide, some in places like the Arctic and Antarctic; this number represents 70 to 75 percent of the originally proposed number of tide gauge stations. In addition, the Franco-American TOPEX/Poseidon satellite, which orbits the Earth, has also added to the data, using a radar altimetry instrument to measure sea surface height to within 5 centimeters (2 inches).

This information is critical, because if global temperatures continue to increase, scientists predict a possible parallel rise in sea level. This would be due to two factors: First, the polar ice caps would melt at increased rates, adding more water to the oceans; second, as ocean waters became warmer, they would expand. Both factors would lead to an increasing sea level. So far, the combined data from tide gauges and spacecraft have enabled scientists to estimate that the global sea level will rise about 20 inches (51 centimeters) by the end of the twenty-first century—much less than the 6.5 to 9.8 foot-increase (2 to 3 meter-increase) that had been estimated before GLOSS was established.

## What would happen if the **sea level rose today**?

If the worldwide sea level rose even a few feet (roughly a meter), low-lying coastal areas would be inundated with water. Millions, perhaps billions, of people who live along the coastlines would be displaced. Island nations, including the Maldives and Marshall Islands, would virtually disappear; extremely populated delta areas, such as those at the mouths of the Mississippi, Bengal, Nile, and Niger rivers, would be inundated with water. There would also be increased coastal erosion, the intrusion of saltwater into river estuaries, the reduction of freshwater supplies, and the flooding of major ports around the world.

## What is **atmospheric pressure**?

Atmospheric, or air, pressure is the weight of the atmosphere over a certain region of the Earth's surface. To put it another way, it is the "squeezing" or pressure of the air around us. As gravity pulls air molecules tightly toward the Earth's surface, the number of molecules increase, and thus, the air pressure increases; conversely, as one travels higher in altitude, the pressure decreases as the number of molecules decrease.

## What is the **atmospheric pressure at sea level**?

The atmospheric (air) pressure is measured in many ways—including pounds per square inch, millibars, hectopascals, and inches of mercury. On the average, the atmospheric pressure at the surface of the Earth (or at sea level) is 14.75 pounds per square inch (1 kilogram per square centimeter); 29.92 inches of mercury; or 1013.25 millibars. Weather maps show atmospheric pressure in millibars, or units equal to a thousandth of a bar. (A bar is a unit of pressure equal to 29.53 inches of mercury in the English system, and 1 million dynes per square centimeters in the metric system.) More commonly now, meteorologists measure air pressure in hectopascals (formerly known as millibars). This measurement allows the scientists to take more accurate and precise readings, with instruments measuring to within 0.1 hectopascals.



## How is **pressure** measured in the **oceans**?

Air pressure decreases with altitude above the surface of the oceans; in contrast, water exerts more pressure than the air. Pressure in the ocean is often measured in "atmospheres" ; 1 atmosphere is equal to the average air pressure at sea level.

## How does the **pressure change** in the **ocean depths**?

As we sink deeper and deeper into the ocean, the pressure of the water around us, known as the hydrostatic pressure, continues to increase. At a given depth, this hydrostatic pressure is based on the mass of the water above that depth. For each 2.25 feet (0.69 meters) we descend, the pressure increases approximately 1 pound per square inch (0.5 kilograms per 6.5 square centimeters); if measured in atmospheres—in which one atmosphere is the air pressure at sea level—pressure increases at the rate of one atmosphere per 33 feet (10 meters) of depth in the oceans.

## What are some **approximate pressures** in the **oceans**?

The average depth of the oceans is about 12,460 feet (3,798 meters) below sea level, with an average pressure of about 388 atmospheres. In the Pacific Ocean's Mariana Trench—which, at 36,198 feet (11,033 meters) below sea level, is the deepest part of the ocean—the pressure is about 1,070 atmospheres. These numbers are only approximations, as pressures can change with temperature and compositional (such as salinity) differences in the seawater.

On a calm day (in 1971), the Pacific Ocean lives up to its name. But, like any other ocean, it can also turn up heavy seas. *NOAA; Lieutenant (JG) Lester B. Smith, NOAA Corps*

# COMPARING THE OCEANS AND SEAS

## What **major oceans** are included in the **world ocean**?

The five large bodies of saltwater that make up the world ocean are called, in order of size, the Pacific, Atlantic, Indian, Antarctic (Southern), and Arctic oceans. Together they are called the "world ocean" because they are interconnected enough to be considered one global ocean.

## What is the **Pacific Ocean**?

The Pacific Ocean is the largest individual ocean on the planet, covering a surface area of approximately 63,980,000 square miles (165,640,000 square kilometers). The average depth is 14,043 feet (4,280 meters), with a maximum depth of 36,203 feet (11,035 meters). The vast Pacific Ocean was first named by the Portuguese explorer Ferdinand Magellan (c. 1480–1521), who mistakenly thought it was peaceful and free of storms—that is, "pacific."

The Pacific Ocean is surrounded by the continental landmasses of South and North America to the east, Asia and Australia to the west, and Antarctica to the south. It is also the home of the Ring of Fire, a belt of seismic activity: In the Pacific, the Earth's tectonic plates are continually moving beneath the continents (in a process called subducting), which leads to the creation of deep trenches, violently-exploding volcanoes, and large earthquakes. Hot spots on the Pacific Ocean floor have also led to the creation of the Hawaiian and Galapagos islands.

The Pacific Ocean can be subdivided into the North and South Pacific; each half has its own circulating currents. In the North Pacific, the currents flow clockwise, with the California Current moving south along the west coast of North America; as it turns westward above the equator, it is

The U.S.S. *Philadelphia* tackles monster waves in the North Atlantic, February 1942. *NOAA; National Archives/Personnel of the U.S.S.* Philadelphia

called the North Equatorial Current; it then turns north and east, becoming the Kuro Siwo (Japan) Current, and flowing by Japan and the Aleutian Islands. In the South Pacific, the currents flow counterclockwise, with the eastward-heading Antarctic Circumpolar Current flowing far to the south near Antarctica, then turning to the north along the west side of South America as the Peru Current. From there, the waters flow to the west just south of the equator, as the South Equatorial Current.

## What is the **Atlantic Ocean**?

The Atlantic Ocean is the second-largest individual ocean on the planet, spanning a surface area of approximately 31,530,000 square miles (81,630,000 square kilometers). It has an average depth of approximately 10,926 feet (3,330 meters), with an approximate maximum depth of 27,495 feet (8,380 meters). The Atlantic Ocean today is bordered on the west by North and South America, on the east by Africa and Europe, Greenland and Iceland in the north, and Antarctica in the South. This body of water was named for Atlas, a Homeric (Greek) mythological figure who held up heaven with great pillars, rising from the sea somewhere beyond the western horizon.

The Atlantic Ocean formed as a result of tectonic activity (movement in the plates that make up the Earth's crust). But unlike the Pacific, where the Earth's plates are moving together, the plates in the Atlantic are spreading apart, leading to large upwellings of lava and the buildup of new crust (into mid-ocean ridges). Most of this activity continues to take place in the depths of the ocean, far from curious human eyes. The spreading of the seafloor along what is known as the Mid-Atlantic Ridge, which stretches the length of the Atlantic Ocean, moved the North American continent away from Europe during the Jurassic period of the Mesozoic era. The seafloor continued to spread, separating South America from Africa during the Cretaceous period. Water filled the gaps, creating the ocean we know today. This spreading continues today at a rate of a less than 1 inch (2.5 centimeters) per year.

The Atlantic can be subdivided into the North and South Atlantic, each with its own circulatory system of currents. In the North Atlantic, the currents flow clockwise, with the Gulf Stream Current flowing north and east along the eastern coast of North America, then heading south and west along Europe and Africa as the Canary Current, and finally moving west above the equator toward North America as the North Equatorial Current. In the South Atlantic, the currents flow counterclockwise; the Benguela Current moves north along the west coast of Africa, travels west just south of the equator as the South Equatorial Current, then south along the east coast of South America as the Brazil Current.

## What is the **Indian Ocean**?

The Indian Ocean is the third-largest individual ocean on Earth, covering a surface area of approximately 28,360,000 square miles (73,420,000 square kilometers), with an average depth of approximately 12,763 feet (3,890 meters) and a maximum depth of 24,443 feet (7,450 meters). This ocean is thought to be originally named after the subcontinent of India, which separated from the supercontinent Gondwanaland and moved north during the Cretaceous period, colliding with the Asian continent during the Cenozoic Era. This great collision created the Himalayas and the high Tibetan plateau beyond. The Indian Ocean is unique because its northern boundary is completely enclosed by a landmass—the Asian continent. On the east lie the Australasian landmasses, such as Australia and Indonesia; to the west is Africa; and far to the south is Antarctica.

The flow of the currents is similar to the flow of currents in the Pacific and Atlantic oceans, but there are some unique differences. Because the northern boundary of the Indian Ocean is totally enclosed with high regions, there are seasonal differences in the winds and currents. The most famous of these is the monsoon, from the Arabic word *mausim,* or "season," which carries large amounts of moisture northward into India during the summer months (and dryness in the winter). Associated winds create a distinct set of currents during summer: The Somali Current flows clockwise, north along the east coast of Africa, then down along the west coast of India; a Southwest Monsoon Current flows east below India. In the southern Indian Ocean, the flow of water runs counterclockwise: The Antarctic Circumpolar Current flows east at the far southern regions, splits with a part running north, just to the west of Australia, then turns to the west as the South Equatorial Current. This current splits north of Madagascar, with part of it becoming the north-flowing Somalia Current and the other part, called the Agulhas Current, flowing south between Africa and Madagascar.

## What is the **Antarctic Ocean**?

The Antarctic Ocean, also known as the Southern Ocean or Antarctic Regions, is the large expanse of water surrounding the continent of Antarctica located at the South Pole of the planet. It can also be thought of as the southernmost parts of the Atlantic, Pacific, and Indian oceans, and has a surface area of approximately 5,000,000 square miles (12,950,000 square kilometers). The greatest depth recorded in these southern waters is 21,043 feet (6,414 meters).

The two major currents in this ocean revolve around the landmass at the bottom of the world: The East Wind Drift flows counterclockwise around Antarctica, close to the landmass in an east-to-west direction. The Antarctic Circumpolar Current revolves in the opposite direction (clockwise), farther north than the East Wind Drift. The Antarctic Circumpolar Current, as well as the winds that drive it, makes passage from the South Atlantic to the South Pacific, around Cape Horn on the tip of

**The frozen tundra of the Arctic: Sheets of jagged ice float near the solid ice sheet, off Canada's Ivvavik National Park in Yukon Territory.** ***CORBIS/Raymond Gehman***

South America, extremely difficult to traverse. These cold, turbulent currents evolved over millions of years as the continents drifted around the globe—the landmasses essentially forcing the currents to go around Antarctica, and the morphology (land configuration) not allowing any inflow of warm water.

## What is the **Arctic Ocean**?

The Arctic Ocean is the least-known and least-understood of the Earth's oceans; it was only recognized as a deep basin about a hundred years ago. Lying at the top of the world, it is almost completely surrounded by the northernmost edges of North America, Europe, and Asia. Unlike the Antarctic Ocean, there is no landmass for currents to flow around—only the seasonally-changing, floating ice packs. The flow of currents in this large basin are in a clockwise direction, with the Greenland Current flowing south past Greenland, and currents entering through the Bering Strait and along the north coast of Norway. Similar in size to the Antarctic, the Arctic Ocean covers a surface area of approximately 5,541,000 square miles (14,350,000 square kilometers). Its maximum depth is 17,848 feet (5,440 meters).



## Are there any **imbalances** between **evaporation** and **precipitation** in the world ocean?

Overall, the world ocean is pretty much in balance, but certain areas do experience more evaporation than precipitation—especially the Atlantic Ocean. But because the world's oceans are connected, there is no true "loss" of water.

## What is the **difference** between an **ocean** and a **sea**?

Oceans are described as continuous bodies of saltwater surrounding the continents. If all the water in the oceans were to disappear, the continents would be surrounded by great depressions. Each of the individual oceans contains shallower areas that differ physically, chemically, or biologically from one another; these are called seas. Geographers define a sea as a division of the ocean that is enclosed or partially enclosed by land. Based on this definition, there are more than 50 seas on Earth, with many of the seas located around Europe and in the western Pacific Ocean. The word sea is also sometimes used to describe the ocean in general, as in the phrase, "sailing on the sea."

Partially enclosed seas are usually found along the margins of the continents, such as the Weddell (off Antarctica) and North Seas. They are more similar to the ocean, as their waters are able to readily mix with the open ocean. Some, such as the South China Sea and Sea of Okhotsk are connected with the ocean by narrow passages between islands. Other seas, such as the Mediterranean Sea and Baltic Sea, are almost completely landlocked, with the only passage to the ocean being through narrow straits. Such nearly enclosed seas also experience a small tidal range or no tides at all.

Still other bodies of saltwater, such as Europe's Caspian Sea, Jordan and Israel's Dead Sea, and Russia's Aral Sea, lack an outlet to the ocean altogether—and are actually lakes. Because of this isolation, such

The Caribbean, one of the largest seas in the world, is famed for its clear turquoise waters, making spots like the Virgin Islands (shown) popular tourist destinations. (For color image, *see* insert.) *CORBIS/Kevin Schafer*

"seas" are distinct in physical, chemical, and biological characteristics from the ocean.

## What are the **Seven Seas**?

The Seven Seas is an ancient reference—dating to before the 15th century—to what were then the seven known major "seas" of the world. They included the Indian Ocean, Black Sea, Caspian Sea, Adriatic Sea, Persian Gulf, Mediterranean Sea, and Red Sea. A more modern interpretation, made popular by the British writer Rudyard Kipling (1865–1936), is the larger known oceans, not seas: the Antarctic, Arctic, Indian, North Atlantic, North Pacific, South Atlantic, and South Pacific oceans.

## What are the **ten largest seas** in the world?

The ten largest seas are the Mediterranean Sea, Bering Sea, Caribbean Sea, Sea of Okhotsk, East China Sea, Sea of Japan, North Sea, Black Sea, Red Sea, and Baltic Sea. (The Gulf of Mexico and Hudson Bay are also large, but do not carry the label "sea.") The largest inland, or landlocked, "seas" are the Caspian and Aral seas, actually both saltwater lakes.

## How **large** are the **Earth's seas**?

The following table details (from largest to smallest) the size of the planet's major seas, including two that are not labeled as seas (the Gulf of Mexico and Hudson Bay), but that still fit the sea criteria.

The National Oceanic and Atmospheric Administration vessel *Discoverer* navigates high waves on the Bering Sea. *NOAA/Commander Richard Behn, NOAA Corps*

| Sea | Size (square miles / square kilometers) |
|---|---|
| Bering Sea | 876,000 / 2,270,000 |
| Caribbean Sea | 749,000 / 1,940,000 |
| Gulf of Mexico | 699,000 / 1,810,000 |
| Sea of Okhotsk | 591,000 / 1,530,000 |
| East China Sea | 483,000 / 1,250,000 |
| Hudson Bay | 475,000 / 1,230,000 |
| Sea of Japan | 405,000 / 1,050,000 |
| North Sea | 224,000 / 580,000 |
| Black Sea | 174,000 / 450,000 |
| Red Sea | 170,000 / 440,000 |
| Baltic Sea | 162,000 / 420,000 |

## How large are the **Caspian** and **Aral seas**?

These saltwater "seas" have no access to the oceans—thus, they are truly lakes. The Caspian Sea, which is bordered by the countries of Kazakhstan, Turkmenistan, Iran, Armenia, Azerbaijan, and Russia, has a surface area of about 143,244 square miles (370,858 square kilometers).

The Aral Sea, which is bordered by Kazakhstan and Uzbekistan, is more difficult to measure. At one time, the sea measured about 24,904 square miles (64,476 square kilometers); today, it is three-fifths that size. Humans are the reason the Aral is shrinking: Within the past half century, waters that once led to the sea were dammed for irrigation, water was drained from the sea to irrigate crops, and the sea became polluted with chemical fertilizers. Today, the ex-Soviet republics surrounding the waters are attempting to restore the area—but it is difficult to say if the Aral will return to its original size.

## What are the **saltiest seas**?

The salinity of seawater varies depending on the region. For example, the salinity of subtropical seas, such as the Mediterranean, is higher because there is more evaporation of water (due to higher atmospheric temperatures); salinity is lower in seas that receive freshwater from rivers or melting ice. Some seas are also super-saline, such as the Red Sea—considered the saltiest in the world—which measures about 41 parts per thousand (or 41 parts of salt per 1,000 parts of water). It is so salty because there are few freshwater rivers flowing into it to dilute the water.

## What **sea** has **volcanic mountains** on both sides of its shores?

The Red Sea, a stretch of water between Africa and Arabia, is actually the water-filled part of the Great Rift Valley, a tectonically active area (an area where the Earth's surface is changing due to the movement of its crustal plates). Coral reefs grow well in the Red Sea, as there are no rivers flowing into it; corals, and the life that grows around reefs, do not like freshwater or the sediments that rivers bring. Off the surrounding coastlines, there are numerous chains of small islands and reefs.

# HOT AND COLD AT SEA

## What is the average **temperature** of the oceans?

The temperature of the oceans varies with depth and latitude on the Earth. For example, in the shallow water of the Persian Gulf, water temperatures can be as high as 104° F (40° C). Overall, there is a definite change in water temperature as you move away from the equator, where the water is about 75° to 85° F (23.9° to 29.4° C), to the polar oceans, where the water is 32° to 40° F (0° to 4.4° C). Deeper ocean waters usually have more constant temperatures. About 87 percent of all the ocean water around the world has an average temperature of 40° F (4.4° C) or less.

## What is a **thermocline**?

A thermocline is a sharp temperature difference in the water. While a steady decrease in water temperature is normal as the depth increases, a thermocline is an area where there is a sudden decrease, much sharper than in neighboring areas. A thermocline can be either permanent or seasonal. Permanent thermoclines are the result of the sinking of cold polar water beneath warmer water layers as it moves toward the equator; seasonal thermoclines result when surface water becomes rapidly warmed by the summer sun. The deeper layers stay colder, not warming until later in the season. In autumn, the seasonal thermocline would disappear.

## Do **all oceans** have a **thermocline**?

No, in the ocean waters close to the north and south poles, the layers of water are relatively the same temperature and no thermocline (sudden decrease in temperature) develops.

## What are the **coldest oceans**?

The Arctic and Antarctic are the coldest oceans on the Earth. A blanket of permanent ice covers both the poles, feeding cold water into the ocean mix—and making these two oceans the coldest. Cold polar weather adds to the ice sheets, making them a permanent feature (though they shift in size) throughout the year.

## What are some of the **coldest seas**?

Most of the coldest seas, of course, are found in the very northern and southern regions of the planet, near to the poles. For example, to the north are the Greenland, Barents, Beaufort, Kara, Laptev, and East Siberian seas; to the south are the Weddell and Ross seas. Midlatitude seas are also cold, such as the brackish waters of the Baltic Sea, which is bordered by such northern countries as Sweden, Finland, Estonia, Latvia, Lithuania, Poland, Germany, and Denmark.

## Does **seawater freeze**?

Yes, seawater does freeze—although most of the world's population lives in temperate, subtropical, or tropical climates, and have never seen the "solid oceans" at the poles. Seawater freezes at about 28° F (-2.2° C), although it really depends on the amount of salt in the water.

Similar to ice in a freshwater pond, oceans generally freeze from the top down. But there is a major difference—and it has to do with salt. The more salt, the lower the temperature required to freeze the seawater (and conversely, with less salt, the seawater will freeze at a higher temperature). When seawater does freeze, freshwater ice crystals form at the surface, leaving salt behind in the surrounding water. This cold, salt-rich water is dense; it sinks and is replaced by deeper water. Because of the deep water's low freezing temperature, it generally freezes only in certain places at the poles. In fact, if seawater acted like most other liquids, our polar seas would be solid ice. Instead the Arctic and Antarctic are a mix of seawater and ice (in various forms).

## What **oceans freeze** over during **winter**?

No oceans freeze over completely during winter, but the Arctic and Antarctic both have sea ice that can reach a thickness of 13 feet (4 meters) in some regions. For example, a blanket of permanent ice covers the central Arctic Ocean all year; the surface layer melts during the summer and refreezes in the Northern Hemisphere's winter. The Antarctic Ocean has a more extensive permanent ice cover. In the Southern Hemisphere's winters and summers, the ice will increase, then shrink, respectively, in response to seasonal changes in conditions.

Sea ice can form from seawater in polar regions. If it is blown together by winds or pushed together by currents it can form a coherent mass, or pack ice. *NOAA/Office of NOAA Corps Operations*

## Why is the water in the **Arctic Ocean saltier** than the water in the **Antarctic Ocean**?

The water in the Arctic Ocean is saltier than the water in the Antarctic Ocean because the ice in the Arctic is formed by water, which freezes, leaving the salt behind. This increases the salinity of the remaining seawater. In Antarctica, snow is responsible for most of the accumulated ice on the continent—though some of the ice is also formed from ocean water. The salinity of these polar oceans does change somewhat during their summer seasons—as both are influenced by the influx of freshwater, which runs into the ocean from melting ice, river drainage, and excessive seasonal rainfall.

## What is **sea ice**?

Sea ice is any form of ice originating from seawater; this excludes icebergs, which form from glaciers (an iceberg is a floating mass of ice that became detached from a glacier). Perennial sea ice can be quite thick, such as the ice in the central Arctic Ocean. In the Antarctic, sea ice only forms in the Weddell Sea and as a narrow strip along the continental edge.

**The National Oceanic and Atmospheric Administration vessel *Surveyor* breaks through the pack ice in the Bering Sea.**
***NOAA/Commander Richard Behn, NOAA Corps***

The reason that many people call any ice in the oceans "sea ice" is that long ago the term was used by mariners for any ice that floated in or drifted into the sea.

## What is **pack ice**?

Pack ice forms when sea ice—frozen seawater in the polar regions—is driven together by waves or wind into a somewhat coherent mass. Ice is considered pack ice when it covers more than half the visible sea surface; unbroken pack ice is when no open water whatsoever is visible, similar to what usually occurs in the central Arctic Ocean. This type of ice is a great hindrance to ships trying to cross the northern ocean, mostly during the winter months. Waves and swells beneath the ice cause cracks and break the ice constantly; the resulting ice floes bump into each other, then refreeze, creating hummocks and deep ridges. Earlier ships exploring the polar region were easily crushed as the ice floes broke apart, then moved together to refreeze. Some modern ships (icebreakers) are structurally sound enough to break through the ice—but it is still very dangerous.

**Most icebergs are white or blue in color—the ice can look blue because it absorbs slightly more red wavelengths than blue. This picture of a bluish arch (for color image, *see* insert) was taken off the Antarctic Peninsula. *NOAA/Commander Richard Behn, NOAA Corps***

## Can **animals** be **trapped** in **pack ice**?

Yes, even animals can get trapped in pack ice. For example, in 1989, three gray whales became trapped by pack ice just off Alaska. Everyone around the world was watching as people opened holes in the ice so the whales could breathe, as the animals worked their way toward the open water. One whale did not survive, but the other two were eventually saved.

## What are **icebergs**?

Icebergs are huge chunks of freshwater ice that break off from a large glacier and fall into the sea; they are actually pieces of glacier ice that find their way to the ocean. Amazingly, icebergs can weigh tons, but still—because of the properties of ice—float easily in the oceans. After falling from the glacier, they are either locally confined by the ocean topography (such as an underwater ridge holding them back), or they float in major ocean currents.

There are about 12,000 to 15,000 icebergs produced each year in the Arctic Ocean, with most of them breaking off of the Greenland ice sheet. In contrast, fewer Antarctic Ocean icebergs break off from the edge of the Antarctic continent's ice shelves. Typically the chunks of ice from the Arctic are shaped like an upside-down cone; Antarctic icebergs usually have flat tops. In addition, Arctic icebergs may rise 250 feet (76 meters) above sea level, while Antarctic icebergs are even larger. Either way, about 90 percent of an iceberg is underwater; the rest is seen "floating" on the surface of the ocean. Thus, when you see an iceberg, you are only seeing a small fragment of the entire mass—the proverbial "tip of the iceberg." For example, if the upper part is more than 200 feet (61 meters) high and several miles (kilometers) long—which is a typical size—the underwater section of the iceberg is immense! In this example, the part that is underwater could extend 2,000 feet (610 meters) and the iceberg might be tens of miles (or kilometers) long.

## How do **glaciers produce icebergs**?

Glaciers produce icebergs during a process called calving. As the glacier flows off the land and into the ocean waters, an ice shelf results—a large mass of ice that floats on the ocean, but is still attached to the coast. Along this shelf, the waters warm the ice, causing cracks and weak spots at the edge of the glacier. Eventually, these weaknesses cause huge chunks of the ice to break away—many falling into the ocean with a thunderous roar.

## Is the **number of icebergs** in the oceans increasing?

No one really knows if the number of icebergs is truly increasing in the oceans. Scientists have found evidence of a periodic increase in some areas. For example, in 1985 scientists noted that one of their predictions had come true: The Columbia Glacier, which enters Prince William Sound near Valdez, Alaska, was beginning to retreat and create more icebergs. In this case, the ice will not float out to sea, as a submerged ridge in the sound stops the ice from escaping. But it is estimated that by the beginning of the twenty-first century, the calving of the glacier will expose a new fjord (a long, narrow inlet).

Icebergs, which form by breaking away (or calving) from glaciers, can take many different forms; this tabular iceberg was photographed in the Antarctic Ocean. *NOAA/Commander Richard Behn, NOAA Corps*

## Can **icebergs** be different **colors**?

Yes, icebergs can be different colors. When we think of an iceberg, we think of the colors blue and white, as most of them calve (are formed) from frozen freshwater glaciers. The ice looks blue because it absorbs slightly more red wavelengths than blue. But icebergs are other colors, too. About one in a thousand icebergs from the Antarctic ice sheet is a green color, occurring when the yellowish-brown remains from dead plankton dissolved in the ocean water are trapped in the ice. The yellow color, along with the blue of the ice, turns the iceberg green. The Antarctic is the only place where green icebergs occur, as the giant ice shelves have remained in contact with the ocean's waters for centuries—and thus have built up the remains of the plankton.

## Why do **icebergs concern mariners**?

Icebergs have always been a concern to anyone who travels by ship in the northern, and sometimes southern, oceans of the world.

One of the most dramatic stories of such a collision occurred in the North Atlantic Ocean on April 12, 1912: The large British luxury liner

**About one in a thousand icebergs from the Antarctic ice sheet is a green color, occurring when the yellowish-brown remains from dead plankton dissolved in the ocean water are trapped in the ice. This greenish iceberg (for color image, *see* insert) was taken in January 1962 off the Antarctic Peninsula. *NOAA/Rear Admiral Harley D. Nygren, NOAA Corps (ret.)***

*Titanic,* in its maiden run, struck an iceberg. A chunk of the ice ripped the ship's hull, sending more than 1,500 people to their deaths in the frigid waters. Soon after the tragedy an international iceberg patrol was established to monitor icebergs and warn ships.

## How **long** do **icebergs** last?

Icebergs can last a long time in the oceans. The larger bergs can drift in ocean currents for many years, the melting and erosion by wave action gradually carving the pieces of ice into fantastic shapes—long ledges, short caves, and tall cliffs and pinnacles. Some icebergs have been known to last for two years or more before they melt.

## Is it possible to use an **iceberg** for **drinking water**?

Using icebergs as a source of freshwater remains a highly debated subject. Some scientists believe it is possible, and point to the larger, more regular icebergs from the Antarctic ice shelf for use.

**This is the killer—the iceberg that sunk the "unsinkable" *Titanic* in the North Atlantic in April 1912. The tragedy, which claimed 1,517 lives, proved a lesson in human arrogance. In its aftermath new safety measures were instituted for all oceangoing ships.** ***CORBIS***

There have been many studies to determine how far an Antarctic iceberg could be towed. But the distances would need to be amazing: About 4,000 miles (6,436 kilometers) from the Antarctic to the southwest coast of Australia; about 9,500 miles (15,286 kilometers) to southern California; and about 12,000 miles (19,308 kilometers) to the Middle East. The technology exists: Iceberg-towing has already been done in connection with oil-drilling activities, in which tugs push Greenland icebergs out of the way of drilling ships or platforms. Other ships have tried to destroy icebergs that come too close to major shipping lanes.

But there are still concerns. Which type of iceberg would be the best for providing drinking water? Who owns an iceberg—and who would get the iceberg? How do you stop the iceberg from significant melting on its way to its destination? Would there be any environmental effects from such an endeavor? We have a long way to go before we can even think about using icebergs as a resource for drinkable water.

## What do **ice cores** taken from the northern and southern ice sheets tell us about the **Earth's past climate**?

Ice cores (thick cylinders of ice pulled out from an ice sheet) drilled in the ice sheets in Greenland and Antarctica, and in the Arctic Ocean have revealed a plethora of information on past climates. For example, Antarctic ice cores reveal a chronology of events: The radioactive fallout from hydrogen bomb tests in the mid-1950s and early 1960s; volcanic eruptions, such as Krakatau (Indonesia, 1883); and trapped air from centuries ago. Other ice cores have shown significant climate changes in the atmosphere during ancient times.

## Which **ocean is warmest**?

One of the warmest oceans is the Indian Ocean—particularly in the northern areas, near the equator. The warmest waters in the world ocean are those that span the Earth's equator.

## Which **sea is warmest**?

The warmest sea is thought to be the Red Sea, a 1,500-mile (2,414-kilometer), narrow body of water between Africa and Arabia; its waters are also thought to be among the clearest in the world. The reasons for its higher temperatures are its proximity to the equator and that it does not mix with any freshwater or colder deep-ocean waters. According to the time of the year, the sea surface temperature ranges from about 68° to 78.8° F (20° to 26° C) in the northern part, and from 77° to 87.8° F (25° to 31° C) in the central and southern parts. The Red Sea also has high salinity—the result of both the hot climate and the lack of a river adding freshwater to the sea.

## What are the **differences** between the **warmest** and **coldest oceans**?

Warm ocean water at the equator will be not only hotter, but less saline than ocean water at the poles. This difference in both temperature and salinity sets up the entire cycling and movement of the global ocean water—which can be seen as a convective pot of swirling waters.

# THE ANCIENT OCEANS

## THE GEOLOGIC TIME SCALE

### What is the **geologic time scale**?

The geologic time scale is a tool scientists developed to represent the Earth's history—from the formation of the planet, about 4.55 billion years ago, to the present. The major divisions on the scale are listed after the Cambrian period (at the beginning of the Paleozoic era). The reason is that major fossils do not show up in the Earth's rock layers until after what is called the great Cambrian Explosion, when life began to flourish in the oceans.

### What does the **geologic time scale** look like?

The following represents one version of the geologic time scale (there are other scales with different dates and names, usually depending on the country).

| Eon | Era | Period | Epoch |
|---|---|---|---|
| Phanerozoic eon (began 544 million years ago) | Cenozoic era | Quaternary period (1.8 million years ago to the present) | Holocene epoch—11,000 years ago to present |
| | | | Pleistocene (glacial) epoch—1.8 million years to 11,000 years ago |

| Eon | Era | Period | Epoch |
|---|---|---|---|
| Phanerozoic eon, (continued) | (Cenozoic era, continued) | Tertiary period (65 to 1.8 million years ago) | Pliocene epoch—5 to 1.8 million years ago |
| | | | Miocene epoch—23 to 5 million years ago |
| | | | Oligocene epoch—38 to 23 million years ago |
| | | | Eocene epoch—54 to 38 million years ago |
| | | | Paleocene epoch—65 to 54 million years ago |
| | Mesozoic era | Cretaceous period (146 to 65 million years ago) | |
| | | Jurassic period (208 to 146 million years ago) | |
| | | Triassic period (245 to 208 million years ago) | |
| | Paleozoic era | Permian period (286 to 245 million years ago) | |
| | | Carboniferous period (360 to 286 million years ago; it can be broken up into the Pennsylvanian period, 325 to 286 million years ago; and Mississippian period, 360 to 325 million years ago) | |
| | | Devonian period (410 to 360 million years ago) | |

| Eon | Era | Period | Epoch |
|---|---|---|---|
| Phanerozoic eon, (continued) | (Paleozoic era, continued) | Silurian period (440 to 410 million years ago) | |
| | | Ordovician period (505 to 440 million years ago) | |
| | | Cambrian period (544 to 505 million years ago) | |
| Precambrian eon (4.55 billion years to 544 million years ago) | | | |

## Who first subdivided Earth's long history?

William Smith (1769–1839), an English canal engineer, was one of the first to subdivide the Earth's long history. His 1815 geological map of England and Wales established a practical system of stratigraphy—the geology of the layers of the Earth's crust; stratigraphy is the basis of the geologic time scale. Smith showed that certain layers of Mesozoic rock in England could be identified by their specific fossils.

An international geologic time scale was drawn up between 1820 and 1870; the era divisions labeled Paleozoic ("ancient life"), Mesozoic ("middle life"), and Cenozoic ("recent life") were established around 1840. By the end of the nineteenth century, the eras were further subdivided into periods, epochs, zones (now usually called ages), and subzones (usually called subages). And by the mid-twentieth century, divisions were more precise, as scientists began using radiometric dating techniques to determine the absolute ages of the rock layers.

## How can a person visualize the extent of the geologic time scale?

The geologic time scale represents billions of years—a time span that is almost beyond human comprehension. One of the best ways to get a handle on how much time the scale actually represents was suggested

The fossil of an early plant, dating from the Devonian period, or between 360 and 410 million years old. *CORBIS/James L. Amos*

by author John McPhee, in his book *Basin and Range*: Stand with your arms held straight out to each side. The extent of the Earth's history, as represented by the geological time scale, is the entire distance from the tip of your fingers on the left hand to the tip of your fingers on the right. If someone were to run a nail file across the fingernail of your right middle finger—and that represented time—it would represent the amount of time humans have been on the planet!

## What are the **time units** used in the **geological time scale**?

There are about six major time units in the geologic time scale. They are not precise, but are merely ways of trying to keep track of the Earth's long history. They are called eons, eras, periods, epochs, ages, and subages. The eon represents the longest geologic unit on the scale; on some scales, it is defined as a division of one billion years. An era is a division of time smaller than the eon, and is normally subdivided into two or more periods. The period is a subdivision of an era; the epoch a subdivision of a period; an age a subdivision of an epoch; and a subage (although it is not often used), the subdivision of an age.

## What do the **divisions** on the **geologic time scale** represent?

The geologic time scale is not an arbitrary listing of the Earth's natural history—each specific boundary between divisions represents a change or event that delineates it from the other divisions. In the majority of cases, a boundary is drawn to represent a major catastrophe or a major evolutionary change in animals or plants (including the evolution of specific species) that lived during that time.

## What is the **Phanerozoic eon**?

The Phanerozoic eon represents the time from 544 million years ago to the present—the time when the fossil record became much richer. The rough translation of Phanerozoic (a word of Greek origins) is "abundant life." The eon comprises the Paleozoic, Mesozoic, and Cenozoic eras.

## How do scientists **divide the Precambrian eon**?

The Precambrian (4.55 billion to 544 million years ago) is broken down into various divisions, often depending on the country in which the scale was made. The following table shows two generally accepted versions of the Precambrian eon.

**Divisions of the Precambrian—version I**

| | |
|---|---|
| Vendian | began 650 million years ago; ended 544 million years ago |
| Proterozoic | began 2.5 billion years ago; ended 650 million years ago |
| Archaean | began 3.8 billion years ago; ended 2.5 billion years ago |
| Hadean | began 4.55 billion years ago; ended 3.8 billion years ago |

**Divisions of the Precambrian—version II**

| | |
|---|---|
| Proterozoic | upper—began 900 million years ago; ended 544 million years ago |
| | middle—began 1.6 billion years ago; ended 900 million years ago |
| | lower—began 2.5 billion years ago; ended 1.6 billion years ago |
| Archean | began 4.55 billion years ago; ended 2.5 billion years ago |

## What does the **Precambrian eon** represent?

The Precambrian is the longest span of time on the geologic time scale, lasting from 4.55 billion years ago to 544 million years ago—about seven-eighths of the Earth's history. There are few fossil clues from this time; thus, generally speaking, scientists lump many events into this division. For example, this time span includes the formation of the Earth, the beginnings of the planet's crust, the formation of the first plates and their movements, the first life on the planet, and the evolution of an oxygen-rich atmosphere. At the very end of the Precambrian, the first multicelled organisms, including the first animals, evolved in the oceans; this event marks the beginning of the Paleozoic era—and of the Phanerozoic eon.

## What does the **Paleozoic era** represent?

The Paleozoic era is delineated by two of the most important events that occurred in the animal world: At the start of this era, multicelled animals experienced an explosion in numbers and diversity (the event is called the Cambrian Explosion); within a few million years, almost all modern animal groups (phyla) appeared. For the duration of the Paleozoic era (which lasted about 300 million years), animals, plants, and fungi began to colonize the land, and insects began to occupy the air. Many of the periods within the Paleozoic era are loosely based on these events. The end of the Paleozoic era (and end of the Permian period) is defined by the largest mass extinction in the planet's history—an event called the Permian extinction. It wiped out the majority of organisms on Earth, including between 90 and 96 percent of all marine animal species.

## What does the **Mesozoic era** represent?

The Mesozoic era began at the end of the Paleozoic era, and lasted for approximately 180 million years. It ended with another mass extinction that eliminated more than 50 percent of the species on the Earth, including the dinosaurs. According to the geological time scale, it was followed by the Cenozoic era.

The Mesozoic was a time when the planet's terrestrial plants changed dramatically. Early in the era, ferns, cycads, ginkgoes, and other unusual plants dominated; in the middle of the era, gymnosperms such as conifers (evergreens) became prolific; toward the end of the era, in the middle Cretaceous period, the earliest flowering plants (called angiosperms) took over from many other plants. This was also a time of great animal diversification: amphibians evolved; then the reptiles evolved from the amphibians; and then mammal-like reptiles evolved from reptiles.

During the Mesozoic era, the dinosaurs became the dominate forms of animal life on land. In the oceans reptiles also dominated, including such creatures as ichthyosaurs and plesiosaurs.

### What does the **Cenozoic era** represent?

The Cenozoic era is the most recent of all the eras, and has spanned only about 65 million years. It began with the mass extinction at the end of the Mesozoic era, and continues through today. This era is sometimes referred to as the Age of Mammals, but in reality, the name could also be the Age of the Birds, Fish, Insects, and Flowering Plants—as all these life forms have diversified and grown in number over the past 65 million years. The Cenozoic is grouped into two main divisions: The Tertiary and Quaternary periods. The present time is part of the Quaternary period.

### What is **relative time**?

Relative time is used as a means of establishing rocks and fossils in a general chronological order. It is based on where a rock layer is located in comparison with other rock layers; thus, it is only relative, not absolute, time. In the nineteenth century, scientists used this method to date rock layers—and to establish and construct the first geologic time scale.

### What is **absolute time**?

Absolute time is the determination of the (approximate) true age of the rock; that is, how long ago the rock layer formed. Absolute time is determined by radiometric means (radiometers are instruments that measure the amount of radiation in the layers of the Earth), and was used to add more precise time spans to the geologic time scale, which is a relative scale. The techniques to determine absolute time were not perfected until after the 1920s.

## FOSSILS

### How do scientists determine the **past history of the life** on our planet?

One of the most useful clues that scientists use to determine the history of our planet—both its life and its changes—are fossils. The fossil record,

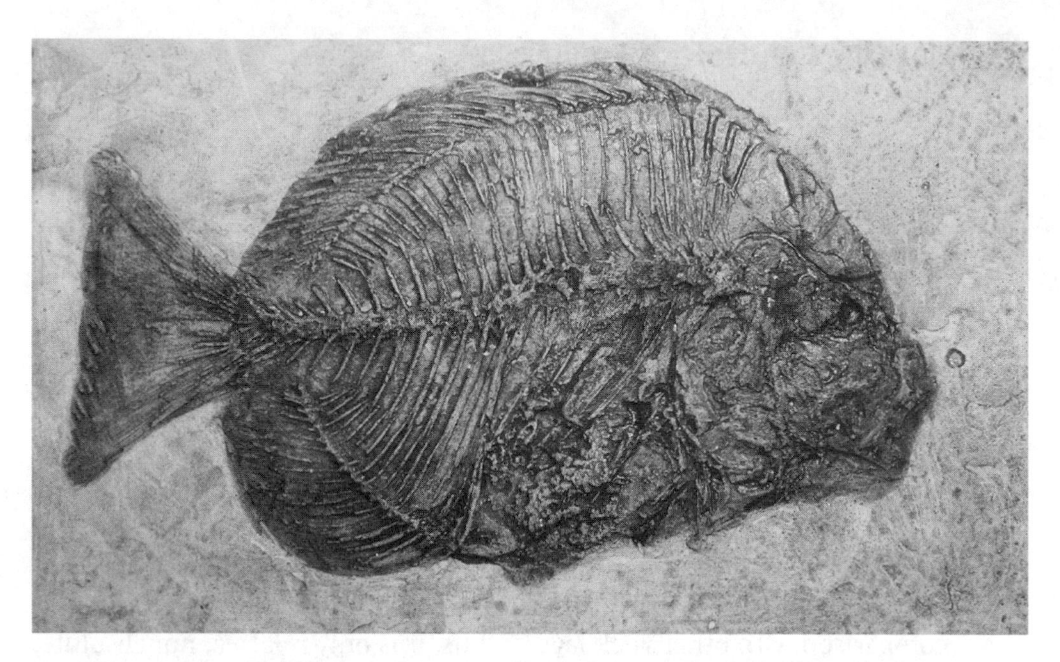

**A 50-million-year-old imprint fossil of an extinct member of the spadefish family. It was found in Verona, Italy.**
***CORBIS/Sally A. Morgan; Ecoscene***

preserved in rock layers over millions of years, allows us to understand our past—and where we are going in the future. Without fossils, our planet's immense and varied past would remain unknown to us.

## What are **fossils**?

Fossils are the remains of plants and animals that have been preserved in the rock layers of the Earth. Fossils are often close to an organism's original shape. The word fossil comes from the Latin *fossilis,* meaning "dug up." There are many different types of fossils: The organism's remains and conditions present at the time the organism died determine what kind of fossil is formed. Most people are familiar with fossils that were formed in rock by the hard parts of an organism, such as teeth, shells, or bones, leaving an imprint. But animals and plants have also been preserved in other materials besides rock: Fossils can be found in ice, tar, peat, and the resin of ancient trees. The process of fossilization continues today—on land and in the oceans—as organisms die and are quickly buried.

## How does a **fossil form**?

A fossil can form in a number of ways, depending on the type of remains and the environment in which the organism died. The general process is for the hard parts of animals (such as bones, teeth, and shells) or the seeds or woody parts of plants to be covered by sediment—sand or mud—either on land or on the ocean floor. Over millions of years, more and more layers of sediment accumulate, burying the remains of organisms deep within layers of sediment. The pressure of the overlying layers causes the sand or mud to eventually turn to stone; this process is called mineralization. The organisms' remains are often chemically altered by mineralization, becoming a form of stone themselves. The same process also produces petrified wood, coprolites (petrified excrement), molds, casts, and trace fossils.

## Are **mineralization** and **fossilization** the same thing?

Not exactly. As far as paleontology is concerned, mineralization is a process of fossilization in which the organic compounds are replaced by inorganic material (minerals).

## What are **molds** and **casts**?

Molds and casts are types of fossils—impressions of an animal's or plant's hard parts (and sometimes soft parts) that are left in the rock after burial and decay. Molds are hollow impressions (cavities) in the rock; if the mold is filled with sediment, it can often harden, forming a corresponding cast.

## What are **trace fossils**?

Not all fossils are hardened bones, teeth, or shells—or even molds or casts. Some are trace fossils: Physical evidence, usually left in soft sediment (such as sand or mud), that creatures once crawled, walked, hopped, burrowed, or ran across the land. For example, small animals in search of food may have bored branching tunnels in the mud of the ocean floor; the tunnels were then filled in by sediment, buried by layers

**Animal hard parts (such as bones and teeth) can become fossilized, as were these eight shark teeth.** ***CORBIS***

of more sediment over millions of years, and eventually solidified, forming trace fossils. Trace fossils also include animal tracks; for example, when dinosaur footprints left in soft mud along a riverbank filled with sediment, the sediment eventually solidified in the form of the footprints. Today we see the results of this long-ago activity as trace fossils. Many originators of trace fossils (in other words, the animals that created them) are unidentifiable since sometimes there are no hard fossils of the creatures left in the area—just the traces of their passing.

## What is **taphomony**?

Taphomony is the study of the way in which an organism was buried and of the origin of plant and animal remains. Through taphomony scientists try to "reconstruct" an animal or plant based on the evidence provided by its fossils. But, for the following reasons, this is a very difficult science that has many variables.

*Scavenging and decay:* After an animal's death, scavengers (birds and other animals) remove the soft flesh from the carcass. The hard parts (bones, teeth, or shells) that are not eaten begin to decay. The rate of decay varies by environment: Decay is faster in a humid climate and slower in an arid climate. The hard parts (the skeleton, shell, or the fibrous structure of the animal) are further reduced by the action of wind, water, sunlight, and chemicals in the surrounding environment. If the hard parts are completely worn away (through the action of wind, water, and sunlight) without being buried, no fossil can form. Often the fossil that forms is the result of a partial wearing-away of an animal's or plant's hard parts, making it difficult for scientists to reconstruct the life form that left the fossil.

*Location:* If an animal dies in a place where it is buried quickly, the organism has a good chance of surviving as a fossil. The ocean provides an ideal environment for fossilization—since organisms living here are very likely to be buried by sediment soon after they die. But if an animal dies in a vulnerable or quickly-changing environment, the bones and other hard parts can be broken or scattered. For example, the action of flash flooding can break up bones and, as the waters recede, the skeletal remains could be carried to various locations. However, flash flooding may also increase the chance of fossilization by moving the bones to a better area for preservation, such as a sandbank in a river.

*Rapid burial:* One of the best ways for a fossil to form is by rapid burial. If an organism's remains are quickly buried by mud or sand, the amount of oxygen is reduced, which slows the rate of decay. Again, the bottom of the ocean is an ideal place for this to occur. But even rapid burial does not ensure the creation of a fossil: As more and more sediment accumulates on top of the bones (or other hard parts of the organism), pressure from these overlying layers may cause damage to the organism's remains; acidic chemicals may also seep into the sediment, causing the bones (or other hard remains) to dissolve.

*Fossilization:* Another factor that figures into the creation of a fossil is a process called fossilization: The sediment (such as mud or sand) that surrounds the organism's remains must be turned into stone—which is accomplished by the pressure exerted by the overlying sediment; there must also be a consequent loss of water in the surrounding sediment. Eventually the sediment grains become cemented into the hard structure we call rock. As the spaces in the rock are filled

A cast fossil of a starfish (or sea star) from the mid-Devonian period, on display at the Black Hills Institute, Hill City, South Dakota. *CORBIS/Layne Kennedy*

with minerals (such as calcium carbonate or pyrite), the remains themselves may also recrystallize (forming a cast of the organism's remains).

*Exposure:* In order to study precious fossils, scientists must first find exposed rocks that contain them. (In other words, some of the fossil record remains buried—hidden from human view.) Fossils may be exposed by the uplift of land (which reveals fossil-containing sedimentary rock at the Earth's surface); by erosion of an area by wind, water, or even an earthquake; or via the development of land (for example, a road cut may expose a fossil-filled rock layer).

## Why are there **gaps** in the **fossil record**?

Gaps in the fossil record are most often the result of erosion. Rock layers, including embedded fossils, can be eroded by the action of wind, water, and ice. Gaps in the fossil record can also be caused by mountain uplift (the process by which masses of land are "pushed up" to create mountains), which destroys fossils. Volcanic activity can also bury fossil evidence, as the hot magma rock physically changes the rock—and thus fossils—it touches.

## How do we determine the **age** of **fossils**?

There are numerous dating techniques used to determine the age of rock and the fossils within the rock. Some of the more common methods use radiometric (isotopic) dating techniques. These techniques use the known rate of decay of radioactive elements into stable isotopes (atoms of a chemical element) or other elements to determine the age of the rock. With these techniques, fossils are not dated directly; the rock around the fossil is dated, and the fossil's age is extrapolated from the data.

Radioactivity within the Earth continuously bombards the atoms in minerals, exciting electrons that become trapped in the crystals' structures. Scientists often use two techniques called electron spin resonance and thermo-luminescence to determine the age of minerals. Both methods measure the number of excited electrons in minerals found in the rock: Spin resonance measures the amount of energy trapped in a crystal, while thermo-luminescence uses heat to free the trapped electrons.

After determining the number of excited electrons present in the minerals—and comparing it with known data representing the actual rate of increase of excited electrons in these minerals—scientists can calculate the time it took for the excited electrons to accumulate. In turn, the data can be used to determine the age of the rock and the fossils within the rock.

Another radiometric technique is uranium-series dating, which measures the amount of thorium-230 present in limestone deposits; these deposits form with uranium present, and have almost no thorium. Because scientists know the decay rate of uranium into thorium-230, the age of limestone rocks—and thus, the fossils found within the rock—can be calculated from the accumulated amount of thorium-230 found within a particular limestone layer.

## What is **carbon dating**?

Carbon-dating techniques are used to determine the age of relatively young organisms (such as wood or bone) as well as the age of very ancient rock. To date younger organisms, the isotope carbon-14 (C-14) is used; it decays into nitrogen-14, and has a half-life (the time it takes for half the isotope to break down) of about 5,730 years. C-14 is produced in the Earth's atmosphere; it then combines with oxygen to form carbon dioxide. All organisms use this gas—as long as they are alive. When an organism dies, C-14 ceases to enter the organism, and the C-14 already present begins to decay. Scientists measure the amount of C-14 present and use its rate of decay to determine the age of the fossilized remains.

Scientists can also use carbon-dating techniques to determine if life existed in very ancient rock. Carbon has two other isotopes (atoms)—carbon-12 (C-12) and carbon-13 (C-13), the first slightly lighter than the second. Living organisms tend to "select" the lighter isotope, as it is easier to absorb (it takes less energy). When sedimentary rock is found to have more than the usual ratio of C-12 to C-13, scientists know that some form of life has altered the amounts—thus there was probably life present in the ancient sediment when it was deposited.

### Why do **fossils form** more readily in the **ocean** than on land?

In the ocean, sediments are continually building up on the bottom, quickly covering the remains of dead organisms and preserving them. This greatly increases the chances of fossilization.

## PALEOCEANOGRAPHY

### What is **paleoceanography**?

Paleoceanography is the study of ancient oceans; it is an interdisciplinary field of study that uses data from geology, biology, physics, and chemistry, among other disciplines. Paleoceanographers can be thought of as detectives, piecing together clues from all types of samples, sources, and data, then using this information to reconstruct the events of the past. One of the central goals of paleoceanography is to reconstruct how the oceans have changed over time. Much of the data is gleaned from ocean sediment samples and is placed in databanks—where it can be accessed and studied by paleoceanographers worldwide.

### How do paleoceanographers know that **oceans** have **changed over time**?

Paleoceanographers know that oceans—and the climate—have changed by studying sediments, including those collected from the deep sea. Over time, each layer of the ocean floor is laid down, one layer on top of the other. Cores drilled deep into the ocean floor show these sediment layers—all of which read like the pages of a book. By knowing which types of sediments formed when, and by analyzing some of the minerals

The *Glomar Challenger* was launched in the mid-1960s. It was the first research vessel to drill into the deep-ocean floor—bringing up invaluable information about the Earth's history. *CORBIS*

or fossils within the sediments, scientists have pieced together a good deal of information about how the ocean has changed over time.

The study of ancient oceans—and how they were created and evolved over millions of years—also gives us an idea of the physical mechanisms at work on our planet today. Paleoceanographers determine how the morphology (land configuration) of the Earth changed in the past, what caused these changes, and what effect such changes had on the oceans, climate, weather, plants, and animals. In this way, scientists can get a feel for what changes will occur in the future—and how those changes will impact our own species, *Homo sapiens sapiens.*

## How do paleoceanographers gather **data about the past**?

Paleoceanographers obtain marine sediment and rock samples from the seafloor to decipher the ocean's past. They also rely on the interpretation of terrestrial rock layers that were once part of shallow and deep oceans.

## How are **sediments obtained** from the **deep-ocean floor**?

Obtaining sediments from the deep-ocean floor is only a recent development. In the mid-1960s, a special ship fitted with a drilling rig, the *Glomar Challenger* was launched; it was named after the company that built it (the Global Marine Company) and after the HMS *Challenger*—the first true ocean research vessel, which sailed the oceans in the 1870s. The *Glomar Challenger* was the first to drill into the deep-ocean floor, bringing up long cylinders of sediment called cores from the bottom of the South Atlantic; the drills brought up material extending about 328 feet (100 meters) into the ocean floor. Improvements in drilling techniques have enabled researchers to penetrate more than 5,000 feet (1,500 meters) into the ocean bottom sediments, bringing up material representing up to 180 million years of the Earth's history.

## What is the **composition** of **ocean sediments**?

There is no way to neatly categorize the composition of ocean sediments. But generally speaking, ocean sediments consist of a multitude of rock sediments and minerals, such as grains of sand, mud, silt, gold (from river deposits), heavy metals, and even ferromanganese (iron-manganese) modules—all depending on where the sediment is deposited. Additionally, sediment can contain small organisms, such as radiolarians (marine protozoans), diatoms (a type of algae), and calcareous nannofossils (minute fossils containing calcium). Material from larger organisms, such as the teeth of fish (such as sharks), corals, broken shells, whole skeletons, and sundry other marine organism remains can also be found in sediment layers.

### What do ocean sediments tell scientists about the conditions in the early oceans?

Ocean sediments containing certain fossils, or of a certain composition, often give scientists clues about the makeup of early oceans—and even information on ancient biological communities. For example, any microfossils in the sediment samples that formed shells of calcium carbonate ($CaCO_3$) are good indicators: The relative amounts of light and heavy isotopes (atoms of a chemical element) are analyzed to obtain information about conditions in the ancient oceans such as the temperature, salinity, nutrient levels—and even to determine the volume of the ice sheets on land. This information can then be integrated with other data, such as paleontological, geochemical, and geophysical findings—and used to reconstruct ancient oceanic circulation patterns and global climate.

## How are the ages of **ocean sediments** determined?

After the sediment samples have been recovered from the ocean floor—by means of a variety of techniques such as drilling or coring—they are prepared in the laboratory, where they are analyzed and their age is determined using techniques such as magnetostratigraphy and fossil identification.

In magnetostratigraphy, iron particles in the sediments are examined to determine their orientation to the Earth's magnetic field. As the sediments were laid down, the iron particles present were oriented with the magnetic field of the Earth at that time. A record of the reversals in the Earth's magnetic field are therefore preserved in the sediment layers by the orientation pattern of their iron particles. This pattern can be matched to magnetic reversal patterns found in the ocean floor basalts—rocks that can be dated by radiometric means (radiometers are instruments that measure the amount of radiation in the layers of the Earth).

Another method of dating ocean sediments is by the fossils and microfossils found in the various rock layers. Scientists use "marker species" found within the core—fossils in the sediments or rocks that have already been dated and the ages determined—to resolve the age of the sediment.

## Are certain types of **ocean sediments** indicative of the **climate**?

Yes, scientists have found that certain sediments are often representative of certain climates. The following list cites the types of sediments that are often used to help determine past climates.

*Ice-rafted sediments (tills):* These sediments are indicative of glaciers and of a very cold climate.

*Organic-rich sediments or phosphorites:* These sediments are indicators of high (or low) productivity levels in the oceans—and could translate into the conditions of the climate.

*Coral reefs:* Coral reefs are indicators of past changes in sea level, since as the sea levels rise and fall, the coral reefs continue to follow the coastlines.

*Pollen within sediments:* In marine sediments, pollens from any vegetation can be used to determine the nature of the climate and changes in vegetation on land.

*Eolian sediments:* Eolian, or wind-blown deposits such as quartz grains, are often indicative of wind intensity and direction.

*Fossil distribution:* The way fossils are distributed in ocean sediments is one of the best indicators of climate. For example, plankton are sensitive to temperature: siliceous organisms typically represent cold climates and carbonates represent warm. The abundance of specific species of Foraminifera or Radiolaria can indicate the climate of a certain locale at a certain time; the discovery of cold (for example pachyderma) and warm species (such as sacculifer) in ocean sediments provide scientists with important clues about ancient climates.

*Fossil chemistry:* The details of the fossil chemistry are also strong indicators of ancient climate conditions. For example, fossil shells record the chemistry of the water; the ratio of stable oxygen isotopes (the oxygen-18 and oxygen-16 atoms) in seawater changes as the volume of ice in the oceans changes.



# THE EARLY OCEANS

## When did the **Earth first form**?

Scientists believe that the Earth formed about 4.55 billion years ago, with the Earth's crust becoming somewhat stable by about 3.9 billion years ago.

## Is there a **connection** between the **Pacific Ocean and the Moon**?

The theory that the Moon came from the Earth was once thought to be foolish by most scientists. This idea was first proposed by British mathematician and astronomer George Darwin (1845–1912) in the early twentieth century. He called it the fission theory, in which a chunk of the rapidly spinning Earth was thrown off into space, and settled into orbit around the Earth. The connection between the ocean and the Moon was simple: The Moon was thought to have been thrown out from the area of the Earth we now call the Pacific Ocean.

The more accepted idea of lunar formation is that the Earth and Moon formed about 4.6 million years ago—as the material from a solar nebula (a collection of interstellar dust and gases) condensed to form our solar system. But amazingly, recent computer models have shown that the Moon may have truly come from the Earth. The theory is that a Mars-sized object struck our planet early after its formation. As the huge object struck the Earth, it tore material from far into the mantle (the part of the Earth that lies below the crust and above the core); this material eventually settled into orbit around our Earth. But the connection to the Pacific Ocean is not part of this more recent theory.

## How did the **first ocean basins form**?

Although it is thought that some of the first "basins" were deep-impact craters on the early Earth's surface, the true ocean basins owe their origins to plate tectonics: The moving crust, driven by the movement of the Earth's mantle (the part of the Earth that lies below the crust and above the core and which is comprised of unconsolidated material), caused the separation of two types of material near the Earth's surface—the lighter and less-dense granitic rock that makes up the continents separated from the heavier and denser basalts that make up the ocean floor.

## Where did **ocean waters come from**?

Scientists believe the Earth had two primary sources of water. First, the gases released from volcanic vents (in a process called "out-gassing") contained water vapor, which created clouds and eventually rain. Second, small (about 30 feet, or 9 meters, in diameter) ice comets, and perhaps frozen asteroid-type bodies, collided with the Earth, providing water for its basins. It is also known that by about 4 billion years ago, a half billion years after the Earth's formation, the planet's surface cooled enough for water to exist primarily as a liquid.

## Have there been any **major changes** in the **oceans** over time?

The composition of the ocean's seawater, including the salts and trace minerals, has remained relatively the same over the Earth's long history. But there have been other major changes: About 2.1 billion years ago, as the planet's atmosphere began to accumulate more oxygen, so did its oceans. And of course, ocean organisms increased in abundance and diversity over time due to a number of factors—especially this increase in oxygen.

## Why have **ocean shapes changed** over time?

The Earth's geologic activity is the main reason ocean shapes have changed over millions of years. Huge continental plates moved (and continue to move) across the planet, resulting in continental drift and plate tectonics (the movement of plates that comprise the Earth's surface). This activity not only changed the position of the continents, but also the shapes, depths, and bottom terrain of the oceans.



## What was the **ocean that covered the entire world**?

Scientists believe that the largest ocean in our planet's history—or the ocean that covered the entire world—formed about 700 million years ago, toward the end of the Precambrian eon and not long before the beginning of the Paleozoic era (which began about 544 million years ago). This ocean, called the Iapetus Ocean, had only one major landmass, referred to as a "supercontinent"; it was called Rodinia.

## What were the **continents and oceans** like during the **Paleozoic era**?

Early in the Paleozoic era, which began 544 million years ago, the supercontinent of Rodinia had already started to break up—forming the southern continent of Gondwanaland, a landmass consisting of parts of Australia, Antarctica, Africa, and South America, plus the Indian subcontinent (all south of the equator); and a northern continent (consisting of present-day North America) as well as some isolated landmasses north of the equator. The major ocean during the Paleozoic was the Iapetus Ocean.

By late in the Paleozoic, the scattered landmasses had joined into two large landmasses—Gondwanaland to the south of the equator and Laurasia to the north of the equator. Toward the end of the Paleozoic, these two continents slowly collided, forming the supercontinent of Pangea, meaning "all Earth."

## What **major catastrophe** happened in the **oceans** at the end of the Paleozoic era?

At the end of the Paleozoic era (or the end of the Permian period, about 245 million years ago), almost all ocean life died out—that is, it became extinct. No one knows why, but there are theories that try to explain this massive global extinction. One theory includes evidence of oxygen starvation in the seas: With very little oxygen around, most living creatures in the seas died. The reason oxygen levels dropped is not known: Some scientists believe massive amounts of carbon dioxide from erupting volcanoes in Siberia warmed the globe, reducing the temperature differences between the poles and the equator. This warming slowed the ocean currents—causing the water to virtually stagnate.

Still another theory of why so many marine creatures died posits that a huge asteroid or comet struck the Earth, wiping out many species. But again, no one can explain why some species did not die, while others did—or where the comet impact occurred.

## What were the **continents and oceans** like during the **Mesozoic era**?

At the beginning of the Mesozoic era (245 million years ago), there was essentially one large expanse of water called the Panthalassa Ocean (today's Pacific Ocean is the remnant of this huge ocean). This ocean surrounded the supercontinent of Pangea, meaning "all Earth." This giant landmass straddled the planet's equator roughly in the form of a **C**; the smaller body of water enclosed by the **C** on the east was known as the Tethys Ocean (or Sea). Only a few scattered bits of continental crust were not attached to Pangea, and lay to the east of the large continent. In addition, the sea level was low, and there was no ice at the polar regions.

Pangea broke apart about 200 million years ago (during the Jurassic period)—splitting again into the great supercontinents of Laurasia (to the north of the equator) and Gondwanaland (to the south). This breakup also saw the initial (east-west) separation of Europe from North America, and of Africa from South America. Thus the North Atlantic Ocean began to open up. Major rises in sea levels flooded many areas, including large parts of what are now Europe and central Asia.

## What **modern continents** were represented in **Laurasia and Gondwanaland**?

The supercontinents of Laurasia and Gondwanaland can be translated into today's continents as follows: Laurasia included North America and

Eurasia (Europe, Siberia, and parts of eastern Asia), and Gondwanaland included South America, Africa, India, Antarctica, and Australia.

## What is the difference between **Gondwanaland** and **Gondwana**?

There is no difference between the terms Gondwanaland and Gondwana. They are synonymous, and the use of the terms appears to be a personal preference.

## What was the **Tethys Ocean**?

The large ocean called the Tethys Ocean (or Tethys Sea) formed about 245 million years ago, during the early Mesozoic era, when the single supercontinent Pangea began to split apart into two separate continents—Laurasia to the north and Gondwanaland to the south of the equator. As the two huge continents split, the gap between them slowly filled with seawater. In the beginning, the Tethys Ocean was very long, narrow, and shallow—but as the continents continued their movements away from each other, the sea became wider and deeper.

The Tethys Ocean was oriented roughly in an east-west direction, and was located just north of the equator for most of the Mesozoic era. This sea was the only body of water on the planet that was distinct from the Panthalassa Ocean, the very large ocean that covered much of the Earth.

## Where is the **seafloor** of the ancient **Tethys Ocean today**?

As the continental movements squeezed the Tethys Ocean into today's remnants (the Mediterranean, Black, Caspian, and Aral seas), the seafloor of this ancient body of water was deformed and uplifted; a process that can be demonstrated by pushing together the ends of a rug. This uplifting produced the European Alps, the Caucasus, and the Himalayas—mountains where fossils of ancient marine life from the Tethys Ocean can still be found. Italian scholar, artist, engineer, and inventor Leonardo da Vinci (1452–1519) was the first European to discover these fossils in the Alps. He correctly assumed how and why these fossil-filled marine rocks were found at such great heights.

## Are there any remnants of the ancient Tethys Ocean still present today?

In a stunning reversal of fortunes, the same continental movements that produced the Tethys Ocean in the Mesozoic era have almost erased any traces of this vast sea today. As the landmasses of Africa and India moved northward during the Cenozoic era, they collided with Europe and Asia, respectively. This movement squeezed shut the Tethys Ocean and only small remnants remain today: the Mediterranean, Black, Caspian, and Aral seas.

## What were the **continents and oceans** like during the **Cenozoic era**?

By the beginning of the Cenozoic era (65 million years ago), the major modern continents had taken shape and they continued to move toward their present positions. The Tethys Ocean was larger, and still separated the northern and southern continents. Laurasia continued to split apart; Gondwanaland no longer existed as South America and Africa had already separated. Australia and India began to move northward from around Antarctica. By about 50 million years ago, the continents were close to their present-day configuration, with India slamming into Asia to create the Himalayas. Also, the Atlantic Ocean continued to open (widen), as North America separated from Europe.

## How will the **oceans** look in the distant **future**?

Though no one really knows for sure, there are some interesting guesses. As the seafloor of the Atlantic Ocean continues to spread, the Atlantic will open wider. In the Pacific, continental plates will continue to subduct (move underneath each other), shrinking the now-largest ocean. And the Red Sea, in eastern Africa's Great Rift Valley, may also develop into a smaller ocean basin, depending on how fast the plates in that location continue to split apart (this activity is similar to the long-ago separation of South America from Africa, and of North America from Europe, which created the Atlantic Ocean). But all these processes will take hundreds of thousands to millions of years to occur.



# LIFE IN THE ANCIENT OCEANS

## When did **life** first **appear** on Earth?

Because the fossil record is incomplete, and the time scale immense, the origins of life on early Earth are highly debated. What scientists do know is that life first appeared in the early oceans, not on land; they estimate that this life began approximately 3.5 to 3.7 billion years ago (during the Precambrian). It's possible that life may have started even earlier—about 4 billion years ago—but scientist currently lack the evidence to support this idea.

## Why is our **knowledge** of **early life** so **limited**?

Although oceans preserve and fossilize organic remains quite efficiently, scientists still lack knowledge of early organisms because of the very nature of the ancient organisms themselves and because of the nature of the Earth. There are three main reasons for the lack of knowledge: First, it is difficult to find 4-billion-year-old rock that has not been changed by heat, pressure, or erosion. Second, because the single-celled organisms (early life forms) are so small, they are difficult to find in rock. And last, because the organisms consisted of soft parts only, they probably decayed after death—leaving no (fossilized) trace of their existence. Without a fossil record, many of the organisms that were present in the ancient oceans will never be known to us.

## Did **life** on Earth **develop continuously**?

Probably not; which is to say, there were "false starts." According to some scientists, life on Earth had to start many times. They theorize that once life began—either around a volcanic ocean vent or in a shallow pond or sea—comets and asteroids bombarded the planet, stamping out the beginnings of life. This may have happened many times over the course of the early Earth—perhaps for millions of years. Eventually, life was able to survive, adapt, and evolve.

## What was the **probable first step** toward **life** on Earth?

The first step toward life on Earth was probably the formation of complex organic molecules (amino acids) from simpler organic molecules. There were many possible origins of these complex organic molecules, ranging from lightning strikes in the primitive atmosphere binding molecules together, to organisms from space—introduced by bombardment of organic-rich comets. Whatever their origin, many scientists believe organic molecules were common and stable on Earth at least 4 billion years ago, due to the absence of large amounts of free oxygen (also called a slightly reducing atmosphere). At some point, these complex organic molecules developed the ability to self-replicate; they also began to synthesize proteins and eventually became compartmentalized, leading to the first cells.

## What was the **possible progression of life** on Earth?

The progression of life on early Earth is hotly debated in the scientific community. What follows is one possible scenario. But it's important to note that the dates are approximate—and even disputed.

*3.7 to 3.5 billion years ago:* The first primitive cells evolve.

*3.5 to 3.2 billion years ago:* Primitive bacteria may have evolved around ancient volcanic hot springs.

*3.45 to 3.55 billion years ago:* Photosynthetic bacteria (cyanobacteria) evolve.

*2.2 to 2.1 billion years ago:* The Earth's atmosphere becomes sufficiently oxygen-rich to support an ozone layer; this layer protects early evolving organisms from the Sun's harmful rays.

*2.1 billion years ago:* The first cells begin to form in greater numbers and develop into simple single-celled (unicellular) organisms.

*1.5 billion years ago:* The first eukaryotic cells form: These organisms have a nucleus, complex internal structures, and are the precursors to protozoa, algae, and all multicellular life.



## What were the **first single-celled organisms** in the ancient **oceans**?

The first single-celled organisms in the ancient oceans were most probably anaerobic heterotrophic bacteria: They did not require free oxygen (anaerobic) and they obtained their food from external sources (heterotrophic). The next type of cells to evolve were probably the anaerobic autotrophic bacteria: They did not require free oxygen (again anaerobic) and they made their own food using the energy from the surrounding environment (autotrophic), such as deep-sea vents. Another term for these organisms is chemoautotrophic, because they made use of chemical energy. But, as chemical energy waned, photoautotrophs developed, which made use of energy from light.

## How were **marine organisms key** to the development of **life** on Earth?

Approximately 3.45 to 3.55 billion years ago (during the Precambrian eon), a certain microscopic organism evolved that would eventually change the world. Today, its living descendants are known as "blue-greens," or cyanobacteria—microorganisms that secrete lime and form stony cushions called stromatolites. Some of the oldest known fossil stromatolites are found at a place called North Pole, Australia; living ones still exist in such places as Shark Bay, Australia.

The blue-greens were photoautotrophs: They used photosynthesis to produce their own food by obtaining energy from the Sun's rays and hydrogen from water; as they removed the hydrogen, they released oxygen. Over time, cyanobacteria released large amounts of oxygen into the atmosphere, most of which became "locked up" by iron on land and in the oceans. Eventually, the amount of oxygen produced was enough to exceed the capture by iron—and by approximately 2.2 billion years ago, the early atmosphere became filled with oxygen.

Stromatolites are stony cushions formed by lime secreted by cyanobacteria. Fossilized stromatolites have provided scientists with important information about the development of life on Earth. *NOAA/OAR National Undersea Research Program*

The presence of oxygen in the atmosphere spurred the evolution of organisms that used this gas; the first organisms to develop aerobic (oxygen-consuming) respiration soon appeared. This aerobic process was more efficient than anaerobic (oxygen-free) respiration—allowing the development of larger cells and multicellular organisms.

## When did the **first soft-bodied animals** appear in the **oceans**?

Fossils show that the first soft-bodied marine animals appeared about 600 million years ago (during the Precambrian). They included a form of jellyfish, sea pens (anthozoans), and segmented worms.

## When did **larger early marine animals** evolve?

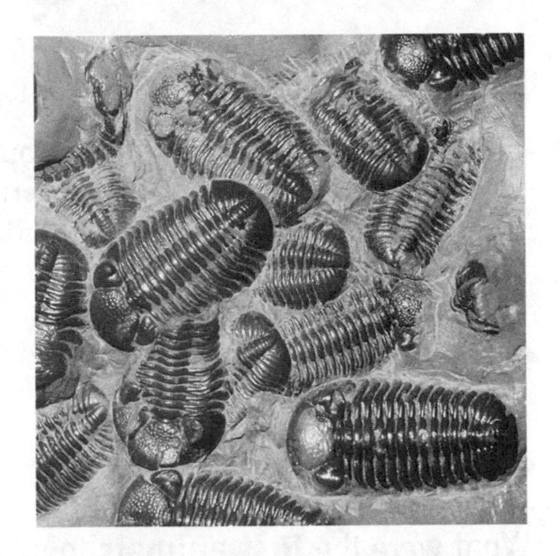

These fossils of trilobites, the first complex animals, are from the mid-Devonian period, or more than 360 million years old. *CORBIS/Kevin Schafer*

Some of the larger early marine animals evolved just after the end of the Precambrian eon, about 544 million years ago during the Cambrian period. At this time—nicknamed the "Cambrian Explosion" or "evolutionary big bang"—a great burst of evolutionary activity began in the world's oceans. New animals appeared rapidly (geologically speaking), filling the oceans with life. Some of these animals included trilobites (the first complex animals), sea pens (anthozoans), jellyfish, and worms.

No one really knows why the animals crowded the seas. One suggestion is that there was a radical change in the climate, allowing the animals to proliferate. Another theory suggests that an overall natural threshold was reached (for example, changes in temperature or oxygen levels), which provided an environment that was conducive to the proliferation of organisms. Whatever the reason, many of the sea creatures evolved from soft-bodied to having hard parts of some kind, including shells and internal and external skeletons. These changes allowed the sea creatures to better survive—and eventually leave evidence of their existence in fossils.

## Why are the **Burgess shale marine fossils** so important?

One of the most significant records of life in the ancient oceans is found at the Burgess shale fossil site. Discovered in 1909 at Burgess Pass in British Columbia, Canada, the shale layer was formed from compressed muds laid down more than 500 million years ago. These muds are thought to have accumulated at the edge of an ancient ocean—and they preserve a variety of organisms that lived during the Cambrian period.

The muds and the nature of the location perfectly preserved the remains—making the Burgess fossils among the best in the world.

Many of the fossilized forms found in the Burgess shale can be recognized as ancestors of our modern-day marine animals, such as jellyfish and starfish. In addition, sea urchins, mollusks, echinoderms, ancient sponges, worms—and even arthropods (of which the trilobites are a member) were discovered there. To date, there have been more than 120 species of invertebrate animals (without a backbone) described from this one location.

## What were the **first animals** to move **from the oceans to the land**?

The first animals to conquer the land may have been arthropods, such as scorpions and spiders. These creatures have been found in Silurian period rock layers, dating from 410 to 440 million years ago.

## When did the **first plants move** from the **oceans to land**?

Fossils show that the first true land plants appeared about 420 million years ago (during the Silurian period), and included flowerless mosses, horsetails, and ferns. They reproduced by throwing out spores—minute organisms carrying the genetic blueprint for the plant.

## Did any **dinosaurs** live in the **oceans**?

No, there were never any dinosaurs in the oceans; the true dinosaurs all lived on land. There were, however, other reptiles that once lived on land—but returned to the ocean to live and feed; these included the ichthyosaur and the plesiosaur. The ichthyosaur was small and had a dolphin-shaped body; the plesiosaur was large (nearly 50 feet, or 15 meters, long), with four paddle-shaped limbs and a long neck for catching fish. These reptiles became extinct approximately 65 million years ago, around the same time as the dinosaurs.

**A form of jellyfish was among the first soft-bodied marine animals to appear in the oceans—about 600 million years ago.** *NOAA/OAR National Undersea Research Program; M. Youngbluth*

## What were some of the **first marine mammals**?

Contrary to popular belief, marine mammals (such as whales and dolphins) did not evolve until a very short time ago—geologically speaking. Mammals evolved on land during the late Triassic period, about 208 million years ago; and they did not proliferate until after the end of the Cretaceous period 65 million years ago. It took even longer for mammals on land to move into the oceans—with some marine mammals evolving during the Miocene epoch about 23 million years ago.

## Where are **prehistoric animals** being discovered below the **ocean surface**?

Not all the animal fossils found in the ocean sediments are from millions of years ago. Recently, some 60 feet (18 meters) below the ocean's surface off the coast of Georgia, scientists uncovered the remnants of prehistoric—about 14,000-year-old—animal life. During this time, the coastline extended out 60 miles (97 kilometers) beyond the present

Georgia shoreline, and the land was populated by enormous animals, such as bison, camels, and mastodons. The area is now completely underwater, and scientists working with Gray's Reef National Marine Sanctuary are uncovering fossils—not only of animals, but of the surrounding plants. So far, they have uncovered mastodon and bison bones, the tooth of a Pleistocene horse, and a marine worm burrow cast dated at about 18,000 years old. Plant fossils, such as pine pollen, and alder and grass seeds have been found—and may give scientists information on the ancient shorelines and possible changes in sea level.

## What is a **living fossil**?

A living fossil is an animal or plant species nearly identical to organisms that lived millions of years ago. Many of these living fossils were first discovered as actual fossils before they were found living on our modern Earth. For example, one species of modern ginkgo survives from the Triassic period about 220 million years ago; the magnolia, one of the earliest true flowering plants, existed during the Cretaceous period, about 125 million years ago. Living fossils of animals include the tuatara, the only living survivor from a reptile group that was abundant during the Triassic period; and the coelacanth, a living fossil fish that was only recently discovered.

Another living fossil is the modern brachiopod *Lingula,* which is barely distinguishable from its Devonian period ancestor. One reason for the longevity of this species, or any other living fossil species, may be due to its ability to adapt and live in stable ecological niches. The *Lingula* live in the intertidal zone, a specialized niche along coastal areas. Even if the sea levels changed, these brachiopods could adapt by changing location to follow the water levels.

## Why is the **coelacanth** famous?

The coelacanth (pronounced SEE-la-kanth) was a primitive fish that first appeared during the Devonian period of the Paleozoic era, almost 400 million years ago. Scientists believe this species of fish was the first to haul itself out of ancient oceans and onto land—and were probably the ancestors of many land animals. Numerous species of coelacanth

**The coelacanth was thought to be extinct until one was caught by a fisherman off the coast of Madagascar in 1938. Others have been found since. Scientists believe this fish lives in the waters deep below the surface, which explains why they are rarely seen.**

have been preserved in the fossil record, but all of them were thought to have gone extinct at the end of the Cretaceous period, about 65 million years ago.

However, in 1938, a fishing boat hauled up a living coelacanth from the waters off South Africa, with other specimens discovered over the years near the Comoros Islands near Madagascar. And in the late 1990s, another specimen was discovered off Indonesia.

The coelacanth is considered to be a living fossil by scientists and of great scientific interest. Fortunately, the modern species does not try to come onto land, where it would be vulnerable to predators and collectors. Instead, today's coelacanth appears to prefer a marine environment; the specimens from Madagascar and Indonesia appear to inhabit similar environments consisting of caves approximately 600 feet (183 meters) below the water's surface, situated along the steep sides of underwater volcanoes.

# UNDERSTANDING THE OCEAN

## THE INTERACTION OF OCEAN AND AIR

### What are the features of **modern oceans**?

Modern oceans are similar to ancient oceans: They have strong currents, which are created by temperature differences (as the Sun does not heat surface waters uniformly) and by the mixing of equatorial (warm) and polar (cold) ocean waters; the oceans interact with the atmosphere; and, since about 600 million years ago, they have teemed with marine life—both fauna (animal life) and flora (vegetation). However, there are also differences between the oceans today and those of ancient times: The most significant difference is the diverse and plentiful marine life in the modern ocean. Another difference can be attributed to the Earth itself: Since our planet is tectonically active (its crustal plates are moving), the shapes and sizes of the oceans have changed over time. New oceans have been created, while others have diminished. It has only been within the last 50 million years that the continents and oceans shifted to their present positions.

### How does the **atmosphere interact** with the **oceans**?

The atmosphere and the oceans could not exist without each other and they interact in a variety of ways to create currents, waves, climate, and weather. In simple terms, a region's climate is the result of the intense radiation from the Sun heating the oceans and atmosphere, creating oceanic and atmospheric circulations—patterns in the flow of water and air. Our weather systems are created by the localized rises and falls in the air, also governed by the oceans and atmosphere (warm air rises, while cool air falls).

## How much of the **Sun's radiation** is **absorbed by the Earth's oceans**?

Over half of the Sun's radiation that reaches the Earth's surface is absorbed by the oceans. Some of this heat causes the evaporation of surface waters; the rest is stored in the ocean's surface layer or is moved downward into deeper waters by the dynamics of currents. The evaporated water ends up in the Earth's atmosphere mainly in the form of water vapor. Dissolved salts from the seawater also enter the atmosphere when the oceans are turbulent—for example, as sea spray during a storm. Each year an estimated 1 billion tons of sea salt particles enter the atmosphere from the ocean—and probably about 90 percent of these salts are eventually carried back to the oceans.

## What is the **ocean's role** in the **water cycle**?

Evaporating water from the oceans and other waterways provides the water cycle (or hydrologic cycle) with vapor. The water cycle begins with the evaporation of water from the surface of oceans (and other waterways) and with evapotranspiration—the loss of water from soil by evaporation and by transpiration (through the membranes and pores) of vegetation. The evaporated water (now as water vapor) rises, reaching the upper atmosphere where it cools and condenses to create clouds. The clouds then produce precipitation. Some of the precipitation (in the form of rain, snow, sleet, etc.) falls into the ocean and adds to the seawater. Precipitation also falls on land, of course, where, if it is in the form of snow or ice, it can be stored on the ground as long as the weather conditions prevail to keep it frozen (but as soon as it melts, it becomes groundwater or runoff). Precipitation that seeps into the ground becomes part of the flow of groundwater, eventually reentering the atmosphere through evapotranspiration. Or, precipitation may remain on the surface (for example, when soil is already saturated or in highly developed areas), eventually running off into lakes, rivers, and streams.

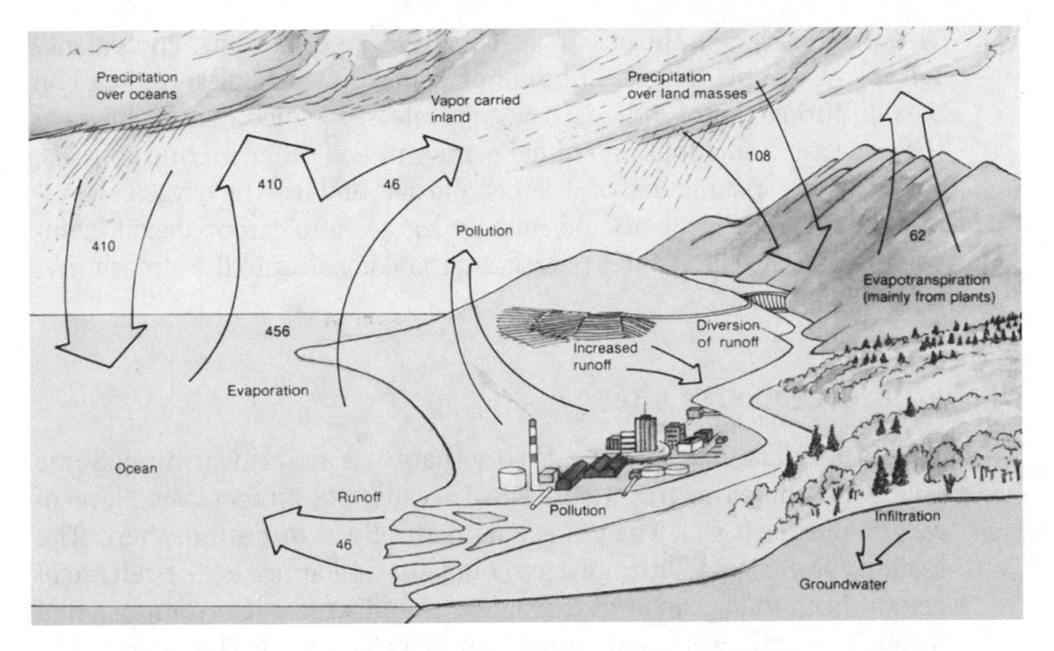


The circulation of water on Earth is shown in this diagram of the hydrologic cycle. *McGraw-Hill*

Eventually, all runoff returns to the oceans where it subsequently evaporates—and the cycle begins again.

## What is the **ocean's role** in the **carbon cycle**?

The Earth's biosphere (the part of the world where life can exist) contains many carbon compounds—including natural gas, limestone, and the shells of certain marine organisms. Carbon is an essential part of the life cycle of all organisms—it is part of the processes of formation, transformation, and decomposition of fauna (animal life) and flora (vegetation).

The Earth's carbon cycle is complex. In very simple terms, carbon dioxide ($CO_2$) originates in volcanoes and carbonate rocks; it is then transmitted to living matter; and eventually, via metabolic processes in organisms, it is regenerated back to carbon dioxide.

The ocean's carbon cycle is somewhat self-contained: Seawater dissolves large quantities of existing carbon dioxide from the atmosphere. The gas also is released in certain processes in or around the oceans, such as carbon dioxide from volcanic eruptions or the dissolution of carbonate rocks.

The major players in the ocean's carbon cycle are organisms: Photoplankton (plant-like plankton that use photosynthesis) "fix" dissolved carbon dioxide during photosynthesis; oxygen is released, which then dissolves in the seawater. Zooplankton (animal plankton) and other marine animals, such as fish, consume the fixed carbon dioxide, and use the oxygen for respiration. And finally, plants and animals degrade into carbon dioxide compounds after they die—thus, releasing carbon dioxide into the atmosphere.

## What is the **greenhouse effect**?

The term refers to a scientific theory that the Earth is warming. Some scientists believe the greenhouse effect is caused by an increasing level of carbon dioxide ($CO_2$). The gas occurs naturally in our atmosphere: The carbon cycle begins with the decay of plants and animals, the release of carbon from the oceans (for example, through volcanic eruptions), and through animal respiration (breathing). But $CO_2$ can also be produced by humans—through the burning of wood and fossil fuels (such as petroleum). Some scientists are concerned that this human production of $CO_2$ could throw off the balance of the gas in the Earth's atmosphere, resulting in an increase in carbon dioxide that would cause the atmosphere to trap heat that would otherwise be radiated back from the Earth—in the same way that the glass of a greenhouse retains heat from the Sun's radiation. Thus, the atmosphere's temperature would increase.

If the atmosphere's temperature were to increase, the seawater's temperatures would also increase—which would cause the water to expand, possibly raising sea levels around the world. In addition, the surface layer of the oceans would not be able to assimilate the increase in carbon dioxide, causing the oceans to become under-saturated with calcium carbonate. In the surface ocean waters, and especially the shallow seas, this would cause a major disruption of the carbon cycle, not to mention the circulation of carbon in general—and could have an enormous effect on world climate, marine life, and life in general.

## What are **atmospheric cells** in the Earth's **atmosphere**?

Atmospheric cells are the major localized circulation features in the Earth's atmosphere—some of which were known even as far back as the eighteenth century. The cells are caused by the interaction of the air and oceans as each is heated by the Sun. The following list describes the Earth's three major atmospheric cells.

*Hadley cells:* As the rays of the Sun strike the Earth's surface (land and ocean) at the equator, humid air rises, creating a low-pressure system. As it rises, it reaches into the troposphere (the lowest layer of the atmosphere) and "splits," spreading north in the Northern Hemisphere and south in the Southern Hemisphere. The air then begins to cool as it heads toward the respective poles. At about 30 degrees north and south latitude, the cooled air sinks toward the surface and heads back toward the equator. This creates the "circulation" of air called the Hadley cells, named after George Hadley (1685–1768), the English physicist and meteorologist who first described them in 1753.

The Hadley cell has a major effect on the equatorial region: As the heated air near the equator rises, it produces thick clouds and heavy rainfall—producing a region of tropical cyclones and rainforests. As the air sinks at 30 degrees latitude, it encounters greater pressures and heat—essentially wringing the moisture out of the clouds, creating a band of dry, sinking air. It is this arid air that maintains the dry conditions found on the Atacama, Australian, Sahara, and Kalahari deserts around these latitudes.

*Ferrel cells:* The Ferrel cells form because not all the air in the Hadley cells flow back toward the equator—some air continues toward the poles along the Earth's surface. As the air reaches about 60 degrees north and south latitude, it runs into the cooler, polar air in each hemisphere, causing the air to rise, some of which heads back toward 30 degrees latitude. These mid-latitude global wind cells are called the Ferrel cells, named after William Ferrel (1817–91), an American meteorologist who first described them in 1856.

*Polar cells:* Finally, some of the rising air at 60 degrees north and south latitude continues to flow toward the respective poles. As it cools, it sinks to the surface—creating Polar circulation cells (sometimes called Hadley polar cells).

## Why are **winds deflected** in specific directions in the Northern and Southern Hemispheres?

The primary (north and south) winds generated by the atmospheric circulation cells are deflected, or flow, in specific directions in each hemisphere: The winds curve to the right (or to the east) in the Northern Hemisphere and to the left (west) in the Southern Hemisphere. This is because of the Coriolis effect. The phenomenon was discovered by French physicist Gaspard-Gustave Coriolis (1792–1843) in 1835; the first person to explain the resulting deflection of the global winds was American meteorologist William Ferrel (1817–91). The Coriolis effect (also called the Coriolis force, although it is not really a force) is the apparent movement of objects when observed from a rotating system. In this case, because the Earth is rotating to the east under the air of the atmosphere, it appears to us that the winds are deflected.

## What are the **major wind bands** in the Earth's atmosphere?

Because of the Coriolis effect, scientists divide the winds in the major circulation cells into six major bands across the planet. From the pole to the equator the winds are labeled as follows (there is one of each type in the Northern and Southern Hemispheres).

*Polar easterlies:* The polar easterlies flow from the east, out of the polar regions; the border between these winds and the westerlies is called the polar front.

*Westerlies:* The westerlies (also called the "roaring forties" and "furious fifties" because of the latitudes where they prevail) flow out of the west. They interact with the cold polar easterlies and the warmer air from the trade winds—a combination that often produces severe storms.

*Trade Winds:* The trade winds (which flow out of the northeast in the Northern Hemisphere and out of the southeast in the Southern Hemisphere) were discovered by English chemist and physicist John Frederic Daniell (1790–1845) in 1823. They are very persistent winds, deviating little from their compass direction. Long ago, the merchant marine put these winds to good use, which lent them their name.

## Do the Earth's features affect the winds?

Even though there are major circulation patterns and wind bands, the planet's features do influence global winds: Mountains, continents, and ocean currents influence surface wind patterns—and produce distinctive climate zones.

## Are there **areas** between the major **wind bands**?

Yes, there are converging areas between the major wind bands (or atmospheric circulation cells). One of these areas, between the Northern and Southern Hemispheres' trade winds, is called the intertropical convergence zone (also called ITCZ). This narrow band of hot and humid air near the equator is punctuated with frequent thunderstorms and squally winds. But at certain times of the year, the trade winds do not come together into a convergence—creating what is called the doldrums. Early mariners labeled this condition the doldrums after sailing ships became caught in its calmness. Sailors could drift aimlessly for days—or even weeks—without any winds.

The bands of the doldrums, ITCZ, and trade winds all change position, swinging across the equator twice a year—accompanied by seasonal changes. For example, during the year, the ITCZ migrates north and south only a few degrees of latitude over the Pacific and Atlantic oceans; in contrast, it changes as much as 20 to 30 degrees of latitude over South America, Africa, and a large region of Southeast Asia and the Indian Ocean.

At higher latitudes, between the trade winds and westerlies, there is a belt called the subtropical high pressure zone, or the horse latitudes. And higher still, between the westerlies and the polar easterlies, there is also a converging area, in which mid-latitude cyclones or frontal depressions occur; this "boundary" is caused by the fierce battle between these two opposing winds.

## How is the **strength of the wind measured** on the oceans?

The strength of the wind on the ocean is determined by a standardized, descriptive scale that was invented in 1806 by English naval office Sir Francis Beaufort (1774–1857). The following scale was originally used to determine the effect of wind on what was called a "man o' war" ship in full sail; it is still used today and is known as the Beaufort scale.

**Beaufort Scale**

| Force | Name | Kilometers per Hour | Miles per Hour | Description |
|---|---|---|---|---|
| 0 | calm | 0–2 | 0–1 | flat, slack sails |
| 1 | light airs | 2–6 | 1–3 | wavelets |
| 2 | light breeze | 7–11 | 4–7 | small waves, sails filling |
| 3 | gentle breeze | 12–18 | 8–12 | distinct waves, filled sails |
| 4 | moderate | 19–30 | 13–18 | wave caps, breeze bending branches on land |
| 5 | fresh breeze | 31–39 | 19–24 | some spray |
| 6 | strong breeze | 40–50 | 25–31 | cresting waves, holding an umbrella is difficult |
| 7 | stiff breeze | 51–62 | 32–38 | high sea, walking is difficult |
| 8 | fresh gale | 63–75 | 39–46 | difficult to stand unaided |
| 9 | strong gale | 76–88 | 47–54 | rolling sea |
| 10 | storm | 89–100 | 55–63 | violent seas, high waves |
| 11 | hurricane-like | 101–117 | 64–72 | blown-foam-covered sea, storm, high sea, impeded visibility |
| 12 | cyclonic | 118+ | 74+ | hurricane or typhoon storm |

## How is **speed measured** at sea?

Nautical speed for ships (it is also used for airplanes) is measured in knots. One knot is equal to one nautical mile per hour; for example, a ship traveling at a 30-knot speed travels 30 nautical miles an hour. Navigators use the nautical mile because of its simple relationship to the degrees and minutes when measuring latitude and longitude. The international nautical mile equals one-sixtieth of one degree, or a minute of arc, of the Earth's circumference. Thus, the international nautical mile equals 1.151 statute miles (1.852 kilometers).

The work "knot" originated long ago when sailing ships carried a speed-measuring device called a log chip and line: A log attached to a rope was heaved overboard—attached to a line that contained knots spaced at intervals of 47 feet, 3 inches (14.4 meters). As the log pulled on the rope and the rope was continually let out, the speed was calculated as the number of knots, counted in a standardized time interval of 28 seconds. (An interval of 28 seconds is to one hour approximately what a distance of 47 feet 3 inches is to 6,077 feet or 1.151 statute miles.) Thus, if the log pulled out 5 knots of line in 28 seconds, the sailing ship was moving at 5 knots or 5 nautical miles an hour.



# WEATHER ON THE OCEANS

## How do **clouds** form over the **oceans**?

Clouds are visible manifestations of collections of water droplets or ice crystals suspended in the air. These droplets are not like the rain we see falling during a thunderstorm on a hot summer's day—they are more than a hundred times smaller. Clouds develop over the oceans or land in the same way: Warmer, humid air, usually from the surface of the oceans or ground heated by the Sun—rises. As it rises into the atmosphere, it is cooled to its dew point (the temperature at which air becomes saturated and can no longer hold any more water vapor); condensation occurs in the form of tiny water droplets, which are suspended in the atmosphere by natural updrafts—what we view as a cloud.

Clouds forming over the oceans—especially in the more hot, humid areas around the equator—have a better chance of growing than do clouds over land, as the oceans continually supply the atmosphere with moisture. This is one of the reasons that tropical storm systems forming around equatorial-warmed waters often gain tremendous strength and size.

## What is a **tropical disturbance**?

A tropical disturbance is an area of low pressure along the tropics; the tropics are the lines of latitude at 23.5 degrees north (called the Tropic of Cancer) and 23.5 degrees south (called the Tropic of Capricorn) of the equator. These are the lines of latitude where the sun is directly overhead during the summer solstice in each hemisphere. Tropical disturbances occur when huge clouds in these regions produce large, clustering thunderstorms. These disturbances can move through the tropics for at least 24 hours, but the winds never create any spinning movements within the clouds. One of the more well-known sources of these disturbances is an area off the west coast of Africa: The disturbances that form here head toward the Caribbean Ocean during the Northern Hemisphere's summer months, when the ocean surface temperatures are warmer. They usually settle down and dissipate—but sometimes, they can be the precursors to tropical depressions.

## What are **tropical depressions** and **tropical storms**?

Tropical depressions form when a tropical disturbance strengthens over warm water—of at least 80° F (27° C); a tropical depression has sustained surface winds of 38 miles (61 kilometers) per hour or less. As the warm air rises, it condenses and releases energy that fuels the storm system. If these storms form at least 5 degrees north or south of the equator, the storms start to spin—counterclockwise in the Northern Hemisphere and clockwise in the Southern Hemisphere.

If the sustained surface winds (taken as a one minute average) increase to between 39 and 74 miles (63 to 119 kilometers) per hour, and the storm continues to rotate, a tropical storm forms.

## What is a **tropical cyclone**?

Tropical cyclones form as tropical storms continue to gather strength from the ocean's warm water. These large-scale circular windstorms travel in the tropics and subtropics, with a well-defined low pressure area called an eye; they have sustained winds (taken as a one minute average) of at least 74 miles (119 kilometers) per hour.

**Satellite image of well-formed tropical cyclones Ione and Kristen in the Pacific Ocean, August 24, 1974.** ***NOAA***

## What are the various **names** for **tropical cyclones** around the world?

Tropical cyclones around the world have a variety of names. For example, in the Indian Ocean, they are called cyclones; in Southeast Asia and the China Sea area, typhoons; off the coast of Australia, willy-willys; and in the Atlantic, Gulf of Mexico, Caribbean, and eastern Pacific, hurricanes.

## How far **north or south** do **tropical storm systems** travel?

Tropical storm systems usually do not travel beyond 30 degrees north or south of the equator; any farther north or south of 30 degrees, there is not enough warm ocean water to either fuel or sustain the storm. In addition, these storms cannot form right at the equator, as there is no real spin of the winds here—and in fact, if such a storm crosses the equator, it would break apart. Of course, there are exceptions to the "warm waters only" rule. For example, tropical storms sometimes develop over the cooler ocean waters of the South Atlantic or southeastern Pacific Oceans.

## What is a **hurricane**?

A hurricane is a tropical cyclone—a system that can become intense and long-lived if the conditions are right—and forms in the Atlantic Ocean, Gulf of Mexico, Caribbean Ocean, and east Pacific Ocean. Hurricanes have sustained winds (taken as a one minute average) of more than 74 miles (119 kilometers) per hour in any part of its weather system; the larger storms can have sustained winds of more than 150 miles (241 kilometers) per hour, with gusts up to 200 miles (322 kilometers) per hour. The storms have a clear, calm area at the center known as the eye, with bands that wrap around an eye extending up to 300 miles (483 kilometers) or more from the center. At the center, the air pressure is very low—the lower it drops, the stronger the winds around it and the stronger the storm.

Hurricane damage is caused not only by extremely high winds, but by torrential rains and storm surges (rises in sea level). Heavy rains can cause extensive flooding in coastal areas, and if a storm lingers, inland areas can also be flooded. If a storm lasts for several days and moves over great distances, the potential for destruction and death is immense. Storm surges can accompany a hurricane; these "bulges" in the water are created by winds and pressures from the system, which push ocean water before the advancing storm. The resulting "dome" of water (which can be as high as 25 feet, or 7.5 meters, above sea level) can be topped by wind-whipped waves. If a storm surge occurs during high tide, the damage to a shoreline can be even worse. Over time, hurricanes can also change the coastline by erosion of the beaches—not only during the storm, but from the accumulated effects of storms that occur close together or by many storms over time.

## What was the **lowest pressure reading** ever recorded for an Atlantic **hurricane**?

So far, the lowest pressure reading for an Atlantic hurricane was 888 millibars, measured during Hurricane Gilbert in 1988. To compare, the usual reading for an average day at sea level is about 1013 millibars. (A millibar is one one-thousandth of a bar, a unit of pressure.)

**Enhanced infrared imagery of Hurricane Hugo, centered off Puerto Rico. The category-4 storm later slammed into South Carolina, where the governor ordered a mandatory evacuation of coastal areas.** ***NOAA***

## How is the **damage potential** of **hurricanes measured**?

Major hurricanes—of which there are approximately a dozen a year in the Atlantic Ocean—have the potential to destroy large areas of coastlines. For example, in October 1998, for a short period of time while over the Caribbean Ocean, Hurricane Mitch was classified as a category-5 hurricane, the most destructive known. Although Mitch struck Central America as a category-2 hurricane, its intensity left its scar on the land, as it dropped several feet of torrential rains on the region—causing more than $1.5 billion dollars in damage and killing more than 10,000 people.

Because of the danger to humans, a scale was developed to keep track of the intensity of hurricanes moving across ocean waters or land. The scale was proposed in 1971 by engineer Herbert Saffir and Dr. Robert Simpson, then the director of the National Hurricane Center. This scale includes wind speeds, estimates of barometric pressure at the hurricane's center, and height of a possible storm surge.

**Saffir-Simpson Hurricane Damage Potential Scale**

| Category | Central Pressure millibars (inches) | Winds per Hour miles (kilometers) | Storm Surge feet (meters) |
|---|---|---|---|
| 1 (minimal) | more than 980 (more than 28.94) | 74–95 (119–153) | less than 6 (less than1.8) |
| 2 (moderate) | 965–979 (28.50–28.91) | 96–110 (154–177) | 6–8 (1.8–2.5) |
| 3 (extensive) | 945–964 (27.91–28.47) | 111–130 (178–209) | 9–12 (2.6–3.7) |
| 4 (extreme) | 920–944 (27.17–27.88) | 131–155 (210–249) | 13–18 (3.8–5.5) |
| 5 (catastrophic) | less than 920 (less than 27.17) | greater than 155 (greater than 249) | greater than 18 (greater than 5.5) |

## What do **hurricane categories** mean in terms of **damage**?

Each hurricane category on the Saffir-Simpson Damage Potential Scale implies a certain amount of damage. The following list describes the kind of damage each level of storm is capable of producing.

Category 1 (minimal): The damage is primarily restricted to shrubbery, trees, and unanchored mobile homes. Other structures probably do not experience much damage. Signs (particularly those that are poorly constructed) may be damaged. Water inundates low-lying roads. Along the coast, there is minor damage to piers; smaller craft in exposed areas are torn from their moorings.

Category 2 (moderate): The damage includes considerable injury to shrubbery and tree foliage, with some trees blown down. There is also major damage to exposed mobile homes, some wreckage to doors, windows, and roofing materials (but not major destruction to buildings), and extensive damage to poorly constructed signs. Rising water about 2 to 4 hours before the hurricane makes landfall causes coastal roads and low-lying escape routes inland to be cut off. Piers are considerably damaged, and marinas flooded; small craft in protected areas are torn from their moorings. In preparation for a category-2 storm, evacuation of some shoreline residences and low-lying areas is required.

Category 3 (extensive): A category-3 hurricane causes foliage to be torn from trees, and larger trees to be blown down. There is some damage to roofing, windows, and doors, some structural devastation

**The aftermath of the Galveston hurricane of September 1900, which devastated coastal Texas. The tropical storm claimed more than 6,000 lives, making it the deadliest hurricane to ever hit the United States.** ***NOAA***

to smaller buildings, and mobile homes are destroyed. There is also serious flooding along the coastline, with many small structures near the coast destroyed; larger coastal structures are damaged by the battering of the waves and by floating debris. Low-lying escape routes inland are cut off by rising waters about 3 to 5 hours before the hurricane makes landfall, and flat terrain that is less than 5 feet (1.5 meters) above sea level is flooded up to 8 miles (13 kilometers) inland. The evacuation of low-lying residences within several blocks of shoreline may be required.

Category 4 (extreme): Shrubs, trees, and all signs are blown down. There is also extensive damage to roofs, windows, and doors, with complete failure of roofs on many smaller residences. Mobile homes are demolished. Any flat terrain less than 10 feet (3 meters) above sea level is flooded inland as far as 6 miles (9.7 kilometers). There is major erosion of the beaches; flooding and battering by waves and floating debris cause major damage to the lower floors of structures near the shore. Low-lying escape routes inland are cut off by rising water about 3 to 5 hours before the hurricane makes landfall. In this case, massive evacuation of all residences within 1,500 feet (457 meters) of the shore may be required, as well as the evacuation of single-story residences in low ground within 2 miles (3.2 kilometers) of the shore.

Category 5 (catastrophic): Trees, shrubs, and all signs are blown down. There is considerable damage to roofs of buildings, with very severe and extensive damage to windows and doors; there is also damage to many residential and industrial buildings, with extensive shattering of glass in the windows and doors. Complete buildings are often destroyed, and smaller buildings are either overturned or blown away, with mobile homes demolished. Damage to lower floors is major, especially for those buildings that are situated less than 15 feet (4.6 meters) above sea level and within 1,500 feet (457 meters) of the shore. Low-lying escape routes inland would be cut off by the rising waters 3 to 5 hours before the hurricane makes landfall, and the beaches would be severely eroded. In this case, massive evacuation of low-lying residential areas within 5 to 10 miles (8 to 16 kilometers) from shore may be required.

## How are **hurricanes named**?

People have named hurricanes for centuries. For example, hurricanes passing through the Caribbean Ocean were often named after the saint's day on which they occurred.

Before 1950, tropical storms were given a number based on when they occurred. The first storm of the season would be called "hurricane number 1"; the second, "hurricane number 2," and so on. For a short time, the military phonetic alphabet was also used as a naming scheme, so that hurricanes had names such as Able, Baker, and Charlie.

By 1953, all tropical storms were given women's names, with each successive hurricane assigned a name that represented the next letter in the alphabet. By 1978, both men's and women's names were used in the eastern North Pacific tropical storm lists; in the Atlantic basin, by 1979, the list expanded to include both male and female names.

More recently, the member nations of the World Meteorological Organization have revised the list to include names common to English, Span-

**1992's Hurricane Andrew, a category-4 storm, devastated coastal areas. The storm's power is amply illustrated here—it was intense enough to toss these boats about like toys.** ***NOAA***

ish, and French-speaking peoples. The order of women's and men's names alternate every year. For example, in 1995, the list of names began with Allison (which was followed by a man's name, Barry); in 1996, the names began with Arthur (which was followed by a woman's name, Bertha). There are six lists of tropical storm names, each comprised of 21 names from the letters A to Z, excluding the letters Q, U, X, Y, and Z. The lists are used on a rotating basis—thus, storm names are recycled.

When a tropical disturbance becomes a tropical storm, the National Hurricane Center in Florida gives the storm its assigned name. But if the storm becomes a hurricane that causes many deaths or extremely heavy damage, the name is retired. For example, on the Atlantic basin list, Andrew, Bob, Camille, David, Elena, Frederic, and Hugo are all retired names.

## What do the various **weather watches** and **warnings** mean in coastal areas?

There are many watches and warnings issued by the weather services to help people understand the severity of a storm and take appropriate precautions. For example, a tropical storm watch is issued by the weather service when tropical storm conditions (winds from 39 to 73 miles per hour, or from 63 to 117 kilometers per hour) pose a possible threat to a specific coastal area within 36 hours; a tropical storm warning is issued when such conditions are expected in a specific coastal area within 24 hours.

In the case of a hurricane, a hurricane watch is issued when such a storm (or a developing hurricane) is a possible threat within 36 hours; a hurricane warning is issued if the same conditions are expected along a specific coastal area within 24 hours. In fact, a hurricane warning can remain in effect when the hurricane has caused dangerously high water, or a combination of high water and exceptionally high waves along the coast—even if the winds have calmed to below hurricane intensity.

## What is a **monsoon**?

A monsoon is a seasonal, prevailing, and large-scale wind system caused by differences in the temperature between the land and the oceans. Monsoons, from the Arabic word for season (*mausim*), were the original names given to winds near Arabia that blew for six months from the southwest (during summer), then six months from the northeast (during winter). These winds also blow over India, southern Asia, northern Australia, and parts of Africa, and, though not as extreme, over North and South America. As the land and oceans "switch places" (in terms of which is warmer than the other), strong winds develop as cooler air rushes into the warmer areas.

Monsoons can be wet and dry. The wet monsoons usually occur in the summer and are accompanied by heavy rains—the moisture-laden winds flowing from the cooler sea to the warmer land. As the warm surface of the land heats the air, the air rises and the cool, moist air from the ocean moves in to take its place—creating the rains.

Dry monsoons have little rain, but still plenty of winds. They usually occur in the winter, with the winds flowing from the land to the sea: As the land quickly cools, it chills the air above its surface; the warmer air above the sea rises, allowing the cool, dry air from the land to push in and take its place, creating a dry monsoon.

## What are **waterspouts**?

Vintage photograph shows the second of the so-called Great Waterspouts that occurred on August 19, 1896, in Vineyard Sound, off Martha's Vineyard, Massachusetts. These tall, whirling columns of air and water vapor can occur on the ocean as well as on lakes and rivers. *NOAA; Monthly Weather Review, July 1906*

Waterspouts are tall, whirling columns of air and water vapor that extend from a cloud to the surface of the ocean (they can also occur on a lake or river). Typically associated with normal rain showers and weak thunderstorms, a waterspout begins the opposite way from a tornado: A strong updraft of air is pulled upward; as the air rises, it rapidly rotates, producing an area of low pressure in its center. The moist air from the water rushes into the low pressure area, is rapidly cooled, condenses, and becomes visible. At the base of the column, seawater is sucked into the waterspout, rising only a few feet in the funnel; closer to the top, most of the water is fresh, formed by condensation.

Most waterspouts are more short-lived and less violent than are the majority of tornadoes. For people who happen to be in the ocean when a waterspout occurs nearby, the danger that is posed is from the winds that the waterspout generates; peak winds are typically between 50 and 100 miles per hour (80 and 160 kilometers per hour). Oceangoing vessels have also reported multiple waterspouts; and still other ships have reported a waterspout passing right over the vessel without causing any damage. Nevertheless, there have also been many reports of waterspouts blowing small vessels and people about, so caution is appropriate.

## What are **land and sea breezes**?

If you've ever been near a sizeable body of water, such as a large river or lake, or an ocean, you've probably experienced land and sea breezes. The interaction of the land and the water are powerful: Sea breezes flow from the cooler waters toward the warming land during the day; land breezes flow from the cooling land to the warmer waters at night. It is

A beach at high tide. *JLM Visuals*

all a matter of air pressure caused by the difference in temperature: Warmer air is lighter and so it rises; cooler air is heavier and it "rushes in" to replace the rising warm air, causing the winds to blow.

# TIDES

## What are **tides**?

Each day, the surface of the ocean—and of those bodies of water that are connected to the ocean (such as gulfs, bays, and deltas)—rises and falls. Most ocean shorelines experience high tide and low tide every day. High tide is often referred to as "when the tide comes in"; and low tide as "when the tide goes out."

## What **causes** the **tides**?

We would not know about tides—the natural up and down movement of the ocean surface—if we had no Moon. Tides are created by the gravita-

The same beach (as the one shown at left) at low tide. Note the measurable difference in the size and extent of the beach.
*JLM Visuals*

tional pull of our only natural satellite, the Moon, and to a lesser extent, our Sun. Because the Moon pulls on the Earth's oceans more than the Sun, the tides usually follow the Moon's cycle.

Tidal flows and cycles are also controlled by the friction of the water against the seafloor surface; the shape of the ocean basins; the presence of landmasses (the continents and islands); and sundry other complexities of ocean currents, winds, and interactions with the climate and weather.

## What is the **tide-raising power** of the **Moon**?

The tide-raising power of the Moon on the oceans is 2.2 times the tidal influence of the Sun. It is estimated that the attraction the Moon exerts on a molecule of water on Earth is six million times smaller than the attraction from gravity—and of course, the Sun has an even lesser effect than the Moon.

## What are the **spring and neap tides**?

The overall effect of the Moon and Sun on the Earth's tides changes over the course of a lunar month. The greatest effect on the Earth's tides results when the Moon and Sun reinforce each other's gravitational pull; spring tides are the result. These tides have nothing to do with spring—the term is from the German word, *springen,* "to jump." Spring tides occur when the Moon and Sun are in line with each other and are either on the same or opposite sides of the Earth. At such a time, the gravitational pull is the strongest and, therefore, the tidal range is the greatest. Neap tides occur when the Moon and Sun are at right angles to each other, with the pull the weakest and the tidal range the smallest.

## What is a **tidal range**?

A tidal range is the difference in the height of the ocean between high tide and the next low tide. The average tidal range around the world is about 6 to 10 feet (2 to 3 meters).

## What was the **greatest tidal range** ever recorded?

The greatest difference ever recorded between high and low tide was at Burntcoat Head, Nova Scotia, Canada. The tidal range measured 53.38 feet (16.27 meters) in the Bay of Fundy's Minas Basin.

## What areas experience the **greatest tidal ranges**?

The greatest tidal ranges occur in bays and estuaries. The following list cites places where the average spring tide is in excess of 16 feet (5 meters).

| Place | Average Tidal Range (feet / meters) |
|---|---|
| Burntcoat Head, Nova Scotia (Bay of Fundy), Canada | 47.5 / 14.5 |
| Rance Estuary, France | 44.3 / 13.5 |
| Anchorage, Alaska | 29.6 / 9.0 |
| Liverpool, England | 27.1 / 8.3 |
| St. John, New Brunswick, Canada | 23.6 / 7.2 |

| Place | Average Tidal Range (feet / meters) |
|---|---|
| Dover, England | 18.6 / 5.7 |
| Cherbourg, France | 18.0 / 5.5 |
| Antwerp, Belgium | 17.8 / 5.4 |
| Rangoon, Burma | 17.0 / 5.2 |
| Juneau, Alaska | 16.6 / 5.1 |
| Panama (Pacific side) | 16.4 / 5.0 |

## Where do **low tidal ranges** occur?

Some of the smallest tidal ranges occur in enclosed seas, such as the Mediterranean Sea, which is virtually tideless.

## How many **tides** occur **per day**?

The number of tides in a tidal cycle is not consistent all over the world. The tides at any given location can be diurnal, semidiurnal, "mixed," or unequal:

**diurnal:** A diurnal tidal cycle is one in which there is just one high tide and one low tide per day. The Caribbean Ocean experiences a diurnal tide.

**semidiurnal:** A semidiurnal tidal cycle is one in which there are two high and two low tides per day. Certain places in southwest Florida experience semidiurnal tides.

**mixed:** A "mixed" tidal cycle is more complex; mixed cycles occur in areas of transition—which are affected by both semidiurnal and diurnal tides. Mixed cycles occur in northwest Florida's Apalachicola Bay (in the Gulf of Mexico) and in certain locations in the Pacific and Indian oceans.

**unequal:** Unequal tides are a condition between the diurnal and semidiurnal tidal cycles; in an unequal cycle, the two high tides are not at the same height on any given day.

## How are **tides tracked**?

There are several ways to track tides. One way is through the use of tidal listings, which predict the time and height of a tide at a particular place. These listings are often used by shipping lines to determine the best times for shipping goods. Recreational sailors also use tidal listings to determine the flow of the currents along shorelines. The listings are based on years of careful records taken of past tidal performance in certain areas.

There are also tidal maps or charts. In one kind of chart, a cotidal chart, all points where high tides occur simultaneously are marked. Another type of cotidal chart connects points where the high tide comes up to the same height above sea level. On these charts, there is one point in a body of water in which all the radiating (or cotidal) lines meet; this is called the "no-tide" (amphidromic) point. There may be more than one such point in a larger body of water, such as the Caribbean Sea.

## What is a **land tide**?

A land tide is just what it sounds like: In most areas, as the surface of the oceans rise and fall, the land responds, too—twice daily. In fact, the continental landmasses can rise and fall as much as 6 inches (15 centimeters) when the Moon is directly overhead and exerts its greatest pull.

## What other effect do **tides** have on the **Earth**?

Tides have an additional effect on the Earth: They have a tendency to slow the planet's spin on its axis (rotation) by fractions of a second annually.

## Do **tides** have an effect on **trees**?

Yes, scientists have found that trees bloat, and then shrink, with the rhythm of the tides. Measurement of the stems of young spruce trees show that their diameters change by several hundredths of a meter during a roughly 25-hour cycle. This cycle has two peaks, one higher than the other—similar to tidal cycles. These patterns may also explain the old folklore of cutting trees before a new moon to get the wood to dry

### What is a storm surge?

A storm surge is an abnormal rise in the sea level along an open coast. This sudden rise in water usually occurs during a storm, as the onshore winds or lower atmospheric pressure pushes and "piles" the water up against the shore. If a high tide occurs at the same time, the storm surge adds to the height of the water—often causing devastation to a populated coast.

faster: New moons produce weaker coastal tides, and it corresponds that plants may take in less water at this time.

## What is the **Earth's tidal bulge**?

The term "tidal bulge" describes the effect of the tide on the side of the Earth closest to the Moon—or where the lunar attraction is greatest.

## Where do the **highest tides** occur?

The highest tides on Earth are found in Nova Scotia's Minas Basin, off the Bay of Fundy, where the water level at high tide can be as much as 52 feet (16 meters) higher than at low tide. The funnel-shaped Bay of Fundy squeezes and pushes the Atlantic Ocean tides to tremendous heights. These high tides occur every 12.5 hours, and an hour later every day, as they follow the changing position of the Moon. Amazingly, Nova Scotia "bends and tilts" when the tide comes in, as about 14 billion tons of seawater spread into the basin twice a day; but you can't really see the land move. In fact, scientists estimate that there is so much water at mid-tide, the flow in the Minas Channel north of Blomidon is equivalent to the combined flow of all the rivers and streams on Earth. And around mid-tide at Cape Split, you can hear the "voice of the Moon"—actually, the roar of the tidal currents as they run by.

The Monkton Tidal Bore occurs on the Pedicodiac River, which flows into Canada's Bay of Fundy (between New Brunswick and Nova Scotia): High tide causes ocean waters to surge upstream; the bore occurs at the point where these waters meet the outflowing river waters. *CORBIS*

## What is a **tidal bore**?

A tidal bore is an abrupt front of high water, caused by the high tide coming in and surging up a river; on the surface of the water, a tidal bore looks like a slight wave trying to move against the flow of the river. In Nova Scotia's Minas Basin, tidal bores surge up several rivers that flow into the basin. These bores also occur in a few other rivers when the high tide begins, including Canada's St. Croix, Meander, Maccan, and Salmon rivers.

## What is a **tidal inlet**?

A tidal inlet usually forms in gaps between barrier islands. Strong currents flow inward and outward through these gaps as the tide rises and falls. The inlets change over time, too: If a major storm reaches the area, it can cause a breach in the island, cutting another inlet; over time, moving sands also can close up a tidal inlet.

## What can happen when the **tide turns**?

When a tide turns (from low to high or high to low), the action can create opposing currents in the water. Those currents meet, and often create a whirlpool, a swirling eddy of water. One of the most violent events that occurs when a tide turns is called a maelstrom, a feature that often forms in the Lofoten Islands off northern Norway.

## Is it possible to harness **tidal energy**?

Yes, it is possible to harness tidal energy. This energy, produced as the ocean waters surge during the rise and fall of the tides, has long been a way to generate electricity. In the mid-1960s, the world's first plant run on tidal energy opened in France—a dam built in an area where the Rance River empties into the English Channel. Unlike regular hydropower (generated by rivers), the reversible turbines in the dam allowed electricity to be generated as the tides came in *and* out. Similar to river-generated hydropower, it is difficult to find sites where tidal power-generation is possible—thus, it is not as widely used as other sources of electric power.

# WAVES

## What is an **ocean wave**?

An ocean wave is the large-scale movement of water molecules. It appears as a ridge or swell on the surface of the ocean. The highest point of a wave is the crest; the lowest point is the trough. The distance between the crest and a neutral point (the water's "at-rest" level) is the amplitude; the distance between the crest and the trough is the wave height; and the distance between one crest and the next (or the distance between two consecutive troughs) is the wavelength. A wave period is the time, in seconds, between the passage of successive wave crests (or troughs) at a stationary point.

There are many different types of waves in the oceans, and they can be extremely large or very small, depending on their source and the area in which they occur. The smallest waves, called ripples, have crests less than

The anatomy of a wave: This shot shows a well-formed crest and trough; the distance between the crest and the trough is the wave height. *CORBIS*

1 inch (2 centimeters) apart. Longer waves are subdivided into shallow-water waves and deep-water waves, depending on the depth of water through which they move. Shallow-water waves include tidal or river bores, in which the water particles move forward and backward in a horizontal plane, with the crests representing "crowds" of water particles. The troughs have very few water particles. Very violent storms at sea or underwater seismic activity will produce fast-moving deep-water waves.

## What is **wave energy**?

The motion of the waves carries an enormous amount of energy; this energy is generated by the source of the waves, such as winds. In deep-water waves, this energy is expended in a lateral motion; in shallower waters, wave energy is released by the water crashing on the shore.

## How do **waves develop** in the ocean?

Most waves on the surface of the ocean are caused by the action of winds; the height determined by the speed of the wind, the length of

The power of ocean waves is manifested here: A French merchant ship labors in heavy seas in the Bay of Biscay, as a huge wave, common in this area of the Atlantic, looms ahead. *NOAA; published in* Mariner's Weather Log, *Fall 1993*

time it blows (duration), and the distance over which it continually acts on the water's surface (called fetch). As wind blows across the surface of the water, it tries to "drag" the water with it. Since the water cannot move as fast as the air, it rises, but gravity pulls the water back. For each wind speed, there is a point at which the waves stop growing; this is because the energy is dissipated by breaking waves at the same rate as the wave energy is added by the wind.

Other types of waves include tide waves, caused by the gravitational attraction of the Moon and Sun; and internal waves, generated within the ocean by numerous causes, including wave interactions, atmospheric disturbances (such as storm surges), and earthquakes.

## What is a **seiche**?

Seiche is the occasional, rhythmic rise and fall of water in a lagoon or bay. This phenomenon is not tidal; in most cases, coastal seiches are the result of interactions between the water and the air—long waves rolling in from the open ocean where they are generated by strong winds, large atmospheric disturbances, or even seismic activity on the coast or underwater. Seiches can be a positive factor along a coast, bringing much-needed nutrients and dissolved gases to the coastal habitat. They can also be very destructive, ramming boats into moorings as the waters rise and fall.

## What is the **tallest open ocean wave** reported so far?

In 1933, the American tanker U.S.S. *Ramapo* was traveling in the Pacific Ocean from Manila, Philippines, to San Diego, California. One large ocean wave, measuring some 112 feet (34 meters) high, was reported during a 68-knot windstorm, but this was not officially recorded.

## Have scientists noted any long-term **changes** in **wave height**?

Over the past 30 to 40 years, the waves in the North Atlantic Ocean have appeared to be getting higher. Less than a century ago, the average wave measured about 7 feet (2 meters); recently, the waves seem to have increased to about 9 to 10 feet (about 3 meters) on average. There is no explanation for the increase, although some scientists believe it may be that the waves have really not grown in height, but rather that the way they're measured today provides more accurate readings.

## What do **swell** and **sea** mean?

Waves within the area of wave generation are called sea; a swell is a long, smooth (crestless), and sometimes massive wave or a succession of long waves, usually occurring outside the area in which they were generated and moving quickly away from the source. As waves move outside the area in which they were generated—such as at a point of a violent storm at sea—they sort themselves out. When wave crests momentarily coincide with each other, each adds its height to the overall wave crest; other times, crests may coincide with troughs, thus canceling each other out and leveling the sea surface. Storms off the Antarctic coast frequently manifest themselves on the beaches of California as long swells.

## What are **rogue waves**?

Rogue waves (also called non-negotiable waves by sailors) are large waves that appear out of nowhere, occurring with little or no warning. Although they are found in all the oceans, certain areas of the globe seem to propagate more of these huge waves than other regions—such as the Cape of Good Hope off South Africa.

These waves are rare, with the majority seen and experienced by crews on ships crossing the oceans. For example, when the H.M.S. *Queen Mary* was pressed into World War II transport service, it was nearly capsized off Scotland by what her captain called, "one freak mountainous wave." And every year a few supertankers or large vessels traveling along a standard route from the Middle East to the United States or Europe experience major structural damage from rogue waves. In 1996, one such cargo ship, with 29 people onboard, sank after being hit by a giant wave.

There are many scientists who do not believe rogue waves exist. One reason is that it is difficult to determine what causes them—after all, not only do they appear out of nowhere, but observers (the witnesses to the event) are often swept away by them. Other scientists believe the gargantuan waves do exist and they suggest several reasons why they occur. The most recent theory, based on a mathematical model, shows that certain ocean currents or large eddies can occasionally concentrate an ocean swell and create a rogue wave. These currents or eddies act like an optical lens or magnifier, focusing the wave action to create huge crests—or the large troughs that mariners have often described as "holes in the sea." One such "focusing" current is the Agulhas Current, which skirts the South African continental shelf. There may be more such currents that form within or near the Gulf Stream in the North Atlantic Ocean, which would explain what the captain of the *Queen Mary* observed off the British Isles during wartime.

## What is a **trapped fetch**?

As the pressure from a fast-moving line of thunderstorms essentially "pushes" the surface ocean water, many small waves get in step with each other and pile up to form one giant ripple, sometimes more than 70 feet (21 meters) tall. This is often called a "trapped fetch"—when a storm moves along with the waves it generates, adding energy to the moving waves.

**Australian windsurfer Luke Hargreaves rides the break of Maui's "Jaws"—the monster waves that form off the island's north shore, making it an international surfing mecca.** ***CORBIS/John Carter***

## What causes the **huge waves** seen near the **island of Maui**?

Off the coast of Maui, one of the Hawaiian Islands, there is an underwater ridge—produced by an ancient lava flow—that acts like a giant magnifying lens, bending and enlarging Pacific Ocean storm swells. These conditions often produce monster surfing waves as high as 70 feet (21 meters). The waves result from the unique shape of the underwater ridge—not from the effects of submerged cliffs, as some observers had previously assumed. This produces "Jaws," a surfing mecca on Maui's northern shore, near the town of Haiku.

At Maui's Jaws site, the ocean depth changes abruptly from 120 feet (37 meters) to just 30 feet (9 meters). Most of the time, the waves are not affected by the ridge. But when swells from stormy weather in the North Pacific Ocean are longer than about 1,000 feet (305 meters), they "trip" over the ridge, wrapping around it. As part of a storm swell passes over the ridge crest, it slows because water travels slower in shallow water; the other parts of the swell travel faster in deeper water, causing the wave to focus on the ridge in a process called refraction. The swell bends inward as it travels on either side of the ridge, focusing its energy on the center of the wave crest—to form the monster wave.



# CURRENTS

## What is a **current**?

A current is the large-scale movement of water in a specific direction. In the oceans, currents can be thought of as the Earth's circulatory system, moving vast quantities of waters of different temperatures around the world, creating climates, providing nutrients for marine life, and influencing weather patterns. They may form permanent circulation systems, as in the Atlantic and Pacific oceans or currents can be relatively short-lived, affecting only a limited area—especially along a coastline.

## What is a **gyre**?

A gyre is the large-scale circular movement of currents—usually a permanent feature—on the surface of the ocean. Gyres normally occur when the surface current, pushed by the Earth's prevailing winds, runs up against a continent. This causes the current to modify its path into a giant, spinning circulation of water. The North Atlantic gyre is formed by the equatorial current, which flows west; it is forced by the North American continent to flow northeast along the coast of the United States as the Gulf Stream Current; this continues to the east toward northern Europe as the North Atlantic Current; then heads southward as the Canaries Current and back to the west, completing the circulation loop.

## What are **tidal currents**?

Tidal currents, a type of subsurface current, are the large-scale horizontal movements of water caused by the gravitational interactions among the Earth, Moon, and Sun. They manifest themselves as localized ocean currents off continental and island shores; in the open ocean, they diminish to the point of being undetectable. The direction of tidal currents changes throughout the tidal cycle; they create a clockwise gyre (large-scale circular movement of upper ocean surface currents) in the Northern Hemisphere and a counterclockwise gyre in the Southern Hemisphere.

## What are **non-tidal currents**?

Non-tidal currents are usually ongoing movements of water—permanent features. They occur at the surface and the subsurface. Non-tidal currents are the circulation system of the oceans; like the atmosphere, the oceanic circulation system is mostly driven by the Sun's radiation and the Earth's rotation (especially the Coriolis effect). Temperature and density differences in the water and temporary winds also drive non-tidal currents, including subsurface gradient currents (which are caused by differences in the density of seawater). Other factors can influence the formation and maintenance of localized non-tidal currents, such as the depth of the water, the size and location of the land, the underwater topography, activity of volcanoes, and discharge from river systems.

## What are **surface currents**?

Surface currents in the ocean are the continuous movements of water found at the surface to just a few feet below. These currents are generated by the planet's prevailing winds—which are a direct result of the Sun's radiation and Earth's rotation. For example, equatorial winds cause westward flowing surface currents north and south of the equator. If there were no continents to redirect their flow, these equatorial currents would continue to flow around the world. They are considered the most important ocean currents in the world, as they circle the ocean basins to either side of the equator, and create specific climate conditions in certain areas.

## How do the **wind** and **rotation** of the Earth create **surface currents**?

The force of sustained winds on the surface layer of water on the ocean causes it to move; this motion is transmitted down through successive layers, but the motion slows with increasing depth due to friction. This

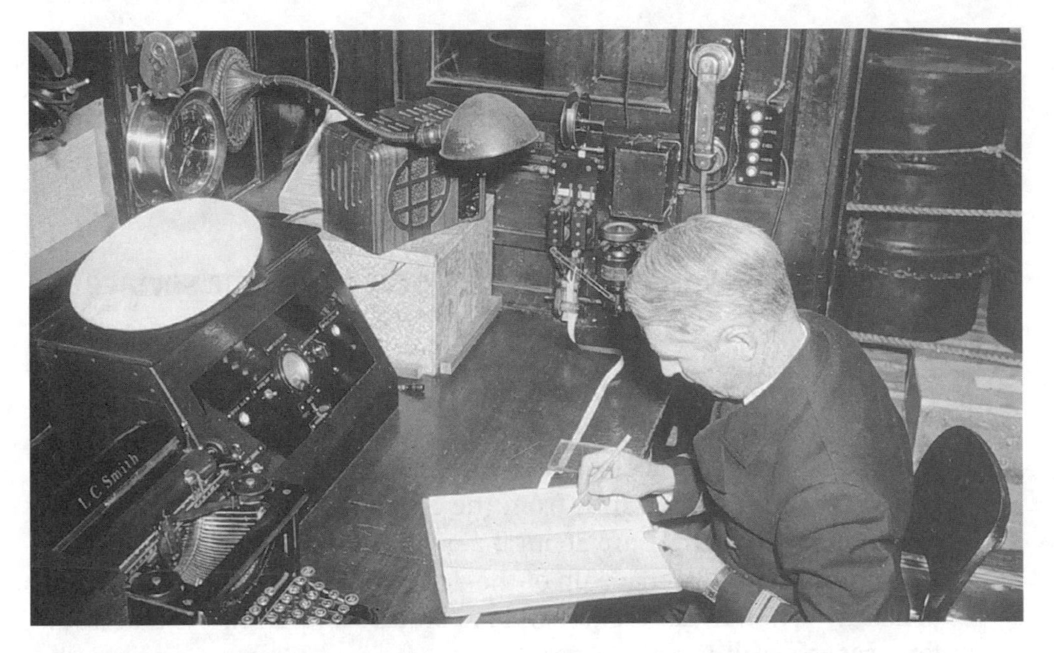


Tapes from recording meters aboard the *Oceanographer* are read off the North Carolina coast in 1940. Because of continued observation and careful record-keeping, scientists have been able to provide the Navy, the merchant marine, and the oceangoing public with better data on the ocean's currents. *NOAA/C&GS Season's Report, Borden 1949*

movement of water is called a wind current, and in order to form, needs a steady wind lasting at least 12 hours.

But this current does not flow in the direction of the wind, which would seem logical. The rotation of the Earth deflects the motion toward the right in the Northern Hemisphere, and toward the left in the Southern, leading to a difference in direction between the winds and the wind-generated currents. This difference in direction can vary between 15 degrees in shallow coastal areas, to 45 degrees in the open ocean.

## How do **surface currents** influence **climate**?

Some surface currents influence climate by their close proximity to land. For example, in the North Atlantic, the Gulf Stream is a fast-moving current carrying warm water in a narrow band from the eastern coast of Florida along the East Coast of the United States, past Newfoundland, eventually arriving at England, where it moderates the climate. To contrast, the cold Labrador Current, which runs southward along the eastern Canadian coast and dips below the Gulf Stream to reach the eastern seaboard of the United States, gives New York, for example, a cool climate—even though the state is at the same latitude as warm southern Italy.

## What **separates** the **waters of the deep ocean** from the **surface ocean waters**?

There is a permanent thermal barrier that separates the warm surface waters from the relatively deep, cold waters; this barrier also prevents the surface and the deep waters from mixing. The surface waters—which receive their warmth from the Sun's radiation and have wind-driven currents—make up about 2 percent of the ocean's volume. The remaining volume is made up of the deep, cooler waters, with currents primarily driven by differences in temperature and salinity—that is, differences in density.

## What are **subsurface currents**?

Subsurface currents in the ocean are those that occur below the domain of the surface currents. They are also referred to as deep-ocean currents. They are caused by density variations in ocean water, which is the result of differences in temperature and salinity. Water that is denser (heavier) will sink downward and move along the seafloor toward the equator—thus creating subsurface, or deep-ocean currents. The movement of these types of currents is called thermohaline circulation.

For example, deep beneath the warm, northeast-flowing Gulf Stream Current, a large cold current flows in the opposite direction, called the North Atlantic Deep Current. This is caused by very cold, salty—thus dense—water from around Greenland, sinking and flowing toward the equator; it is also pushed up against the western edge of the Atlantic Ocean's basin by the Earth's rotation. As it flows toward the equator at approximately 0.5 miles (0.8 kilometers) per hour, it hugs the ocean bottom, displacing the warmer water upward—and creates a subsurface current. When it reaches south of the equator, it is often undermined by another deep-ocean current: The Antarctic Bottom Current, which flows northward toward the equator—and under the North Atlantic Deep Current.

## Who **first discovered** ocean **currents**?

Probably the first currents were discovered around an area that had plenty of famous civilizations—not to mention shipbuilders: The Mediterranean Sea. Ships were used by the Egyptians, Athenians, the "sea kings" of Crete, Phoenicians, Romans, and others to carry on trade. And although there is not a true written record of currents, many ships left the confines of the Mediterranean to travel the Atlantic Ocean, reaching as far as Iceland and perhaps beyond. Some of the earliest records of currents were recorded by the Polynesians more than 2,000 years ago in the Pacific Ocean . They used their knowledge of the weather, winds, migrating seabirds, and ocean currents to navigate between their widely separated islands.

## Why is it difficult to **measure ocean currents**?

One of the major reasons it is difficult to measure ocean currents has to do with the size of the oceans: It is difficult to collect a massive amount of data over such a large area. In addition, the ocean floor has huge, river-like channels—similar to those on the land surface, but not as easy to find. The currents are often dragged through these elusive channels, changing the overall current flow.

## How have **athletic shoes and rubber toys** been used to chart **ocean currents**?

In 1990 a Korean container ship bound for the United States ran into problems, accidentally dropping 80,000 Nike athletic shoes into the Pacific Ocean. Though oceanographers could not follow the sneakers across the ocean, they did retrieve many of the shoes on beaches in Alaska, Oregon, and Hawaii. The shoe landings confirmed and added to several computer models of Pacific Ocean currents.

Other accidental ocean research "tools" were some 29,000 bathtub toys. During a Pacific storm in January 1992, the toys (rubber turtles, frogs, ducks, and beavers), which were being transported aboard a container ship, were swept overboard near the International Dateline. Like the athletic shoes, the toys washed up on various shores, helping confirm and add data to computer models of the currents.

## What are **convergences** and **divergences**?

A convergence is where different currents come together; in the majority of cases, this causes the water to sink. A divergence is where different currents pull apart, causing the water to rise. Often the best evidence of a convergence or divergence is to watch the debris on the oceans—logs, seaweed, garbage, and even shipwrecks can gather at these points. Because this floating material is found in the middle of the oceans, there is often a large concentration of birds and marine animals that frequent these areas.

## What are some of the **major surface currents** in the oceans?

There are many surface currents in the Earth's oceans. They include the North Equatorial Current (warm), the Equatorial Countercurrent (warm), South Equatorial Current (warm), Oyashio Current (cold), Kuro Siwo Current (warm), California Current (cold), Labrador Current (cold), Gulf Stream Current (warm), Benguela Current (cold), Canaries Current (cold), Peru (Humboldt) Current (cold), North Atlantic Drift (warm), and the Antarctic Circumpolar Current (cold).

## What is the **North Equatorial Current**?

The North Equatorial Current (or Equatorial Current) includes a warm, east-to-west-moving Pacific Ocean current and a warm east-to-west-moving Atlantic Ocean current driven by the northeast trade winds that blow over the tropical oceans of the Northern Hemisphere. It eventually splits in two—as the opposite-flowing current called the Equatorial Countercurrent and the Kuro Siwo Current of Japan.

## What is the **Kuro Siwo Current**?

The Kuro Siwo ("black stream") Current is a strong, fast-moving warm current that flows northeastward from the northern Philippines, up the east coast of Japan. It is a continuation of the North Equatorial Current, and carries warmer waters into the North Pacific Ocean. The approximately 50-mile- (80- kilometer-) wide and 1,300-foot- (400-meter-) deep current actually appears dark from far away—thus its name.

**American statesman Benjamin Franklin drew this map of the Gulf Stream (c. 1782) to explain why ships returning across the Atlantic traveled slower if they dipped too far south, where their westward progress was impeded by the northeast-moving current of the Gulf Stream waters.** ***NOAA Central Library***

## What is the **California Current**?

The cold, shallow California Current flows southeastward along the west coast of North America; it then meets the North Equatorial Current around the equator. The fog banks of San Francisco are caused by the California Current, as moist, relatively warmer air moves over the cold current's waters.

## Why is the **Labrador Current** important?

The cold, slow-moving Labrador Current travels in a southward direction; its cold waters dip below the northeastward-flowing Gulf Stream Current. In the warmer, summer months in the Northern Hemisphere, the current often extends down to about Cape Cod, Massachusetts; in the winter months, it reaches down to about Virginia. The Labrador Current is also the current that carries icebergs into the major shipping lanes between North America and Europe during the winter.

## Why is the **Gulf Stream Current** important?

The warm, fast-moving Gulf Stream Current is located just off the East Coast of the United States, and brings a temperate climate to the coastlines it passes; it is also part of the major Atlantic gyre (large-scale circular movement of currents). The Gulf Stream moves north from the

Caribbean Ocean; the warm water then reaches the Gulf of Mexico, where it is warmed even more. It then moves through the Florida Straits, where it becomes one of the fastest currents known. From there, it travels up the East Coast of the United States, traveling about 80 miles (130 kilometers) a day—until it reaches the colder Labrador Current off the coast of northeastern Canada.

## What is the **Benguela Current**?

The Benguela Current is part of the South Atlantic Ocean gyre (large-scale circular movement of currents). It is a cold, slow-moving, broad, shallow current that flows northward along the west coast of South Africa. There is a great deal of upwelling ("pushing" of water) along this current, forming where currents part (diverge). These divergences provide food for many plants and animals, and are known for their dense marine life.

## What is the **Canaries Current**?

The cool Canaries Current is a split off from the North Atlantic Drift, and tempers the climate of the Canary Islands—which would be tropical if the Canaries Current did not flow past the islands. As this cool current moves south, it meets the warm winds and waters of the Iberian peninsula, creating the famous fogs along the western edge of the Pyrenees Mountains between Spain and France.

## What is the **Humboldt Current**?

The Humboldt Current, now called the Peru Current, is a cold, shallow, and slow-moving current that flows north and then northwest along the western coast of South America. It is a well-known area of upwelling (pushing of water caused by temperature differences or diverging currents), which encourages the growth of small plankton (nutrients). This, in turn, attracts a wealth of other marine animals that feed on the plankton, and other animals that feed on those animals. Thus, it is a tremendously productive fishing ground. If the Peru Current changes course, it inhibits the upwelling, affecting the marine life in the area. This change in course usually occurs during El Niño years, as warm waters move south, displacing the cold, northward flowing Peru Current.



## Why is the **North Atlantic Drift** important?

The warm, shallow, and slow-moving North Atlantic Drift (or Current) is a branch of the Gulf Stream Current; it flows to the northeast, across the North Atlantic, and onward toward the Arctic Ocean. Because it is a split from the Gulf Stream, the warmer waters moderate the climate of western Europe and England, resulting in moderate winters for the latitude, and keeping Norwegian ports free of ice during the winter.

## What are some of the **major subsurface currents** in the oceans?

There are many subsurface currents in the Earth's oceans. They include the Deep Western Boundary Current (cold), Cromwell Current (cold), Weddell Sea Bottom Water (cold), and North Atlantic Deep Water (cold).

## What is the **Deep Western Boundary Current**?

The Deep Western Boundary Current is the world's largest deep-ocean current—a fast, cold current in the southwest Pacific Ocean. It is about 100 times the size of the Amazon River, and a part of the global ocean circulation system. In fact, this current channels about 40 percent of the world's cold, deep water throughout the oceans.

The current is also responsible for carrying a great deal of sediment into the oceans: As it travels north from the Antarctic Ocean to the Pacific, it hits the landmass of New Zealand, where the mountains provide an abundance of eroded sediment. Fine-grained muds drift into the current to form huge deep-sea drifts of sediment, some of which are several hundred miles long. In fact, scientists are using this buildup of sediments to find out how the climate has changed in the Southern Hemisphere and to determine the changes in the strength of the Deep Western Boundary Current over time.

## What is a rip tide?

The so-called "rip tide," also called the undertow, is actually a strong, localized current that flows quickly outward from the shoreline. These currents form in a variety of ways: They can be caused by the rapid escape of water, by means of waves, from the shore through a gap in a reef; by waves that break obliquely (at an angle) across a longshore current (a current flowing parallel to the shore); by the head-on meeting of two tidal currents; or by a strong tidal stream that enters shallow water (a stream of moving water caused by tides that, for example, moves through an inlet). The rapid movements of rip currents return a substantial amount of water to the oceans and they often carry a great deal of sediment as they tear through the shoreline. They are also responsible for many deaths, as swimmers can become caught in the strong, moving waters, which carry them toward the open ocean.

## What is the **Cromwell Current**?

Named after Townsend Cromwell (?–1958), the oceanographer who discovered it in 1952, this subsurface current flows swiftly from west to east along the equator in the Pacific Ocean. A cold current, the Cromwell is only about 200 miles (322 kilometers) wide and about 700 feet (213 meters) deep, and flows at about 3.5 knots (nautical miles) per hour.

## What is the **fastest ocean current**?

The fastest ocean current known is the warm Florida current, part of the Gulf Stream system. It flows from the southern tip of Florida to the north, traveling past Miami, and onward toward Cape Hatteras, North Carolina. It can often reach speeds of 2 to 4 knots, covering 3.3 to 6.6 feet (1 to 2 meters) per second. Some estimates gauge it at 5 knots. Another contender for the fastest current is the Kuro Siwo in the Pacific Ocean.

## What are the **slowest currents** in the Earth's oceans?

The slowest currents in the Earth's oceans are those of the deep ocean. The average speed of these currents is about 1 knot (nautical mile), covering 1.6 feet (0.5 meters) per second.



# OTHER OCEAN PATTERNS

## What is an **upwelling**?

An upwelling is a push of water caused by water temperature differences. Subsurface water, which is colder and denser, wells up to reach the surface—most often replacing wind-displaced surface water. Upwellings also form where currents diverge (split apart) and in places where water is blown away from the coast by strong winds. Upwellings occur most often off the western coasts of continents and in the Antarctic Ocean. For example, the cold Peru (formerly called the Humboldt) Current flows north along the western coast of South America. This causes an upwelling of water off the coast of Peru. Another area of upwelling is the California coast near San Francisco.

## What constitutes **nutrient-rich water**?

When water upwells, it often drags along with it nutrient-rich water, or water that contains chemical substances necessary for the maintenance of life in the ocean. The nutrients include materials necessary for the growth of plants, such as nitrites, nitrates, phosphates, ammonia, and silicates. Nutrient-rich water, especially in the cold upwelling waters, often has enough chemical substances to allow plankton to bloom in great numbers at the surface. This in turn feeds the fish and bird population, and creates a fertile breeding ground for all sorts of animals that live in the upwelling area. Consequently, upwellings are prime areas for fishing.

This false-color image of El Niño was taken by the U.S./French TOPEX/Poseidon satellite in mid-September 1997. It shows (in white) the heated surface waters near the equator—a mass that at this time was one-and-a-half times the size of the continental United States. (Also *see* the color insert, where the warmed waters appear in red and white.) *AP Photo/Jet Propulsion Laboratory*

## What is **El Niño**?

The phenomenon of El Niño is the unusual warming of a mass of tropical Pacific Ocean surface waters just above the equator and off the western coasts of Peru and Ecuador, South America. El Niño is a Spanish term meaning "the Christ child," or "the little boy," so named because the event seems to peak around Christmastime. The temperature of the ocean waters only rises by a few degrees more than usual, but the increase nevertheless causes major climate changes in pockets around the world. The reason for the phenomenon is unclear. Scientists believe some natural events may be exacerbated by the atmospheric pollutants emitted during larger volcanic events, such as the eruption of Mount Pinatubo (in the Philippines) in June 1991.

It was once thought that an El Niño appears every 15 years—but now it is known to occur every 3.5 to 7 years, and typically lasts 12 to 18 months. Over the past 50 years, El Niño events have begun in 1952, 1958, 1964, 1966, 1969, 1972, 1976, 1982, 1987, 1991, 1994, and 1997. In more recent years, scientists have been able to predict the occurrence of El Niños up to one year in advance. This is possible because of global climate models that were developed to understand the events. But no one yet has been able to predict the intensity of an El Niño.

## How do **El Niños impact** the world weather?

Occurring off the western coast of South America, El Niño was once thought to be a localized phenomenon. But it is now understood that El Niño events have a great effect on weather around the world. The following list describes several El Niño events and their effects.

1972: As the waters warmed in the Pacific Ocean, the anchovy population declined, wreaking havoc on the local fishing industry.

1976–77: During this event, severe cold struck the central and eastern United States, and a major drought struck California.

1982–83: This El Niño event caused drought in Australia—the worst one in 200 years. It also worsened a drought that was already underway in Africa. It sent torrential rains to Peru and Ecuador. The United States Pacific coast was inundated with winter storms, while the first typhoon in 75 years hit French Polynesia.

1991–95: This long El Niño changed rainfall patterns around the world. In California, an entire year's precipitation fell in January 1995. The worst floods in a century occurred in Germany, Holland, Belgium, and France. Major droughts occurred in Australia, Indonesia, and southern Africa.

1997–98: This El Niño was classified as a Type 1 (the strongest), involving a high sea-surface temperature. It was also huge—extending from about 160 degrees east to 80 degrees west longitude. What made this El Niño different than the eight other Type 1 events that occurred between 1949 and 1993 was its rapid maturity and growth.

## What was the **longest El Niño** ever recorded?

The longest El Niño so far began in 1991 and lasted into 1995, a time when worldwide weather disasters increased. The long-term event may have been helped along by the June 1991 eruption of Mount Pinatubo in the Philippines. Scientists estimate that such long El Niño events occur only every several thousand years—but we don't know enough about past El Niños to know for certain.

## What was the **strongest El Niño** every recorded?

The strongest El Niño occurred in 1997 and lasted into 1998; the area of warm surface waters off the coast of South America was also the largest on record. The ocean temperatures averaged almost 11° F (5° C) warmer than usual: temperatures were about 8° F above normal off the South American coast, and 5° F above normal off the coast of Baja, California.

## What **problems** can a **strong El Niño** cause?

It is difficult to name all the problems that have occurred during strong El Niño years. For example, in 1982 to 1983, El Niño caused tens of thousands of animals to die off the Galapagos Islands; the droughts in Australia and northern Brazil also caused hardships in those areas. The 1997 to 1998 El Niño probably caused many hurricanes to reach the highest category of hurricane intensity (category 5)—and to move farther north than usual.

## What theories are there about **El Niño** events and the end of the **Ice Age**?

Scientists have found that there may be an amazing connection between El Niño occurrences at the end of the Ice Age and glaciers in the Great Lakes. It is known that Lake Huron was once covered by thick ice sheets. When the Ice Ages ended, the glaciers melted and sent water surging through all the Great Lakes; it may be that El Niños contributed to that melting. Evidence from Lake Huron glacial sediments shows that there were several episodes of rapid warming in 10- to 20-year intervals. Scientists believe this is evidence of two events we know about today: El Niño Southern Oscillation (ENSO) and something called the Quasi-Biennial Oscillation (QBO), which alters wind speeds in the upper levels of the atmosphere (the stratosphere) every 2 to 2.5 years.

## What is **El Niño Southern Oscillation**?

El Niño Southern Oscillation (ENSO) is a term describing the back-and-forth movement between periodic warming (El Niño) and periodic cooling (called La Niña) in the Pacific Ocean. In other words, the oscillation accounts for the repeated occurrence of both events.

## What is **La Niña**?

La Niña is the opposite of El Niño, a time when the surface water of the Pacific Ocean off the coast of South America becomes cooler; in addition, the Earth's trade winds are stronger than normal. Similar to an El

Niño, La Niña changes global weather patterns. Most La Niña years lead to a weaker winter jet stream over the central and eastern Pacific, with fewer storms along coastal California. La Niña is also associated with less moisture in the air, which usually causes less rain along the west coasts of North and South America. But over Southeast Asia and Africa, the usual summer monsoons intensify, and Australia is usually deluged with rains as a result of La Niña.

## How do scientists tell whether there is an **El Niño** or **La Niña** in the Pacific Ocean?

Thanks to the launch of special satellites within the past few decades, scientists now have a way to determine El Niño and La Niña events—also detecting if one is getting weaker and the other one stronger. For example, sea surface height readings, which are an indicator of the heat content of the ocean, have been taken by the *TOPEX/Poseidon* satellite. From 1997 to 1998, the rapid cooling occurring in the central tropical Pacific Ocean (indicative of a transition from an El Niño to a possible La Niña) slowed, with the area of lower sea level (or colder water) of La Niña decreasing in size and strength. Thus, scientists could not tell whether or not the La Niña would occur—but the satellite continued to send data to Earth and scientists continued to keep watch. In 1999 most media reported that a La Niña event, albeit a relatively weak one, had unfolded in the Pacific Ocean. But some scientists disputed this conclusion.

## What are **"roving blobs"**?

Scientists may talk about "roving blobs" in the Atlantic Ocean; these are massive temperature anomalies, which are thought to be similar to the well-known El Niño and La Niña in the Pacific Ocean. But in this case, these giant patches of warm or cold water drift much more slowly around the North Atlantic—and may affect European weather.

The "roving blob" phenomenon, first described in 1996, was found by studying sea surface temperature records from 1948 to 1992. The "blobs" measure hundreds—or maybe thousands—of miles across and typically last 3 to 10 years; to compare, an El Niño warming persists for 12 to 18 months. These temperature anomalies appear to drift along the path of prevailing ocean currents, but they move at only one-third to one-half the speed of the actual currents. In addition, some scientists believe these "blobs" are also drifting around other ocean basins.

Scientists have traced some of these patches. For example, a cold anomaly developed off the coast of Florida in 1968. It drifted eastward; by 1971, it elongated and hit the coast of Africa. It then traveled south, then west across the tropical Atlantic, reaching the coast of South America in 1975—and dying out within the next two years.

Scientists have yet to determine the cause or details of these massive patches, or if they truly affect climate and weather. For example, some scientists believe a warm patch in the Atlantic Ocean in the late 1950s helped to prolong a Scandinavian drought—but they still need to conduct more research on the phenomenon to determine its effects.

# THE SHORE

## THE COAST DEFINED

### What is a **coast**?

A coast is an area of land extending from the shore (where the land meets the water) inland. On the landward side, the coast ends at the first major change in terrain. The actual width and morphology (configuration of rocks and other surface materials) of a coast varies from place to place, as does the shape of a coast. Some coasts are smooth and straight, while others are irregular and rugged. And still others consist of coral reefs or low-lying wetlands and marshes.

### What is a **coastline**?

A coastline is the line that forms the boundary between the shore and the ocean (or another body of water). The term may also refer to the shape of a coast.

### How many **types of coasts** are there?

Coasts are described by their special geological features or by the geological event that formed them. Scientists have classified eleven types of coasts around the world: deltas, cuspate forelands (triangular accumula-

**Fjords were carved out of granite by Ice Age glaciers. The coast of Norway, along the Norwegian Sea and the Arctic Ocean, is characterized by deep troughs such as this one.** ***CORBIS/Paul A. Souders***

tions of sand projecting seaward from the shoreline), wave-erosion (irregular and straight) coasts, drowned rivers (valleys), fjords, drumlins (elongated or oval hills), volcanic coasts, fault coasts, mangroves, barrier islands, and coral atolls. In everyday vernacular coasts are referred to as rocky, like the sheer volcanic cliffs of the Hawaiian Islands or the cobbles of the Maine coast, or as sandy (which are more common than rocky), like those along the North Carolina shore.

## How do **coasts form**?

Many factors combine to form coasts. These include the fluctuation in sea levels over time; glacial action; weathering and erosion; and the type of rock along the coast. For example, a delta coast is created as rivers drop sediment at the mouth of a river; cuspate forelands (triangular accumulations of sands) are also formed through the action of deposition. The movement of sediments by the action of waves creates wave-erosion coasts. Fjord coasts are deep troughs that were formed by glacial action (during the Ice Age); drowned rivers and drumlins were also formed by glaciers. A volcanic coast forms from the eruption of volcanoes and lava flows. A fault coast occurs when a shoreline rests against the edge of a fault scarp (a slope occurring at the boundary between two plates or at a fracture in a crustal plate). Mangrove coasts form as mangrove trees overtake a coastal area; their extensive root systems trap sediments to "build" the coast. Barrier islands are made of deposits of unconsolidated sediments and shells that build up parallel to the shore and above high tide. Atoll reefs form after a volcanic island rises above the ocean's surface and then is worn away by the action of the ocean

**Today, the relatively narrow coastal belt is home to approximately two-thirds of the world's population — not surprising when you consider the major seaport cities around the globe. Manhattan (shown) alone is home to more than 1.5 million people. *NOAA***

## Why are **coastal areas** important to **plants and animals**?

A coastal area is important to all resident organisms, mainly because it offers diverse habitats, as well as plenty of food and water.

## Why are **coastal areas** important to **humans**?

Throughout history coastal areas have been important to humans in many ways—including as a source of water, edible plants, and animals. Coasts are also gateways for trade (via shipping). Major cities have developed along seacoasts, or along lakes and rivers that are connected by a waterway with the ocean. Today, the relatively narrow coastal belt is home to approximately two-thirds of the world's population—and some 6.3 billion people are expected to be living in coastal areas by the year 2025.

In the United States today, it is estimated that nearly 50 percent of the population lives within 100 miles (161 kilometers) of the coastal areas.

Some scientists believe this figure may be low. It is also estimated that by 2010, approximately 127 million people will live in U.S. coastal areas. Currently seasonal visitors to the coast number almost 180 million more, with the average American spending 10 vacation days on the coast every year. Coastal recreational activities alone total about $75 billion annually.

## What **manmade problems** affect **coasts today**?

The large—and ever-growing—number of people living along or visiting the world's coasts has led to over-development and to the deterioration of coastal environments. For example, water contamination from non-point-source pollution (pollution from non-specific places, such as runoff from parking lots) has resulted in outbreaks of toxic algae-blooms and the killing of fish. Many coastal wetlands, important as buffers against storms and as food sources for humans, have been lost due to encroaching commercial and residential development. This encroachment, which has destroyed natural habitats, has also reduced the number and diversity of coastal plants and animals, which are important in the food chain.

## What **natural conditions** adversely **affect the coasts**?

Many natural conditions create problems in coastal areas. These include low-lying land (for example, an island that is less than 5 feet, or 1.5 meters, above sea level could easily be flooded); the absence of dunes to hold back water; an erosion rate of more than 3 feet (1 meter) per year; a lack of vegetation to hold the sand or silt; and poor drainage during rainy seasons. In addition, beaches are naturally modified by winds, waves, tides, and storm surges, which constantly change the nature of coastal areas.

## What is a **sea cliff**?

A sea cliff is a high, steep cliff or slight slope formed by wave erosion. Sea cliffs sit at the seaward edge of a coast, or on the landward side of a wave-cut plateau along the coast.

Developments such as this one, perched on a spit of land in the Gulf of Mexico at St. Petersburg Beach, Florida, provide much sought-after coastal housing. But peril could accompany pleasure: Such low-lying areas risk damage from storm surges, erosion, and hurricanes. *CORBIS/Jonathan Blair*

## How are **arches** and **sea stacks formed**?

Arches and sea stacks are coastal rock formations that are created by the action of waves breaking against a cliff. Because sea cliffs are made of various types of rock, the ocean waves differentially erode the cliffs (in other words, the erosion is irregular rather than uniform). Over time, the crashing of the water can erode the rock so much that it creates holes in the rock; these holes can eventually develop into caves; if water breaks through the cave, it can eventually form an arch, around which the water rushes; and if water continues to pound away at the arch, the top may fall, creating tall rocky columns called sea stacks.

## What is a **coastal plain**?

A coastal plain is a low, usually broad, flat land that slopes gently toward the shore. Coastal plains form in three ways. Some formed as the land rose, pushed upward from underwater by forces within the Earth. Others formed as sediment from rivers was carried to the nearby sea; over time, the layers of sediment built up along the coast to form a plain. The third and most familiar way is through sea level changes over time. The coastal plain of the southeastern United States, called the Atlantic Coastal Plain, extends from Maine to Florida, and west to the foothills of the Appalachian Mountains. Most of this area has gone through a succession of emergence and submergence as sea levels changed over the past hundreds of millions of years. In fact, during the last Ice Age, water taken up by the ice sheets caused sea levels to fall hundreds of feet, meaning that our coastal areas used to be much larger and broader than they are today.

## What is the **Coast Range**?

The Coast Range consists of mountains along the Pacific coast of North America, stretching from northwestern Mexico and Baja California north into southern Alaska. Although not all the mountains in the range are geologically similar, they are classified together (as a discontinuous mountain range) because of their close proximity to each other and the ocean. In Canada they are called the Coast Mountains.

**Sea stacks, such as these off the Oregon coast (west of Astoria), are formed by the action of waves breaking against cliffs.**
*NOAA/Commander Grady Tuell, NOAA Corps*

Hurricane Beulah produced coastal flooding in Texas in September 1967. *NOAA*

## How do **changes in sea level** affect a **coast**?

Long- and short-term sea level changes can affect a coast, mostly in terms of inundation or loss of water. In the short term, sea level rises—from such things as tsunamis (giant seismic waves) or storm surges (often from hurricanes)—can affect a coast. In the long term, the rise or fall of global sea levels can cause a coast to emerge from the water, exposing new land—or sink under water by flooding the land. Changes in sea level, whether short-lived or more lasting, can also impact the coastal communities—the flora (vegetation) and fauna (animal-life) dwelling there.

## Can **erosion affect** a **coast**?

Yes, like land erosion, coastal erosion of sand and sediment can chronically affect a coast. Short-term changes, such as those brought about by a major hurricane, can erode a beach area, undercutting homes and other manmade structures along a coast. Residential and commercial development also affect erosion rates by disrupting or altering the flow of water, changing the natural regime along the coasts. Long-term erosional changes occur as a result of common—and naturally occurring—coastal processes, such as the rise and fall of the sea level.

Approximately one quarter of the coasts in the United States are greatly affected by erosion. The populated coasts of the Atlantic Ocean and the Great Lakes are especially affected by erosion. In California, sea cliffs along the coast are eroded by breaking waves. Along the Gulf Coast of Texas, coastal areas, including Padre Island and Sargent Beach, have shown extensive erosion since the 1800s.

Beach erosion can adversely affect near-shore structures such as this one on the Outer Banks of North Carolina; this erosion was caused not by a hurricane, but by a Noreaster—a low-pressure system that can form over the eastern U.S., where moist Atlantic air feeds the storm to create high levels of precipitation and strong winds. *NOAA/Richard B. Mieremet, Senior Advisor, NOAA OSDIA*

# COASTAL FEATURES

## What is an **estuary**?

An estuary is a wide, funnel-shaped area near the tidal mouth of a river—where freshwater meets saltwater. Here, the flow of freshwater dilutes the saltwater. Tides also periodically force saltwater upriver—how far up river is determined by the strength of the tide, the amount of freshwater in the river, any offshore currents, and the surrounding land formations (what scientists call the local geological formations). Estuaries are protected from the full action of the ocean's waves, storms, and winds by their seaward boundaries, which can consist of barrier islands, reefs, or fingers of land, mud, or sand.

Estuaries come in all shapes and sizes. Depending on where you live, they can be called marshes, swamps, lagoons, sounds, sloughs, bays, or inlets.

## Why are **estuaries important**?

Estuaries are among the most complex habitats of the planet's ecosystem—and are therefore important for a number of reasons. They provide nursery grounds for almost two-thirds of the fish and shellfish consumed by humans, and are prime habitats for thousands of birds and other wildlife. Estuaries can produce four to ten times the weight of organic matter than a comparable-sized cultivated corn field. These areas are also important for water filtration. Sediments and nutrients drained from the land are carried by rivers, then filtered out as the water flows through dense meshes of marsh grass and peat, resulting in clean water. Estuaries also serve as flood control. They act as natural buffers between the land and ocean, protecting billions of dollars of real estate—especially the estuarine soils and grasses that absorb flood waters and dissipate storm surges.

## Do **estuaries** have **problems**?

Although they are now recognized as vital parts of the Earth's ecosystem, estuaries were long thought to be useless land. As such, they have been filled in, dredged, and built on—and polluted by a variety of chemicals. And they continue to be threatened, as the human population in coastal regions continues to grow.

## What is the **most productive estuary** in the **Northern Hemisphere**?

The Apalachicola Bay National Estuarine Research Reserve in the Florida panhandle is the most productive estuary in the Northern Hemisphere. The habitats within the reserve include beaches, oyster bars, salt and freshwater marshes, forested floodplains, and sandhills. More than 1,300 species live in the Apalachicola River's drainage basin (at the Gulf of Mexico), including 315 species of birds, 87 species of reptiles, and 57 species of mammals. Threatened and endangered species include the Florida black bear and West Indian manatee, respectively. Over 90 percent of all oysters harvested in Florida come from the Apalachicola estuary, which represents about 10 percent of the United States' overall harvest.

## Are there **problems** in the **San Francisco Bay estuary**?

Yes, California's San Francisco Bay estuary, located at the confluence of the Sacramento and San Joaquin rivers, has encountered problems related to increasing human activity. More than 95 percent of the tidal marshes have been leveed and filled, destroying fish and wildlife habitat. Contaminants, from municipal and industrial sewage, along with runoff from the surrounding farms, ranches, and cities, have entered and polluted the estuary. Dredging altered the water flow patterns and subsequent salinity (salt) distribution, and the flow of freshwater into the system has been reduced by diversion for irrigation.

## What is a **coastal marshland**?

A coastal marshland is usually a very large, flat tract of land that is protected from waves by its distance from the ocean. Although the land is not affected by waves, it is strongly affected by high and low tides. Coastal marshlands are divided into several types, including salt marshes and tidal flats, and are prime wetland habitats.

## What is a **salt marsh**?

A salt marsh is a flat, poorly drained area that forms in sheltered areas and is periodically inundated with brackish to salty water. Salt marshes contain some of the most diverse and prolific habitats on Earth. The main reason for this is the natural cycle of the salt marsh itself: Salt marsh grasses rot; grass is decomposed by bacteria; organisms die and decay; and the tides mix the rich organic remains, spreading them (like fertilizer) to the rest of the estuary.

## What is a **mud flat**?

A mud flat—also often called a tidal mud flat, tidal flat, or low marsh—is a relatively level, usually barren area of very fine-grained silt and sand situated along a sheltered estuary or island. Flats are alternately covered and uncovered by the tides; some are always covered by shallow waters. They are usually devoid of any vegetation, but they do attract shorebirds

It's exactly what it sounds like: A mud flat is a relatively level, usually barren area of very fine-grained silt and sand. Here, National Oceanic and Atmospheric Administration personnel attempt the mud flats along Kalgin Island, Cook Inlet, Alaska. *NOAA/C&GS Season's Report, 1967*

that feed on small animals living within the muds. Many tidal flats are also associated with deltas (extensive deposits at the mouth of a river).

## What is a **tidal marsh**?

A tidal marsh is found on the landward side of a salt marsh or a tidal mud flat.

## What is a **delta** and how does it form?

A delta is the sediment—clay, silt, mud, gravel and/or sand—that often collects at the mouth of a river. Deltas form as flat, low-lying plains that are broken into numerous channels. Deltas are largest at the mouth of the biggest rivers, such as the Mississippi River and the Nile River. The river slows as it flows into a larger body of water (in this case, the ocean) and sediment drops out of the water to form the delta.

The buildup of a delta depends on the amount of sediment carried by the river. For example, the Mississippi River carries more than 200 million tons of sediment annually through New Orleans and into the Gulf of Mexico. But not every river forms a delta. Some, such as the St. Lawrence River on the border between the United States and Canada, does not carry enough sediment to form a delta. Other rivers, such as the Columbia River in southwestern Canada and the northwestern United States, carry a great deal of sediment, but strong ocean currents do not allow sediment deposits to form.

## Are there **different types of deltas**?

Yes, there are three major types: bird-foot, estuarine, and fan-shaped (arcuate) deltas. The bird-foot deltas have many channels that branch

The longest river in the world, the Nile flows north from east Africa to the Mediterranean Sea, where it forms a fan-shaped delta. (This view, taken from the space shuttle *Columbia,* looks south, or landward; the Mediterranean is at the bottom of the picture.) *CORBIS/Courtesy of NASA*

The Mississippi River Delta, where the mighty river flows into the Gulf of Mexico, is a bird-foot delta. *NOAA*

out from the river's main channel. The Mississippi is an example of a bird-foot delta. An estuarine delta forms as an extensive plain of marshlands, and follows the (nonspecific) shape of the estuary. France's Rhone River Delta is an estuarine delta. The fan-shaped is one of the most familiar deltas; the area where the river meets the ocean is triangular (the origin of the word "delta" is from the Greek alphabet, in which the letter delta, or *d,* is designated by a triangle). Egypt's famous Nile River Delta is a classic example of the fan-shaped type.

## What is a **mangrove**?

A mangrove is a tropical tree or shrub that can tolerate low oxygen and can grow in seawater, where its roots are surrounded by muddy silt. The leaves and trunks of mangroves resemble those of trees found on land. The roots, too, are similar, except the root systems of mangroves are exposed—as if the soil has been dug out from the bottom of the trees. Sediment becomes trapped in this tangle of roots. As more sediment accumulates, it forms a dense mass, eventually growing seaward. Mangroves (sometimes called mangrove islands, since they form what

Mangrove coasts form as mangrove trees overtake a coastal area; their extensive root systems (shown) trap sediments to "build" the coast. *NOAA*

appear to be treed landmasses) are found in or near sheltered bays and estuaries. Some of the best examples of mangroves in the United States are found along Florida's southern and western coasts, including the Florida Keys. Others are found in the Philippines, Ecuador, India, Bangladesh, and Indonesia. In these tropical locations, mangroves are the primary coastal vegetation.

# THE SHORE DEFINED

## What is the **shore**?

The shore is the narrow strip of land that immediately borders the ocean; it is also called the seashore. This area is in constant contact with the water, and is alternately covered and uncovered by the waves and tides. The width of the shore is defined on the landward side by the high-water line and on the seaward side by the low-water line. Though sometimes used interchangeably with "coast," the shore is actually a narrow—and well-defined—part of the coast.

## What is the **shoreline**?

The shoreline is a changing boundary on the shore, defined by the edge of a specific condition, such as the average high-water mark or the average low-water mark. This boundary moves with changes in the tide and the sea level. It is literally where the land meets the sea.

## What is the **shore complex**?

The shore complex is a narrow, landward area that parallels a coastline. It usually crosses many diverse landforms including beaches, beach ridges, over-wash fans, dunes, barrier islands, sea cliffs, and sea stacks.

## What are **shoreline zones**?

For study, the shoreline is broken into various zones that are dependent on the tides.

## What is the **intertidal zone**?

The intertidal, or littoral, zone is the easiest shoreline habitat for humans to observe. It is the part of the ocean shore that is periodically covered by the highest tides and exposed by the lowest tides. Seaward of the intertidal is the subtidal zone; landward of the intertidal is the splash zone.

An intertidal zone is defined by the type of land or ocean animals it contains, since the life found in the zone depends on how long the section of water is exposed to the air.

## What are the major **zones within** the **intertidal zone**?

The zones within the intertidal are easily seen in many areas along an ocean (tidal) shore during the low spring tides. Not all ocean shores, however, contain every division of the intertidal zone, since the patterns rely on the rock, sand, and soil that make up the shoreline. The intertidal can include the supralittoral zone, supralittoral fringe, midlittoral zone, infralittoral fringe, and infralittoral zone. Each is described in the following table.

| Intertidal Zone | Description |
|---|---|
| supralittoral zone | above the high-tide mark; receives both the splash of the waves and seawater misting |
| supralittoral fringe | the upper level of the high-tide zone; receives regular splashing of the waves when the high tides are in |
| midlittoral zone | the major part of the intertidal zone below the high-tide mark |
| infralittoral fringe | the lowest level exposed by extreme spring tides; the base of the fringe marks the point of the lowest tide (or the beginning of the marine environment below the tides) |
| infralittoral zone | the marine environment below the tides |

## What **types of shores** are there?

There are various types of shores around the world—they are all dependent on the prevailing type of rock. For example, there are muddy shores, such as those found along the Louisiana coastline, where rivers drop silt and sand along the shoreline; stony or rocky shores, such as those along the Maine and Oregon coasts, where hard rock breaks apart and tumbles in the ocean waves, leaving rounded cobblestones along the shore; and sandy shores—the most well-known and well-liked—where the action of the waves has refined the sediments to form smooth beaches, perfect for walking.

## What is a **beach**?

A beach is a specific type of shore washed by the waves and tides. Most beaches slope gently, typically with a concave profile. They normally

**Evidence of the human love affair with the ocean—the summer crowd at a Santa Monica, California, beach.** ***CORBIS/Cydney Conger***

consist of an expanse of sand, pebbles, or shells, and extend between the extremes of the low- and high-tide water levels or to a line of permanent vegetation. Beaches are, like all features of the coastal regions, in a constant state of change, influenced by the tides, storms, seasons, and waves. In the long-term, geological processes such as changes in sea level, have a decided influence on a beach.

## How are **beaches created** and **destroyed**?

The geologic processes of erosion and deposition (the accumulation of sedimentary deposits) help create and destroy beaches. The agents of these processes are waves, currents, tides, and winds. Some beaches are formed by the deposition of sediment, which comes from nearby rivers and the transportation (of the sediment) by waves. Others are formed by the process of erosion, as the action of wind and waves reduces existing rock to sand and pebbles. In the short-term, during the winter or extreme storms when waves and winds are more energetic, erosion can reduce the width of the beach. Such storms and energetic waves can also deposit sediment in places, covering over structures. For this reason, wooden or concrete barriers called groins are often erected along a beach to curb the effects of these processes.



## What is the **composition** of **beach sand**?

In general, the sand found on beaches consists of light-colored quartz and feldspar sand grains; the erosion and weathering of rocks such as granite produces these fine grains. While some beach sand may be produced directly by the shoreline erosional process, much of it is produced by the action of inland rivers, eroding rock and producing sediment that eventually reaches the sea.

There are many other types of beach sand. Some sands are comprised of very small pieces of clam shells or other marine animals—rounded and smoothed over time by ocean waves, which act as a huge rock tumbler. Some tropical beaches are made up of coral fragments. Still other beaches are composed of fine sands: black sand, the result of eroded volcanic rock; bright pink sand, from the fine breakup of shells and corals; and green sand, from the weathering of a green mineral called olivine from volcanic rock.

## What **types of beaches** exist along the east and west **coasts of the United States**?

Along the western coast of the United States, in Oregon and northern California, the beaches are usually a narrow stretch of sand between rocky cliffs and rolling surf. In these areas, waves crash on hard bedrock; there is little sediment available for deposition. The finer sand is washed out to sea, while the back and forth motion of the waves tends to carry pebbles ashore, creating a rocky beach. In contrast, many of the beaches along the eastern, or Atlantic, coast of the United States are much broader and comprised of finer sand. They tend to be created by the action of waves over a long time.

## How are **beaches divided**?

Although the beach appears, to our eyes, as one expanse of sand or pebbles, it is comprised of three distinct zones:

**The surf breaking on the rocky, volcanic coastline of Lanai, Hawaii.** ***CORBIS/Douglas Peebles***

The **offshore zone** is the area where incoming waves begin to be affected by the bottom and begin to curl over to form surf or breakers.

The **foreshore zone** is the area of the beach that is regularly exposed to the high and low tides; on the landward side of the foreshore there may be a scarp, which is a rise of several feet that is the result of the eroding action of strong waves.

The **backshore** is the area of the beach from the waterline inward to where the ocean no longer influences the plants. It is usually horizontal or slopes gently landward. In this zone a berm marks the limit of the high tide; the width of this berm is dependent on the conditions of the surf. Some beaches, depending on conditions, may have two berms. This area is exposed to waves or covered by water only during exceptionally severe storms or very high tides.

## What is a **sand dune**?

A sand dune is loosely piled sand deposited by the wind. Dunes are common along sandy beaches, usually along the low-lying seashores above the high-tide level. These sand dunes move in the same way as desert sand dunes, although they are smaller in size: As the winds blow the sand, the particles race ahead of the winds, essentially jumping from place to place. This is why dunes seem to "move" over time.

## What is a **foredune**?

Landward of many sandy beaches (or at the landward limit of the highest tides) are narrow belts of dunes, usually irregularly shaped hills and depressions, called the foredunes. They are stable dunes or dune ridges usually running parallel to the shoreline. They are usually covered with beach grasses and other types of plants, which act as a baffle, trapping sand that tries to move landward from the beach. Thus, the foredunes can rise many feet above the high tide level. For example, the dunes on the Landes coast of France measure up to 300 feet (91 meters) high and 6 miles (10 kilometers) wide.

## Why are **foredunes important**?

The foredunes are often carved (reduced) in more extreme storms, but they usually rebuild in a short time. However, if the plant cover is depleted due to human intervention—such as building or vehicular and foot traffic—a blowout will occur. This blowout, or cavity, can allow sands and water to spread past the foredunes and into the tidal marshes beyond, causing extensive environmental damage. Such a release of water could also cause extensive property damage as well.

## What is the **strandline**?

The strandline is the high-tide mark on a beach. It is often visible even after high tide, since the tidal waters will deposit all types of debris here and when the tide recedes, the debris is left behind.

## What **problems** can result from squeaky **clean beaches**?

When beaches are cleaned using heavy machinery, as is normal in resort areas, the ecosystem along the strandline is disrupted, leading to a loss of marine creatures such as flies, beetles, centipedes, crustaceans—and the birds that prey on them. The removal of the strandline also contributes to the loss of sand dunes and increases beach erosion.

## What is a **barrier beach**?

A barrier beach is a long (a few miles or kilometers), narrow, and coarsely textured ridge that rises slightly above the high-tide level. It usually runs parallel to the shore and is separated from the shore by a lagoon or marsh.

## Have **nesting loggerhead sea turtles** helped to **preserve** a shore?

Yes, in the southeastern United States, from Florida to the Carolinas, nesting loggerhead sea turtles have helped to preserve the shore. Recent studies show that the eggs laid by the endangered sea turtle bring much-needed nutrients to the nesting beaches. The eggs contain elements such as nitrogen, phosphorus, and lipids, which are brought to the nutrient-starved beaches by the turtles from their rich feeding grounds such as the Bahamas, Cuba, the Dominican Republic, and the Florida Keys.

The eggs, whether they hatch or not, eventually deposit their nutrients along the dunes. This makes it possible for vegetation to become established and stabilize the dune system, preventing erosion. The vegetation, in turn, supports a variety of insects and other plant-eating animals, keeping the ecosystem in balance.

# SHORE WAVES AND CURRENTS

## Do **waves lose energy** and **change form** as they reach the shore?

Yes, as waves reach a beach and enter shallow water, they lose energy and change form. Most waves (which are usually caused by wind) travel across the oceans, losing little energy. But as they reach shallow water, the drag of the bottom slows the wave, causing it to steepen. The wave becomes higher and its wavelength (the distance between one crest and the next or between two consecutive troughs) shortens as it slows. As the wave steepens and the top of the wave narrows, it becomes unstable—eventually breaking on the shore.

## What are **breakers**?

Breakers are the finale of the wave's interaction with the beach, when the crest of the growing wave exceeds the speed of the wave itself. The wave becomes unstable and collapses, creating a frothy surf, filled with foam and bubbles. This also results in the distinctive, and often relaxing, sound so characteristic of the shore.

### How is the force of a wave measured?

The actual force of a wave on a beach has been accurately measured by an instrument called a spring dynamometer. This instrument records the amount of pressure exerted against a specific area. For example, at a seawall in Dunbar, Scotland, a dynamometer registered a force of 3.5 tons per square foot, which was generated by 20-foot (6-meter) waves.

## What are the **types of breakers**?

There are three major types of breakers (as most surfers can tell you), which are defined by the slope of the beach before it reaches the shallow water:

*Spilling breakers* occur on gently sloping beaches, and gradually break a considerable distance from the shore.

*Plunging breakers* are found on a moderately steep beach; these breakers curl over to form the characteristic tunnel most surfers like to ride.

*Surging breakers* don't break at all—they merely roll up on a beach without plunging forward. These "breakers" are found on very steep beaches and are considered the most destructive of all the breakers.

## What is the **surf**?

The surf is the wave activity seen between the shoreline and the outermost limit of the breakers. The surf zone is the area between the landward limit of the wave rushing onto the shore and the farthest seaward breakers.

## How can a **heavy surf** contribute to a **rainy day**?

Heavy surf off Washington's Olympic Peninsula, near La Push. *CORBIS/Joel W. Rogers*

As strange as it may seem, a heavy surf can contribute to a rainy day far from the shore. All raindrops have a nucleus, which can be a particle of dust—or even salt from any ocean. In a heavy surf, the breaking waves crash onto the shore, producing a wash of seawater and foam. The air bubbles in this surf, usually less than a fraction of an inch, burst and send salty particles into the air. Some of these particles are captured by the air currents in the atmosphere. They drift upward to become the nucleus of a raindrop or snowflake.

## What is the **swash**?

The swash are the waves that occur after a breaker collapses, forming a foamy, turbulent sheet of water that runs up the beach's slope. Swash is mostly responsible for pushing sand and gravel landward—up the beach.

## What is the **swash zone**?

The swash zone is the sloping part of the beach that is alternately covered and uncovered by the ocean waves.

## What is **backwash**?

Backwash is the return flow of the swash, or water that returns toward the ocean. It is felt by surfers and waders as a strong seaward current in the breaker zone (the zone where the waves break). Backwash is responsible for pushing sand and gravel seaward on a beach.

## What is **beach drift**?

The onshore-offshore shifting of sand and gravel that occurs through the action of swash and backwash also creates a sideways movement called beach drift. This causes the sand to move in a series of arched paths—moving particles long distances along the shore.

## What are **longshore currents**?

Longshore currents are currents in the surf zone that flow parallel to the shore. They form when the incoming wind-generated waves strike the beach at an angle. As the waves enter the shallow water near the shoreline, the leading edges slow before the rest of the wave, bending the wave as it moves ashore. This shoreward movement results in longshore currents, which are more prevalent along long and relatively straight coastlines. The strength of this type of current is variable and tends to be stronger in the winter months. The speed depends on a number of factors, including wave height, period, the angle of the wave to the beach, and the beach slope.

## Why are **longshore currents important**?

Longshore currents, by their nature, carry large amounts of suspended sediment and sand. They can erode a beach in one area and build it up in another. When a longshore current encounters deeper water, its forward movement slows, allowing the sediment and sand to settle to the bottom. One consequence of this process is the formation of sandbars in areas where longshore currents are common. These currents also deposit large amounts of sediment in harbors and channels, requiring repeated dredging to keep these areas open for shipping.

## What is **longshore drift**?

Longshore drift occurs when the longshore current and wind combine to move sand along the sea bottom in a direction parallel to the shore.

## What is **littoral drift**?

Littoral drift is the sediment moved along the shoreline under the influence of waves and currents. Beach drift and longshore drift move particles of sand in the same direction for a given set of onshore winds to create littoral drift. The two actions complement each other, and help shape shorelines. A straight or broadly curved shoreline usually means that the littoral drift is moving the sand in one direction for a certain set of prevailing winds. Rugged coastlines develop wave-cut cliffs; the sediment is carried away by the littoral drift moving along the sides of a bay. This action often forms crescent-shaped, "pocket" beaches.

## What is a **sand spit**?

A sand spit is a small point or low tongue of land made of sand or gravel, attached at one end to the mainland and having the other end in open water (usually a bay). This finger-like extension of the beach is deposited by the process of littoral drift, in which particles of sand are pushed by currents and onshore winds that move in the same direction. A long, narrow part of a coral reef extending from shore is also often called a spit.

## What is a **bar**?

As a sand spit grows from the littoral drift, it forms a barrier called a bar (or sand bar or sand reef)—a low ridge of sand bordering the shore. It is built up near the water surface by wave action along the shore and usually forms across the mouth of a bay.

## What are **tidal currents**?

Tidal currents are caused by the rising and lowering of tides in bays and estuaries. When a tide begins to fall, it creates an ebb current that runs toward the sea. When the tide is at its lowest point, the flow stops. And as the tide begins to rise again, a landward current develops, called the flood current.

## What is a tsunami?

The term tsunami is from the Japanese *tsu* (harbor) and *nami* (wave); it is often incorrectly called a "tidal wave." A tsunami is a gravity or deep-ocean wave, or succession of waves, generated by seismic activity—such as an earthquake—on the ocean floor. Some tsunamis may be caused by volcanic activity or underwater landslides; unlike earthquake-generated tsunamis, those that are the result of volcanic activity or landslides cannot be detected by current technology. No matter how they are generated, all tsunamis are influenced by the water depth and the shape of the shoreline. Because of this, it is often difficult to determine if a tsunami will adversely affect a shoreline or coast.

A tsunami generated by an earthquake can generate huge waves in the open oceans. The distance between crest to crest can be several miles, and the height from trough to crest a few feet. These waves cannot be felt on ships in deep water. Thus, in the open ocean, tsunami waves are harmless; it is when they reach shore that they become destructive and deadly.

## Why are **tidal currents** important to **seashores**?

Tidal currents are important to the shallow waters off the seashores of continents and islands because they erode and deposit sediment. The direction of the usually strong tidal currents changes throughout the tidal cycle. These changes make a clockwise path of water, or gyre, in the Northern Hemisphere, and a counterclockwise gyre in the Southern Hemisphere. In the open oceans, tidal currents are virtually undetectable.

## Why do **tsunamis move** so **quickly**?

The tsunami waves move so fast because of their energy. In a tsunami, as in other wave systems, the atmosphere interacts with the ocean. This relationship creates an increase in atmospheric pressure, which, in turn, causes the ocean to become shallower. This "squeezes" the waves, allowing them to move faster than they normally would in deeper water. This effect results in an increase in the speed of tsunami waves, with some estimates of waves exceeding 620 miles (1,000 kilometers) per hour.

## How can a **tsunami affect a coastal area**?

Tsunamis are feared along coasts because of their potential for destruction. If the underwater earthquake is small, the resulting tsunami is usually small, often dissipating before the waves reach a coast. If the earthquake is strong enough, it can generate a tsunami so large that its waves can be up to 200 feet (61 meters) high, and travel at speeds of up to 150 miles (241 kilometers) per hour—the tsunami slamming into the shore and surging onto the coast.

As a tsunami gets closer to shore, the velocity of its waves slows, while the wave height increases. In shallow coastal waters, tsunamis—often with crest heights of more than 98 to 164 feet (30 to 50 meters)—become a threat to life and property as they hit the coast with a devastating force. These waves can cause major damage, and even wipe out entire villages along the shore, by the energy released by the monster waves.

The dissipation of the tsunamis varies. For example, the Unimak, Alaska, tsunami of April 1, 1946, consisted of a succession of crests 2 feet (0.6 meters) high, with 122 miles (196 kilometers) between the crests in the open ocean. Once these waves entered shallower waters, they manifested themselves as 55-foot- (17-meter-) high waves, which hit the Pololu Valley on the coast of Hawaii at 15-minute intervals—causing major destruction. By the time the tsunami reached Bikini Atoll in the Marshall Islands, the deep-water wave height had dropped from 2 to 1.5 feet (0.6 to 0.4 meters), resulting in somewhat less powerful onshore waves.

## Which **coastal areas** are at **risk** for **tsunamis**?

About 80 percent of tsunamis occur in the Pacific Ocean (mainly because of the ocean's size and the significant earthquake and volcanic activity on its floor); 10 percent in the Atlantic; and 10 percent in other oceans. These figures seem to indicate that landmasses that ring the Pacific are most likely to experience tsunamis.

Recent studies have shown, however, that smaller earthquakes, formerly thought to be too weak to generate tsunamis, can trigger massive underwater landslides that create large, local tsunamis. This put more coastal regions at risk than previously thought. The Cascadia Subduction Zone off Washington State is this type of area (where smaller earthquakes occur and generate landslides and localized tsunamis), putting the British Columbia, Washington, Oregon, and northern California coasts at risk. Even southern California, because of its offshore topography and population density, faces the risk of tsunamis.

But it isn't just the west coast of North America that can experience these locally generated tsunamis; Puerto Rico and the Virgin Islands have also experienced these destructive waves. These islands lie on the eastern edge of the Caribbean tectonic plate, where even small earthquakes can cause underwater landslides resulting in local tsunamis. The Virgin Islands were hit by tsunamis in 1867 and 1868, while Puerto Rico experienced these waves in 1867 and 1918.

## Is there a way to **limit destruction** from a **tsunami**?

The largest amount of destruction from a tsunami comes from the action of the waves on land structures, and from the debris that is created and hurled by these waves. These effects are greatest when the affected land is relatively flat and the structures are flimsy. In order to limit the amount of destruction in tsunami-prone areas, scientists are recommending the construction of sea walls to break up and absorb some of the wave's energy; they also recommend a ban on construction in areas where the risk of tsunamis is high. These suggestions will be very hard to implement in the poorer, developing countries of the world that don't have the money necessary to build sea walls or to adequately control development in coastal areas.

## What are the **problems** with **tsunami prediction**?

There are two major problems with predicting tsunamis and warning people about them. First, since scientists cannot predict earthquakes on the ocean floor (and earthquakes generate tsunamis), we do not know when a tsunami will occur. Second, even when a tsunami is detected, because of all the variables involved—such as how deep the ocean is where the tsunami was generated and the shape of the shoreline it eventually hits—it is difficult to predict whether a tsunami will cause any problems when it does reach land. However, there are efforts under way to develop better systems for detecting tsunamis and, when possible, warning people about them (for more on this, *see* page 451, in the chapter on Global Ocean Conditions).

# BEYOND THE SHORE

## THE EDGES OF THE CONTINENTAL MARGINS

### What is the **continental margin**?

The continental margin is the part of a continental landmass that extends from the lowest low-tide line on a shore (or from the edge of the coastal plain) seaward to where it descends steeply to the ocean floor. Most of this area is underwater, and for much of recorded history was unknown to humans. Today, we know the dominant feature of the continental margin is a plain (or shelf) covered by sediment transported there by inland waterways.

The continental margin can be divided into specific areas or zones. Jutting out from a continent, and lying under relatively shallow water, is the zone called the continental shelf. It begins at the shoreline and slopes gently downward to a point called the shelf edge. Continuing seaward, at the shelf edge the underwater land begins to slope more steeply downward; this steeper area is called the continental slope. The point where this slope reaches the more level bottom of the ocean is called the base—both of the slope and the continent. Beyond this there is an area of sediment buildup called the continental rise, which extends to the ocean basin floor (or the abyssal plains; the deep ocean).

## What **features** comprise the **continental margin**?

The exact shape of the underwater landscape depends on whether the area is close to shore or farther out toward the deep ocean. In any case, the land beneath the waves is not uniformly smooth, like a beach; rather, the underwater landscape has a richly varied shape, rivaling the features found on the continents. Many of these features are similar to those found on dry land, such as valleys; others are unique, such as deep, sinuous canyons—the result of the ever-changing underwater environment.

## Have **continental margins always** been **underwater**?

No, scientists have determined that parts of the continental margins were alternately exposed and submerged, over and over again, during the past millions of years. This especially occurred during the Ice Ages, as the water in the oceans was alternately taken up by huge ice sheets and then released when the ice sheets melted during warming periods.

## What is the **continental shelf**?

The continental shelf (also called the continental platform) is the outward extension of land from the continental landmasses, covered by relatively shallow water. The continental shelves are typically not smooth beaches; they can have hills, deep gorges, and other topographical features.

The average angle of the continental shelf is about 0.1 degree, the shelf gradually increasing in depth as it extends into the ocean. In some areas, the slope descends 5 feet (1.5 meters) for every mile (1.6 kilometers)—a rate of decent that would go almost unnoticed on dry land. On average, the depth of the continental shelf is approximately 400 feet (122 meters) below sea level. The edge of the continental shelf is marked by an abrupt change in slope; this area is known as the shelf break or shelf edge. Landward of the continental shelf lies the shoreline and then the coastal plain. Seaward of the coastal shelf lies the continental slope.

## How is **sediment deposited** on the **continental shelf**?

Most continental shelves (also called embankments or aprons of debris) consist of recently deposited continental sediment on top of older layers of sedimentary rock. These layers form as sediment is washed away from the continents in rivers and streams.



## What is the **importance** of the **continental shelf**?

The continental shelves are extremely important areas. They are shallow enough to allow sunlight to reach the bottom, and are the recipients of nutrient-rich sediments from the adjacent continents. This combination has led to an almost infinite amount and variety of life: These areas are home to almost 80 percent of Earth's flora (vegetation) and fauna (animal life), and yield approximately 90 percent of the world's harvest of fish and shellfish. In some localized areas, the continental shelf is also a source of oil, natural gas, and mineral deposits—all packed into an area that is only about 8 percent of the total global seafloor.

## How big is the **Earth's total continental area** when the **continental shelves** are figured in?

Scientists often consider the continental shelves (and the islands on the continental shelves) as part of the continents. If the continental shelves are taken into account, the total continental area would be 35 percent of the Earth's surface; without the continental shelves, the total continental area is about 29 percent of the Earth's surface.

## What is the **average width** of the **continental shelves**?

On average, the global continental shelf is approximately 45 miles (72 kilometers) wide. But it varies greatly, and can range from less than 1 mile (1.6 kilometers) wide off parts of the western shores of North and South America, to more than 900 miles (1,448 kilometers) wide off the Arctic shore of Siberia.

## Is the **continental shelf** off the **United States shores** uniform in **width**?

No, the widths of the continental shelves off the United States shores are not uniform, varying in size from coast to coast. The width is almost

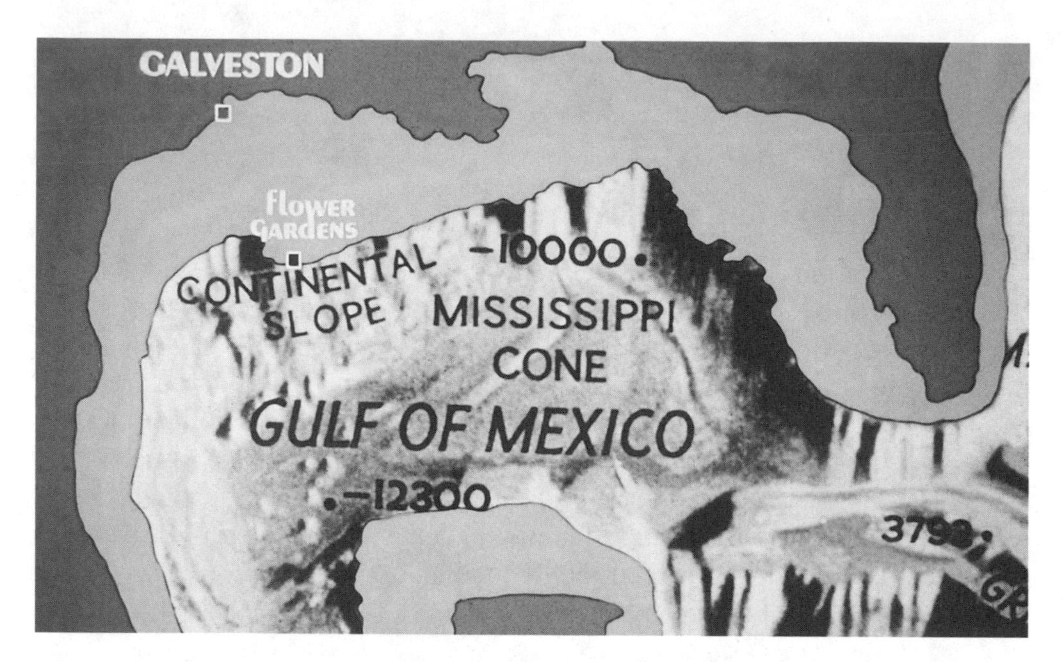


The underwater topography of the Gulf of Mexico is varied: Off the western coast of Florida the continental shelf extends about 200 miles (322 kilometers) before the continental slope (where the land rapidly descends toward the ocean floor) forms an almost vertical drop. *NOAA/OAR National Undersea Research Program*

negligible off the coast of southern California and southeast Florida. But on Florida's western coast, the continental shelf extends about 200 miles (322 kilometers) into the Gulf of Mexico.

## What is the **shape of the continental shelves** off the shores of the **United States**?

The continental shelves vary along the shores of the United States, but generally speaking, they approximate the features of the adjacent coast. For example, off the mountainous coast of southern California, the continental shelf has a similar, rugged look; the continental shelves along the Atlantic and Gulf coasts resemble the onshore low hills and plains of those regions; and the continental shelf off the mouths of large rivers, such as the Mississippi, tend to be broad and have large, muddy bases made up of sediments that were washed away from the continents by the action of the river water.

### What do continental shelves look like off other continents around the world?

The continental shelves around the world also greatly vary in size and area. For example, approximately one-third of the Arctic Ocean is underlain by continental shelves. It includes a large, broad shelf north of Europe and Asia and the narrower shelves of North America (above Canada) and Greenland. In fact, the average depth of the Arctic Ocean is only about 3,407 feet (1,038 meters) because of the large shallow expanse of the continental shelf.

## What is the **size of the continental shelves** off the shores of the **United States**?

The continental shelves off the shores of the United States cover a total surface area of approximately 891,000 square miles (2,307,000 square kilometers); at 591,000 square miles (1,530,000 square kilometers), the Alaskan continental shelf accounts for half of that area. The relatively small shelf off California, Oregon, and Washington covers approximately 25,000 square miles (65,000 square kilometers), while the Atlantic continental shelf has an area of approximately 140,000 square miles (362,000 square kilometers)—equivalent to the surface area of New York, New Jersey, Ohio, and Pennsylvania combined. And in the Gulf of Mexico, the continental shelf covers nearly 135,000 square miles (350,000 square kilometers), about the combined size of Nebraska and Iowa.

## How do **sand ridges** form on the **continental shelves**?

Sand ridges are distinct patterns on the ocean bottom. These crests of sand rise to a height of a few yards above the continental shelf, and extend in length for hundreds of yards. These sand ridges, which include longshore bars and barrier beaches, can occasionally emerge from the water as they build up sand, or they can remain submerged. They are found where there is an abundance of sand and they can encompass large areas of the continental shelf. Often, there are row upon row of these ridges, arranged in tidy, parallel lines, and separated from each other by a distance of a few hundred yards. Some of these ridges remain stable over long periods of time, while others move and change shape quickly. Still others swiftly appear then disappear.

For many years, scientists puzzled over the origin of these regular patterns, believing they were collections of sand formed by the actions of tidal or ocean currents, or by waves induced by storms. A recent model suggests that certain types of ocean waves—those with long wavelengths (distances between two consecutive crests or troughs) and large amplitudes (distances between wave crests and a neutral point, usually the water's "at-rest" level)—are largely, though not completely, responsible for the formation of these sand ridge patterns. This type of wave is thought to ruffle the ocean surface and create similar movements in the water close to the shelf's sandy floor. This action moves sand out of one area and deposits it in another, creating a ridge. But this model must still be verified in the actual ocean environment before it can be accepted as the final word in sand ridge formation.

## What is the **continental slope**?

The continental slope is part of the continental margin in which the land takes a much more rapid descent toward the ocean bottom. It characteristically has a slope of 3 to 6 degrees and is the true edge between the continental landmass and the ocean basin.

The continental slope is relatively narrow, ranging from 6 to 60 miles (10 to 100 kilometers) wide. This plunge to the bottom can be a steep slope or, in some areas, a precipitous cliff. For example, off western Florida in the Gulf of Mexico, the continental slope is an almost vertical wall; in one area, there is an almost perpendicular drop of 1 mile (1.6 kilometers). The continental slope off North Carolina is generally steeper than the slope to the north, as it is carved by the strong south-flowing Western Boundary Undercurrent. Its slope averages about 7 degrees, with a maximum of 16 degrees off Cape Hatteras, and reaching about 3.5 degrees at Blake Ridge. Off Santiago, Chile, South America, the continental slope drops steeply and uninterrupted from the edge of the relatively shallow continental shelf to a depth of 20,000 feet (6,096 meters).

This three-dimensional image (from 1989) shows the continental slope off the coast of Oregon, south of the Columbia River entrance. *NOAA; Captain Albert E. Theberge, NOAA Corps (ret.)*

The final depth at the bottom of the continental slope varies from ocean to ocean. Along the Atlantic coast the slope stops at about 9,843 feet (3,000 meters); on the Pacific side of North and South America, the slope continues into deep-ocean trenches, reaching depths up to 26,248 feet (8,000 meters).

The true origin of continental slopes is unknown. Scientists believe some slopes developed as continental landmasses pulled apart along rifts and faults (fractures in the Earth's crust) to create the ocean basins. Others may have been created by uplifting of land at the edge of the continental shelf (for example, by volcanic activity), or by coral reef build-up.

## What is a **submarine canyon**?

The inclined landscape of the continental slope is frequently cut by spectacularly steep valleys called submarine, or underwater, canyons. These canyons are thought to be either ancient river channels or geologic extensions of canyons on land. The narrow, V-shaped valleys of submarine canyons are located in all the oceans, and generally originate near the mouths of large rivers (or near the sites of former river mouths). They are similar to canyons on land, and even have tributaries that cut into the sides of the main canyon. These deep, steep-sided valleys are up to 1 mile (1.6 kilometers) deep; the slope of the channels range from only 1 degree inclination to 30 degrees inclination. Submarine canyons extend from the continental slope and generally open onto the continental rise, where the slope "drops" sediment on the ocean floor.

## What are the **types of submarine canyons**?

There are several types of submarine canyons, including the steep-sided, narrow, V-shaped valleys and the broad, flat-bottomed valleys.

## What is a **V-shaped valley**?

A V-shaped valley, on land or in the ocean, is just as it sounds: a valley in the shape of a V, with steep sides and short tributaries entering into the main valley. They are formed by streams or ocean currents that cut in a downward motion, narrowing the valley into its characteristic shape. Over time, the V-shape may become broader, forming a V-shaped valley.

## Are there **submarine canyons** off the **eastern United States**?

Yes, there are several submarine canyons along the eastern United States. These underwater canyons usually begin near the shore and extend along the wide continental shelf in an almost straight line, gradually getting deeper and deeper as they approach the continental slope.

One of the best examples of a submarine canyon is the Hudson Canyon—a major feature of the mid-Atlantic Ocean, and the largest submarine canyon on the eastern United States continental margin. First identified in 1864, this canyon originates as a shallow valley crossing the continental shelf. The Hudson Canyon channel extends from the mouth of the Hudson River at New York City, as the Hudson Shelf Valley; seaward of the continental slope, it cuts across the continental rise as the Hudson Valley; it continues to the floor of the ocean some 100 miles

(161 kilometers) away and approximately 12,000 feet (3,658 meters) deep, where it forms the Hudson Fan. The canyon, measured at the continental slope, is some 3,600 feet (1,097 meters) deep and approximately 5.5 miles (8.9 kilometers) wide. And this is not the largest canyon in the world cutting through the continental slope!



There are numerous other submarine canyons off the eastern United States, including the Hendrickson, Washington, Baltimore, Hatteras, and Pamlico Canyons—deep valleys that extend across the upper continental rise and stop on the lower rise where they form submarine fans.

## Are there **submarine canyons** off the **western United States**?

Yes, the submarine canyons off the western United States are somewhat different from their eastern counterparts. Here, the canyons are extremely rugged and twisted. There is no gradually deepening, shallow valley. The continental shelf is very narrow here, so the deep canyons begin less than 1,000 feet (305 meters) off the shore.

The Monterey Canyon off the coast of northern California is one of the most spectacular canyons in the world—on land or underwater. It is the largest and deepest submarine canyon on the west coast—beginning right at Moss Landing, California, where the Salinas River flows into the ocean, and at a depth of 50 feet (15 meters). It extends out to sea for more than 60 miles (97 kilometers), terminating at a depth of approximately 10,000 feet (3,048 meters), and has wall heights up to 6,000 feet (1,829 meters). At the end of the canyon is the Monterey Canyon Fan, which is more than 100 miles (161 kilometers) from land. The head of the canyon acts as a giant sediment trap: North-to-south moving longshore currents carry sediment into the canyon where turbidity currents (currents caused by the force of gravity "pulling" the water and sediment down the sides of the canyon) are created; these currents then deposit sediment onto the fan.

## Where else in the **world** can **submarine canyons** be found?

Some of the largest and most well-known submarine canyons in the world are found at the mouths of mighty rivers, including at the mouths of the Amazon River (off Brazil, in the Atlantic Ocean), the Indus River

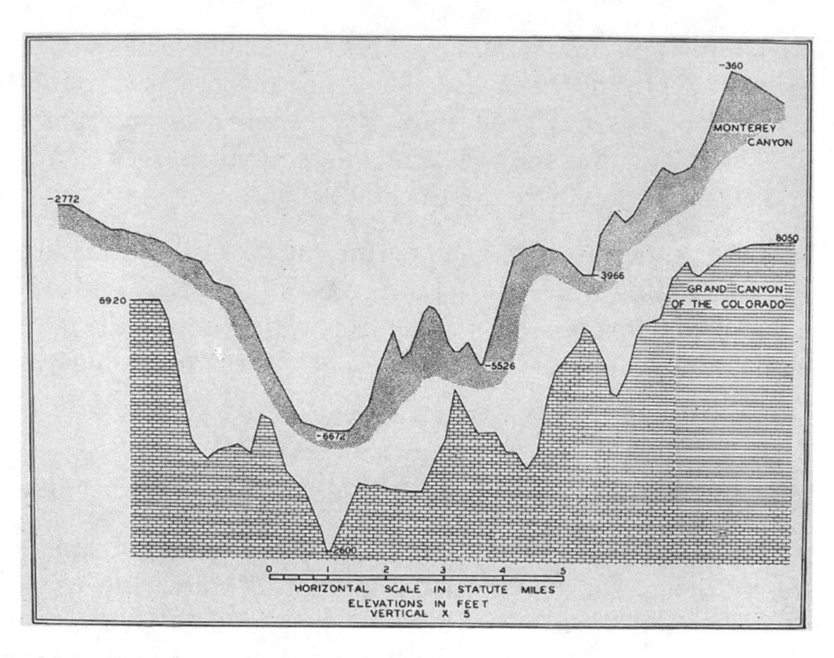


A graphic (from *Submarine Canyons* by Shepard and Dill, 1966) showing how the underwater Monterey Canyon (top) compares with the Grand Canyon (bottom). *NOAA Central Library*

(off Pakistan, in the Arabian Sea), the Ganges River Delta (off India, in the Bay of Bengal), and the Congo River (off western equatorial Africa, in the Atlantic Ocean). Many of these underwater canyons are similar in size and proportion to the Grand Canyon of Arizona. For example, the Congo Canyon extends 500 miles (805 kilometers) right up to the mouth of the Congo River, and river sediment bypasses the continental shelf to travel directly down the canyon to the continental rise. Closer to home, there is a well-known canyon where the St. Lawrence River empties into the Atlantic Ocean, off Canada's Gaspe Peninsula.

## How did **submarine canyons form**?

There is still considerable debate concerning the origin of the submarine canyons. Most of the canyons were originally thought to have formed during the last Ice Age (the last ice sheet retreated only about 11,000 years ago). Scientists believe these valleys were extensions of large rivers that were later submerged when sea levels rose. But if this is true, the oceans would have had to rise (or land would have had to fall) more than 2.5 miles (4 kilometers) worldwide—which is much greater than the estimate of the rise in sea level that occurred after the Ice Ages, which is just a few hundred feet (a hundred or so meters).

The best current theory states that the canyons formed—and continue to evolve—from intense erosion caused by rapidly flowing, sediment-carrying currents called turbidity currents.



## What are **turbidity currents**?

Turbidity currents are a mix of water and suspended sediment that gravity causes to flow down a slope, usually a submarine canyon. No one has ever seen a forming turbidity current—or at least there are no confirmed reports. But some divers have observed slow, steady currents of sand flowing down canyon walls, and a submersible that hit a canyon wall apparently started an avalanche of sediment and water. And there is indirect evidence of the existence of turbidity currents—the sediments deposited by them.

Scientists think the turbidity currents affect submarine canyons as follows: Sediment is eroded from the continents, then carried by currents and waves to the edge of the continental shelf. An indirect "event" such as an earthquake, strong ocean current, flood, violent ocean storm, or the sediment's own weight causes the sediment to "break off" of the edge. This triggers rapid water movements—an underwater version of storm surge—and a dense slurry (a suspension of sedimentary particles in the water) moves rapidly down the slope. As the turbidity current flows down the canyon, it carves out more of the valley, eroding and transporting the sediment as it travels—thus starting a new valley or deepening an existing canyon by cutting into its walls.

## How were the first **turbidity currents detected**?

The theory of the connection between turbidity currents and submarine canyons was first put forward in the mid-twentieth century. The first true evidence was gathered in 1929, after a major earthquake occurred on the Grand Banks off Newfoundland, Canada. There several submarine telegraph cables broke in sequence—from the shallow to deeper waters. Scientists decided that strong currents raced down the continental slope traveling from 25 to 50 miles (40 to 80 kilometers) per hour. They further deduced that turbidity currents, strong currents of water and sediment, were responsible for the cable breaks. Since that time, other cables have broken in the same manner. For example, in 1935 a cable broke in a canyon off the Magdalena River, South America. This was further evidence that turbidity currents carve out an underwater canyon: Twisted around the cable at a depth of about 1 mile (1.61 kilometers) was a mass of shallow-water marsh grass.

## What is a **turbidite**?

The graded bed of sediment left by a turbidity current—usually at the mouth of a submarine canyon—is called a turbidite. Coarse particles settle out first, and are found at the bottom of turbidite deposits; finer particles settle toward the top of the turbidites.

## What is a **submarine fan** or **delta**?

A submarine fan or delta is a cone- or fan-shaped continental sediment deposit located seaward of large rivers and submarine canyons. They are usually associated with turbidites (the graded bed of sediment left by a turbidity current). The sediment travels from the rivers to the oceans, and as the water slows, the sediment drops onto the ocean floor in the shape of a fan. If the sediment is carried farther down the continental shelf and slope, it slows as it reaches the base of the slope. It then spreads out, putting down the sediment load in the form of a "delta" on the continental rise—or even as far out as the abyssal plains (the depths of the ocean). Submarine fans are also called submarine cones, abyssal cones or fans, subsea aprons, and deep-sea or sea fans. Some of the fans are created by the deposit of the sediment normally present in the water; but turbidity currents contribute to the accumulation of some fans.

## What is the **largest underwater turbidite** discovered to date?

The largest underwater turbidite (the graded bed of sediment deposited by a turbidity current) discovered to date was found in 1998; the vast blanket of sand and silt was larger than any other ever witnessed. The

26- to 33-foot- (8- to 10-meter-) thick deposit sits at a depth of 9,184 feet (2,800 meters) between France and Algeria in the Mediterranean Sea. The volume of sediment is estimated to be enough to bury the island of Manhattan under a pile nearly the height of Mount Everest! It is estimated to have formed about 22,000 years ago, at the peak of the last Ice Age.



## Why are scientists interested in **ancient, gigantic turbidites**?

There are other gigantic turbidites under the ocean besides the one in the Mediterranean, but there is little information on these sites. Some scientists believe these underwater features may be important in interpreting the world's ancient climate. One such climate-turbidite theory states that as the sea levels naturally rose and fell over time, drops in ocean levels would have reduced the amount of pressure on the seabed. This could have caused buried deposits of methane hydrates (deposits containing methane gas) to break down, giving way to underwater landslides. If this is true, methane—a greenhouse gas—could have been released into the atmosphere and warmed the climate. But so far, ice cores taken deep under the ice sheets of the polar regions have shown no evidence of huge methane releases.

## Are **turbidites** found anywhere **on land**?

Turbidites can be found on land, including rock layers in several mountain ranges, such as the European Alps. Such deposits indicate that these areas were once underwater—situated on the continental rise, at the mouth of a submarine canyon.

## Is there a connection between **whales** and **turbidity currents**?

At one time, some scientists believed that turbidity currents could be started by whales ramming against the side of a canyon wall. Scientists believe that this rarely (if ever) occurs, as there is no direct evidence—and whales are usually found farther out, in the open oceans.

## What is the **continental rise**?

The continental rise is the transition area between the steeply pitched continental slope and the ocean basin floor. The width of this area varies between approximately 62 and 621 miles (100 and 1,000 kilometers). In this region, sediment that originated on land and was transported down the continental shelf and slope accumulates. Continental rises are uncommon in the Pacific, but are well developed in the Atlantic and Indian oceans, as well as around Antarctica.

## Are all **continental shelves, slopes, and rises** similar around all the continents?

No, just like the continents themselves, the continental shelves, slopes, and rises vary from ocean to ocean. For example, the continents ringing the Pacific Ocean have very narrow continental shelves, measuring about 12 to 24 miles (19 to 39 kilometers) wide. Many areas in the ocean also have very steep continental slopes that drop into an ocean trench. Most landmasses around the Atlantic Ocean, and much of the Indian Ocean, have much wider continental shelves and slopes, with many valleys and pronounced continental rises.

# REEFS

## What is a **reef**?

A reef is a permanent ridge, mound, or mass of material located in a waterway, standing above the sediment and rock on which it was deposited. Reefs can be found in shallow water, just offshore, on the continental shelf or farther out, on the continental rise. Shallow-water reefs are strong and mostly wave-resistant; because of this, reefs are often a danger to navigation.

## How do **reefs form**?

Reefs form for a variety of reasons. Some are really submerged landforms and are therefore composed of the same rock as the nearby (and sometimes adjoining) landmass. Other reefs are composed of previously deposited masses of sand, gravel, or shells. Still other reefs are artificial, created by humans to encourage the growth of plants and animals. Various materials have been used to created these artificial reefs, ranging from old tires to ships sunken on purpose. But the most familiar reefs are those created by living organisms, such as coralline algae and polyp corals.

## What is **coralline algae**?

Coralline algae is a red algae that is encased in a transparent calcium-containing shell. They are colonial, with numerous coralline algae building up hard structures that resemble polyp-built reefs. Coralline algae reefs are not, however, as prevalent as polyp-built coral reefs. In 1984 off the coast of the Bahamas, scientists found a coralline algae at 884 feet (270 meters) below the water surface. This was a surprising place to find coralline algae, an organism that relies on photosynthesis; after all, it was previously assumed that the photosynthesis process, which requires sunlight (radiant energy), could not take place at such depths.

## What are **corals**?

Corals are sedentary, calcareous marine organisms. They are part of the phylum (animal group or classification) Cnidaria (or Coelenterata), whose members are somewhat symmetrical and have tentacles. These animals live attached to the seabed or to other hard objects.

## What is a **coral reef**?

Coral reefs, often called the "rainforests of the oceans," are the richest and most colorful marine habitats in the oceans. Not only do coral reefs contain some of the most diverse marine plants and animals, but are themselves also animals (and the remnants of animals). And although there are many types of corals—such as sea fans or sea whips—the ones

**Coral can take many different shapes—including a dome-shape that resembles the brain.** ***NOAA/OAR National Undersea Research Program***

that build reefs are known as hard, or stony, corals. Since not all corals are alike, each reef has its own characteristic growth patterns (which depend on the coral), and its own characteristic population.

## Where are **coral reefs found**?

Corals usually grow in shallow, clear tropical waters; they thrive in waters of about 70° F (21° C). They are found in many regions of the world, extending north and south of the equator where there are warm currents, such as along the coasts of Florida, Japan, and Australia. The reefs usually form as the warm surface currents move close to the eastern side of the continents located near the equator. Along the western edges of the continent, deep polar upwellings move toward the equator—and are not conducive to growing coral reefs.

## Are **coral reefs** always **islands**?

No, coral reefs are not always islands. They can also be attached to a mainland, from which they extend into the ocean. The Florida Keys are one example of this kind of reef. But most coral reefs, including Australia's Great Barrier Reef, are low-lying island structures built up on basaltic lava flows under the ocean's surface. Or they can be islands of coral called an uplifted coral platform, which are usually slightly larger than the low coral islands. Banaba Island (formerly Ocean Island) and Makatea Island of French Polynesia are examples of uplifted coral platforms.

**Sea fan coral is made up of polyps attached in a fan-like pattern to a central internal skeleton.** ***NOAA; Linda Wade***

## What are the major **types of coral reefs**?

There are three main types of coral reefs: Fringing, barrier, and atoll.

Fringing reefs are usually found close to the shorelines of continents (or along the continental shelf) or around islands. They form as long bars attached to the land, usually parallel to the continental shoreline along the eastern side of the landforms. They grow outward from the continent to the edge of the continental shelf. The Florida Keys are excellent examples of fringing reefs.

Barrier reefs are found at some distance from the shore. They are separated from the shore by a lagoon, and are situated in ocean shallows on the eastern side of continents. They are usually shaped as multiple bars separated by deep channels. Their rate of growth changes over time, often growing faster in the shallows. The best example of a barrier reef is the Great Barrier Reef off eastern Australia.

An atoll reef actually forms after a volcanic island rises above the ocean's surface, and then is worn away by the action of the ocean—as the volcanic island "sinks" (subsides), the reef grows. They are roughly circular

**Fringing reefs, one of three types of coral reefs, form as long bars attached to the land. This reef (the outline is visible in the water) formed adjacent to a high volcanic island in the Western Caroline Islands (western Pacific Ocean).** ***NOAA/Dr. James P. McVey, NOAA Sea Grant Program***

in shape, often dotted with small, sandy islands on top, and are surrounded by the open ocean. They grow slowly, getting larger and rising above sea level to form the small coral reefs. The atoll reef usually encircles a lagoon—a warm, shallow, quiet body of water. One good example is the Bikini Atoll in the Pacific Ocean, famous for being the site where the United States tested the atomic bomb in the 1940s.

## How do **coral reefs grow**?

Coral reefs grow though the action of tiny animals, called coral polyps, which can be as small as a pinhead and no larger than a human fingernail. They are either found separately or in a colony (most coral reefs are animals living in colonies). These animals are mostly made up of soft tissue, but they also secrete a stony-like, cup-shaped skeleton (usually calcareous) around them. The lower part of the coral reef is actually the skeletons of the coral animals. The reefs form as small coral animals divide and create new animals, surrounding themselves with hard, outer skeletons; as new animals are created, they lie on the skeletons of dead animals.

There is a great variety of polyp life. Some polyps build corals that look like branching trees or bushes, domes, brains, fans, or even deer antlers. The living polyps themselves range in color from pink, yellow, green, blue, and purple—causing the reef to resemble an underwater flower garden.



## How **fast** does a **coral reef grow**?

Coral reefs grow very slowly, usually only a few fractions of an inch (millimeters) per year. For example, brain corals grow very slowly and have annual growth rings that resemble those of trees. Other, larger coral colonies are extremely old—dating between 800 and 1,000 years. Some very large reefs have been growing for millions of years and measure hundreds of meters (or yards) thick.

## How do **corals reproduce**?

Corals reproduce in one of two ways: by budding or from eggs. Most corals use the process of budding, in which the polyp forms small buds that develop into new polyps. Sometimes, corals reproduce from eggs, the larva hatching from the egg of an adult in a process called spawning.

## How do **corals spawn**?

Corals spawn using a process called broadcast spawning. With this method, timing is all: Not only does it only happen once a year, it has to be synchronized perfectly with other spawning corals. In many species of coral, the males release sperm into the water; then a short time after, the females release egg bundles (collectively called gametes). With such a flurry of activity in the water, larger animals are attracted to the site for food—and many of the eggs do not survive. The sheer number of eggs and sperm, however, ensure the survival of the species.

If a gamete (egg bundle) survives, it reaches the planktonic larval stage. The organism floats with the ocean currents for several days—or several months—making it highly vulnerable to predators. But if it survives to this state and finds a suitable surface—often part of an existing coral reef—the larva attaches itself to the spot by secreting calcium carbonate

The coral atoll of Tetiaroa, Society Islands, French Polynesia. *CORBIS/Douglas Peebles*

(it absorbs calcium from seawater to produce the calcium carbonate). From there, it continues the process of building up the existing reef.

## How do **coral polyps eat**?

In order to eat, the polyps reach tentacles from their cup-shaped housing, catching microscopic food particles as they float by in the water. Corals also eat tiny single-celled plants that live inside the coral's tissues. Corals and the single-celled plants (algae) they eat have a symbiotic relationship—or each depends on the other to survive. The corals need to eat the algae, while the corals build the reef where the algae can grow.

## Why are **coral reefs** so important?

Coral reefs are the richest marine habitats in the oceans, and attract more species of fish than any other habitat. Many of those fish have familiar names, such as butterfly fish, clown fish, and gobies. The fish in and around the reefs feed on sea grasses and on tiny planktonic organ-

**Since coral reefs can sustain a diversity of life rivaling that of the rainforests, preserving these underwater habitats is crucial.** *NOAA/OAR National Undersea Research Program; B. Walden*

isms—both phytoplankton (tiny plant organisms) and zooplankton (tiny animal organisms). In return, the fish excrete nitrogen and phosphorus-containing materials. These materials are recycled—used by the coral or other organisms living in and around the reef.

Coral reefs are also important, if not vital, to humans. They represent millions of dollars in tourism (reefs make excellent spots for recreational snorkelers and divers to investigate underwater life); they create a food-rich spawning ground for commercial fish species; they act as barriers, protecting coastal areas from extreme storms; and more recently, they have become a source of ingredients for many life-saving medicines.

## What is the **keystone species** of a **coral reef**?

The keystone species of a coral reef is the dominant species of the reef. In most cases, it is a single species that controls the population of a reef, such as a predator preventing the possible intrusion of another species. For example, damselfish eat algae and expose the coral, but they also keep other coral predators, such as parrotfish, from chewing up the coral.

## Which **reef** contains the **most plant and animal species**?

There are conflicting reports about which of the world's reefs teems with the most life. It seems to be a toss-up between two: The coral reef region off southeast Asia and the Great Barrier Reef are each estimated to be home to more than 3,000 plant and animal species.

## What is the **largest coral reef**?

The Great Barrier Reef, in the Coral Sea off the northeastern coast of Australia, is the largest coral reef on Earth—and the largest thing on Earth built by living beings. The 1,200 mile- (1,931 kilometer-) long coral reef is in a perfect place: The area has clear, warm water for the coral to grow in, and receives very little silt in runoff from the continent. Most of the area was designated a marine park in 1983.

## Is the **Coral Sea made of coral**?

The Coral Sea is not a sea filled with coral. It is a body of water that is part of the Pacific Ocean—bounded by Australia's Great Barrier Reef to the west, Papua New Guinea to the north, and the New Hebrides (Vanuatu) and New Caledonian island groups to the east. The reason for the name is the sea's proximity to the Great Barrier Reef. There are also two other major coral reefs in the Coral Sea.

## Is the **Great Barrier Reef** in **trouble**?

Yes, the Great Barrier Reef is in trouble. Its microorganisms are dying—and no one knows why. One suggestion is that manmade pollutants on land flush out to the sea via nearby rivers, killing off the organisms. Another idea is that global warming is causing the reef to die. In this scenario, the warmer air temperatures would increase the shallow water temperatures in which the reef exists, killing off the organisms. Or global warming may be causing the sea level to rise too high; too much water prevents the reef from growing.

## What causes **destruction** to **coral reefs**?

Coral reefs can be destroyed in many ways; they can be damaged by physical or biological means. Physically, the action of waves or storm surges can destroy or weaken the platforms that hold a reef. Biologically, organisms prey on live corals. For example, crown-of-thorns starfish and parrotfish eat live coral, and can cause extensive damage to reefs.

Humans also cause problems for coral reefs, directly and indirectly. Some reefs have been over-fished, destroying the balance of life, and others have been damaged by commercial fishing practices that use poisons and explosives. Reefs have been nearly buried by an over-abundance of silt and sediments that have washed out to sea from land—the result of bad forestry, farming, and construction practices. Coral reefs have also become polluted by sewage and other environmental contaminants. In addition, corals have been broken by ships and divers; sometimes the corals have been altogether removed, along with their accompanying fish species, for use in aquariums.

Global warming has also been blamed for a recent die-off of coral. Almost two-thirds of the world's coral reefs are in decline or threatened. High sea temperatures—the highest on record—were determined to be the direct cause of the large-scale death of some reefs. The hardest-hit areas were in the Indian Ocean, such as the Seychelles, Mauritius, and the Maldives, where the die-off was between 70 and 90 percent. In the western Pacific, from Vietnam to the Philippines and Indonesia, thousands of miles of corals have died-off or have been bleached. So far, the only large areas of coral that have escaped this massive devastation are located in the atolls of the central Pacific.

## What is **coral bleaching**?

Coral bleaching occurs when the coral tissues expel zooxanthellae, a type of algae that resides in the coral structure; the coral then turns white and dies. When in balance, the algae and coral live harmoniously in a symbiotic relationship: The algae gives the color to the otherwise white coral skeleton and provides carbon compounds that nourish the coral. In return, the coral gives the algae a place to live, and the nitrogen and phosphorous the algae need to survive. But if there is a change in salinity or an increase in temperatures, ultraviolet exposure, or pollu-

**The aerial view of Australia's Great Barrier Reef, Queensland. Its microorganisms are dying, endangering the reef.**
***CORBIS/Robert Holmes***

tion, coral bleaching can occur. If any one of these conditions persists for very long, the corals may never recover.

## Is there a **connection** between **climate change** and **coral bleaching**?

Yes, while some of the recent destruction of corals might be attributable to stressors such as substantial freshwater runoff or human use of the area around the reefs, many scientists believe there is a connection between changes in global climate patterns and coral bleaching. For example, during the strong El Niño event of 1982–83, in which the sea surface warmed about 1 to 2 degrees F higher than the seasonal average, there was widespread, permanent coral bleaching in the equatorial Pacific Ocean from Japan to Panama. During the 1997–98 El Niño, global sea temperatures were again high and it was also the worst year ever recorded for coral bleaching. The bleaching occurred in waters around the globe, but close to home, the Florida Keys were heavily damaged.

While scientists debate whether global warming is a function of human activity and if El Niño is a product of global warming, the facts show a correlation between high ocean temperatures and coral bleaching. If this pattern continues, it also follows that warming will diminish marine biodiversity.



## What **fungus** is responsible for the **death of sea fan coral**?

The fungus *Aspergillus sydowii,* first described in science literature in 1913, has been responsible for the devastating loss of sea fan coral over the last 15 years. The fungus was identified by genetic studies of DNA taken from infected corals at different sites, including the Bahamas, the Florida Keys, the Virgin Islands, Puerto Rico, Mexico, Panama, Venezuela, and many other coral reef locations.

Sea fan coral is made up of polyps attached in a fan-like pattern to a central internal skeleton. The skeleton supports all the branches of the colony and older colonies can reach up to 5 feet (1.5 meters). The fungus disease causes a characteristic receding of the polyps, revealing the dead internal skeleton.

## What has been done in the **Florida Keys** to **overcome coral reef problems**?

Over-fishing and pollution have put a strain on the coral reefs in the Florida Keys. It has been difficult to curb the pollution problems, but there is a plan to combat over-fishing, called Tortugas 2000. This plan establishes no-take zones, areas in which no plants or animals can be removed, although diving is still allowed. It is hoped that this will relieve some of the pressure on the Keys' coral reef system.

## What **plants** may help **protect coral reefs**?

A recent experiment showed that Sargassum, a kind of seaweed, may protect certain coral reefs. For example, many inshore reefs of the Great Barrier Reef were experiencing coral bleaching. Scientists didn't know if Sargassum, a large, brown seaweed, was causing the bleaching problems or not. To test if the seaweed beds had any effect on the corals, researchers studied several parts of the reef, some with 3- to 6-foot- (1- to 2-meter-) thick canopies of the seaweed, and other parts with no covering of seaweed at all. After many months, they compared the data and found that the average percentage of bleached corals was significantly higher in plots without the Sargassum canopy than in plots where the seaweed canopy was intact. Overall, 19.6 percent of corals were bleached under "normal" conditions for these reefs—but 36.4 percent were bleached when the Sargassum canopy had been removed.

Based on this experiment, scientists suggest that the coral reef with the canopy was healthier because the seaweed cover acted as a natural umbrella—decreasing the corals' exposure to damaging high temperatures and ultraviolet light. Or the seaweed may have reduced the mixing of water, thus maintaining a more constant salinity around the coral reef.

# BARRIER AND CONTINENTAL ISLANDS

## What are **barrier islands**?

Barrier islands are long, narrow, sandy islands that separate the open ocean from the mainland; in between the barrier islands and the mainland lies a lagoon or embayment. The islands are made of recent (geologically speaking) deposits of unconsolidated sediments and shells that build up parallel to the shore and above high tide. Thus they are exposed land just offshore. Barrier islands commonly have dunes on the ocean side, and vegetation and swampy areas (such as tidal flats) on the lagoon/mainland side.

## How do **barrier islands form**?

Barrier islands form as waves build up a ridge of sand just offshore the mainland. The islands increase in height as sand dunes on the islands grow. Behind the barrier islands, a lagoon forms, a broad area of water (often several miles wide), which is also shallow, the bottom filled in with tidal sediments. The islands may also be cut by tidal inlets—gaps through which strong currents flow seaward and landward as the tide rises and falls.

Each barrier island chain is unique, with its own rate of erosion and deposition. For example, the barrier islands off North Carolina began after the last Ice Age, when sea levels changed in response to the melting ice sheets. These islands have been carved into sharply pointed headlands—or capes—with a very broad inner embayment made up of mostly shallow water areas and salt marshes.



## Do **barrier islands change** over time?

Yes, barrier islands change—even over a short period of time. In fact, they are often referred to as nomadic landforms, as they are constantly changing in shape, size, height, and location—all in response to the forces of wind, waves, currents, sea level changes, storms, and the amount of sediment carried there from land and sea. For example, the Virginia barrier islands off the East Coast of the United States have retreated landward in the past 100 years—and many islands no longer exist (they have become part of the mainland).

## Why are **barrier islands important**?

Barrier islands are physically important to a coast: They protect the bays and mainland behind them from the constant battering of ocean waves; during major storms, they act as a buffer from erosion and damage that are caused by intense wave activity; and ecologically, they provide wildlife habitat, including dunes for vegetation and tidal flats for shorebirds and other near-shore marine animals.

## Where are **barrier islands** found?

Barrier islands are found all over the world. Off the East Coast of the United States lie the Outer Banks of North Carolina. Off the Gulf coast lie several barrier islands, including the Padre and Matagorda islands on the east coast of Texas—which form one of the longest unbroken sec-

The bird's eye view of North Carolina's barrier islands; the mainland appears in the upper left corner. *JLM Visuals*

tions of barrier islands, measuring more than 100 miles (161 kilometers) long. In the Adriatic Sea, Lido Island, off the coast of Italy, protects the city of Venice. And in the North Sea, the Frisian Islands protect farming regions of coastal Germany and Denmark.

## What are **continental islands**?

As the name implies, continental islands are islands that were once part of a major continental landmass; the island and continent still share the same continental shelf. New Guinea in the western Pacific, California's Channel Islands, the islands of New Zealand, and the Philippines are all continental islands—they are structurally associated with nearby continents.

## How do **continental islands** form?

Most continental islands separated from the mainland because of rising sea levels after huge glaciers melted thousands of years ago (at the end of the last Ice Age), cutting off higher ground from the mainland as the lowlands filled with water. For example, the British Isles formed in this way, as rising water cut the islands off from the European mainland.

Southern California's Channel Islands, five landmasses (including the Anacapas, shown here) are continental islands—they were once part of the nearby continental landmass. *CORBIS/George Huey*

Other continental islands formed as the crustal movements of the Earth separated them from the original land mass. For example, Greenland was separated from North America when the crust moved apart.

Still others form by the weathering and erosion of a link of land that once connected the island to the mainland. For example, as the Orinoco River flowed into the ocean, it cut through the land linking the South American mainland and what is now the island of Trinidad, off the coast of Venezuela.

## How do scientists know which **islands** are of **continental origin**?

In most cases, continental islands are distinguished by being close to the neighboring continent. They can sometimes be verified as continental if the two bodies are made of the same types of geologic structures and/or rock layers (usually granites).

## When did certain **continental islands separate** from a **parent continent**?

In most cases, it takes millions of years for a continental island to separate from its parent continent. For example, the Greater Antilles (West Indies, in the Caribbean Ocean) separated from the North American continent about 80 million years ago; Madagascar from Africa about 100 million years ago; New Zealand from Australia approximately 80 to 90 million years ago; and the Seychelles from Africa 65 million years ago.

### Which continental island is the largest?

The world's largest subcontinental island is Greenland. This can be verified by comparing the landmasses' major rock units: Rocks from Greenland can be matched with those from North America and northern Europe—all the same age and composition.

Many of today's islands were connected to a mainland during the Pleistocene epoch—during the Ice Ages—but are no longer connected due to the melting of huge global ice sheets, which caused sea levels to rise, flooding the lands around the islands. The British Isles, the Sunda Isles, the Falklands, and the Philippines were all formed in this way, as were the islands of Ceylon, Japan, Newfoundland, Greenland, New Guinea, Taiwan, Tasmania, and Kodiak.

# THE OCEAN FLOOR

## What is the **ocean floor**?

The ocean floor is the very bottom of an ocean basin. The ocean floor is a multi-level structure that "floats" on top of the denser and more viscous mantle (the part of the Earth that lies below the crust and above the core and which is comprised mostly of unconsolidated material). The layer resting directly on top of the mantle is the oceanic crust, an approximately 3-mile- (5-kilometer-) thick layer of volcanic rock known as basalt; this is the youngest rock on the planet. The next layer, on top of the oceanic crust, is made of lava, and based on its composition, scientists divide it into two different areas: The bottom area, directly on top of the crust, is about 1.2 miles (2 kilometers) thick, and has feeder dikes or lava sheets intruded into other rock; the top area is approximately 0.3 miles (0.5 km) thick, and is made of pillow lava. On top of all of this is a layer of unconsolidated (loose) sediment that was transported there from dry land. This uppermost layer is approximately 1.2 miles (2 km) thick.

## What is the **composition** of **oceanic crust**?

The oceanic crust is composed of basalt, a dark gray to black igneous rock that is created by volcanic activity. Basalt can be found at the mid-ocean ridges, where new crust is slowly being created: At the mid-ocean ridges, or spreading centers, the Earth's plates are spreading apart, creating large upwellings of lava, which accumulates to form the ridges. Basalt has a microcrystalline structure, the result of its rapid cooling. It

## Are there basins on land?

Yes, there are basins on land. But land basins are much smaller than ocean basins and, of course, they are surrounded by land, with rivers flowing directly into them. Inland, also called interior, basins are usually salty: Rivers pour dissolved salts into the basin and since there is no outflowing river to drain the basin, salts accumulate. For example, the Great Salt Lake in Utah is the low point of the Great Basin, and its water is four times as salty as seawater. The Dead Sea, between Jordan and Israel, is the low point of an interior basin and it is the saltiest lake in the world. Its water is seven times saltier than seawater.

is composed of approximately 50 percent silicon dioxide ($SiO_2$), 16 percent aluminum oxide ($Al_2O_3$), and 9 percent calcium oxide (CaO), with many minerals making up the remaining 25 percent.

## Where is the **ocean floor**?

Many people think of the ocean floor as all the land beneath the water's surface—including the continental shelf, slope, and rise. However, the true definition of the ocean floor is the land that makes up the ocean basins (in other words, the deep-ocean floor); it is bordered by, but does not include, the continental margins (shelves, slopes, and rises). For example, the North Atlantic Ocean floor does *not* extend from the European and African shorelines to the North American shoreline; instead, it extends, in an east-west direction, from the edge of the European and African continental margins to the edge of the North American continental margin.

## What are **ocean basins**?

Ocean basins are the roughly circular or oval-shaped areas of the ocean floor that do not include the continental margins or the structures of the mid-ocean ridges. In other words, the basins stretch from the outer margins of the continents to the mid-ocean ridges. Some are large, such as the Pacific Ocean basin, and measure around 2 to 3 miles (3 to 5 kilometers) deep. Others are smaller, such as the Natal Basin, east of South Africa and south of the island of Madagascar, in the Indian Ocean.

The term "basin" is also used to describe the entire depression that holds the global oceans or it may be used to describe smaller areas of the ocean floor that are surrounded by higher terrain—such as the Gulf of Mexico basin.



## Are there any **inland basins below sea level**?

Yes, there are many inland basins below sea level. For example, Death Valley in California is 282 feet (86 meters) below sea level. The Dead Sea is the deepest interior basin, and reaches 1,339 feet (408 meters) below sea level (as of 1996). Because Israel and Jordan gather water from the Jordan River before it flows into the Dead Sea, the water level in this basin drops about 1.5 feet (0.5 meters) annually.

## Why don't the **ocean basins fill with sediment** from the continents?

It does seem that over time the ocean basins should fill with sediments from the continents. After all, rivers carry a great deal of sediment from the landmasses to the oceans. But the oceans differ, each accumulating deposits at its own rate. Plus, the accumulation is very slow—not only because the oceans are far from many river sources, but because of the tremendous size of the oceans, it takes a very long time for sediments to accumulate appreciably.

## How did the **ocean basins form**?

Scientists believe that the process of plate tectonics (the slow movement of the several plates that comprise the Earth's surface) was responsible for the formation of the ocean basins. The continents are granitic and "float" on a denser mantle rock of silica/magnesium. As a mid-ocean ridge opened, the heavier material from the mantle rose and became seafloor. This material eventually "dove" under another plate at a specific place called a subduction zone (an area where one continental crustal plate moves underneath another); when the mantle remelted (sinking low enough to become molten), it released the lighter rock at the surface as volcanic islands—and eventually formed continents. Therefore, ocean basins—or the places between the continents—have extremely thin crusts when compared with the continental plates; for this reason, they are sometimes referred to as "crustless," or as the "naked plates."

### What are some **major features** on the **ocean floor**?

Many people believe, erroneously, that the ocean floor is relatively flat and featureless. The truth is, the land beneath the waves is more diverse than on the continents—in other words, more varied than the land formations we see around us. The ocean floor has ridges, rises, chasms (extending deeper than Mount Everest is high), rolling hills, flat (abyssal) plains, fracture zones, and volcanic islands—and these are just some of the geologic features found under the water.

### What caused the **ocean floor's geological features**?

The mountains, trenches, and other features found on the ocean floor were the result of the creation, movement, and loss of the planet's crust. Over billions of years, this ongoing process, which is the result of plate tectonics (the slow movement of the several plates that comprise the Earth's surface), has gradually shaped and changed the face of the Earth—and the ocean floor.

## CONTINENTAL DRIFT AND PLATE TECTONICS

### When did scientists first realize the **Earth's crust moves**?

In 1620, English philosopher, statesman, and scientist Sir Francis Bacon (1561–1626) noted the similarity between the west coast of Africa and the east coast of South America; he was the first to do so. But his ideas were ignored until the late 1800s and early 1900s. Before that time, the scientific community's prevailing view was that the Earth's surface was static, having been in the same place since the crust cooled.

In 1858, French geographer Antonio Snider-Pelligrini (also seen spelled Pellegrini) first proposed the modern continents had fit together like a gigantic jigsaw puzzle. In his work "Creation and Its Mysteries Revealed," he showed the most obvious fit—the corresponding shape between the west coast of Africa and the east coast of South America. He also believed that the formerly linked continents were split by the Noachian flood (the great flood or deluge mentioned in the Old Testament of the Bible, in the story of Noah's Ark).

The theory that today's continents were once connected was brought up again over the years, but it was ignored by most scientists. In 1912 Alfred Wegener (1880–1930), a German meteorologist and geologist, published a book titled *The Origins of Continents and Oceans.* The book expanded on the theory put forth by Snider-Pelligrini and others, proposing that all the continents had once fit together in a large supercontinent, which Wegener called Pangea (or Pangaea, from the Greek *pangaia,* meaning "all Earth"). He believed Pangea had broken up into smaller landmasses, with the shape and location of today's continents a result of their movement across the Earth's surface. He called this process continental drift—a term that is believed to have been inspired by the way ice floes move across the surface of the water. However, Wegener could not explain the mechanisms driving this movement and his theory was not taken seriously. By the 1950s and 1960s, scientists finally accepted Wegener's general theory, since by this time enough evidence had accumulated to support the theory and scientists had worked out a mechanism, called plate tectonics, to explain the movement of the continental plates.

## What are **continental drift** and **plate tectonics**?

The most accepted theory of continental movement is called continental drift; the theory of its mechanism is called plate tectonics. The term tectonic comes from the Greek word *tekton,* meaning "builder," a reference to the building of mountains from the movement of continental plates.

The continental plates move laterally across the face of the planet. Boundaries between plates include the mid-ocean ridges (or spreading centers), subduction zones (areas where one continental crustal plate moves underneath another), and areas in which plates slip by—or pass—one another.

The theory of continental drift was first suggested by German geophysicist Alfred Wegener (1880–1930). The evidence for this movement has been collected by observation satellites, which allow scientists to use sophisticated laser-ranging instruments to track minute movements in the plates that make up the Earth's crust. These plates move just fractions of an inch to inches each year.

No one can truly explain the reason (or reasons) why the continental plates, and thus the continents, move (or "drift") across the face of the Earth. But scientists are certain that the plates do move—which has prompted numerous theories to explain why they move. One theory suggests that the continents are situated on large tabular blocks of the Earth's crust. The mechanism that drives the movement of these blocks is the fluid mantle (the part of the Earth that lies below the crust and above the core, and is comprised mostly of unconsolidated materials) underneath. The interactions between the plates, which occurs at or near their boundaries, give rise to many of the features found on land and the ocean floor.

## What evidence led to the **acceptance** of **continental drift** and **plate tectonics**?

Evidence accumulated over time and by the 1960s there was a general acceptance of the theories of continental drift and plate tectonics. Satellite data, fossil findings, and seafloor spreading indicated to the scientific community that the Earth's plates had moved and continue to do so. The theory was strengthened by further evidence—the discovery of magnetic anomalies in ocean rocks.

And although no one can, at this stage, definitively explain the reason (or reasons) why continental plates—and thus the continents—move across the face of the Earth, scientists are certain the plates do move.

## How did fossils help prove the theory of plate tectonics?

Similar fossils were found on widely separated continents, indicating that separate species either developed identically across the far-flung continents (a notion that is highly unlikely), or the continents were in contact millions of years ago when these species developed. For example, fossils found in South America were found to be related to those in Australia and Antarctica; others discovered in Antarctica were related to those in Australia. These findings seemed to indicate that the landmasses had once been joined together: Species roamed freely across the surface, died, were buried by sediment, and became fossilized. As the continents drifted apart, they carried the organisms' remains—leading to the finding of widely separated, but identical, fossils.

## How do **satellites** help **prove the theory of plate tectonics**?

Observation satellites, which use sophisticated laser-ranging instruments, allow scientists to track minute (extremely small) movements in plates. Without the data sent back to Earth by these satellites, the almost infinitesimally small movements of the plates could not be detected.

## What are **active** and **passive margins**?

Active and passive margins describe the different characteristics of continental margins that are a result of plate tectonics (the slow movement of the plates). For example, the continental margin (or edge) of the Atlantic Ocean is a passive margin, with a well-developed continental shelf, slope, and rise. This type of margin developed as continents split and drifted apart, and an ocean basin was created between them. A passive margin is the "trailing edge" of a plate; these margins are also called "constructive margins" because sediment deposited from the landmass builds the margin outward—toward the sea.

An example of an active margin can be found on the western (Pacific) side of the South American plate. Here, the continental shelves are narrow, the slopes are deep and narrow, and the rises are frequently missing. These features are characteristic of colliding plates. Active margins are areas of frequent geological activity, with numerous volcanoes, earthquakes, and mountains. An active margin, in other words, is the "leading edge" of a plate.

## What is **plume tectonics**?

Plume tectonics is how scientists describe the behavior of the magma (the hot, liquid rock) in the Earth's mantle—the below-the-surface activity that causes the plates to move. They use seismic tomography, a technology that scans the Earth's interior with seismic waves, much like a CT scan examines the human body, to track the movement in the magma. Based on this and other studies, mantle convection (heating) occurs in association with "hot plumes" and "super-plumes" that move up from the boundary between the core and mantle, and "cold plumes" that drop toward the boundary. For example, beneath Africa and Tahiti, there are hot super-plumes rising, creating volcanoes and moving plates; beneath Japan, a cold plume falls, creating a subduction zone (an area where one plate moves beneath another).

## What are the **major plates** of the **Earth's crust**?

The Earth's crust is divided into 13 major plates: the North American, Eurasian, African, South American, Pacific, Nazca, Cocos, Somali, Australian, Bering, Philippine, Arabian, and Antarctic. Each plate boundary has a specific direction of movement. There are also other, smaller plates within the larger plates.

## What **happens** at the **plate boundaries**?

There are three main events that occur at the boundaries of moving plates: 1) The plates may separate or pull apart; 2) they may collide with other plates, with one plate subducting (moving underneath) the other—usually forming deep trenches or building mountains in the process; or 3) two plates can slip past one another, moving two pieces of land in opposite directions.

## What happens when **two of the Earth's plates separate**?

At the boundary where two plates separate, molten rock from the mantle (the part of the Earth that lies below the crust and above the core) rises from deep within the Earth. As it flows out, it accumulates to create a mid-ocean ridge; and as the plates continue to separate, new oceanic crust is continually created. One well-known example of this geological process is the Mid-Atlantic Ridge, a long chain of volcanic mountains cutting through the center of the Atlantic Ocean. On land, the boundaries where plates are separating are called rift valleys. Eastern Africa's Great Rift Valley is one example.

## What happens when **two of the Earth's plates collide**?

At the boundary between two colliding plates, one plate can be forced under the other—causing the oceanic crust to sink into the mantle (the part of the Earth that lies below the crust and above the core). This process is called subduction, and the area where it occurs is called a subduction zone. These areas generate many earthquakes and volcanoes, and form deep-ocean trenches. One well-known subduction zone lies along the west coast of South America. This is where the Nazca plate is subducting (moving underneath) the South American plate, creating a subduction zone. This activity once created the volcanic Andes Mountains and still creates frequent earthquakes.

## What happens when **crustal plates slip** by one another?

At some boundaries, the plates are neither separating nor colliding—sometimes they move by each other (side by side), in opposite directions, in a process called slip. When these moving plates get stuck, a large amount of energy builds up along the boundary; when they let go or move slightly, earthquakes result. The most famous example of a boundary between two slipping plates is the San Andreas Fault, which runs through California; here a part of the North American plate slides by the Pacific plate.

## How was the process of **seafloor spreading** discovered?

In the mid-twentieth century, American educator and geologist Harry Hess (1906–69) took part in an ocean-drilling project called MOHOLE. Examining material from the ocean's bottom layers, Hess discovered that ocean floor rocks are younger than those found on the continental landmasses. He also found that the ages of the ocean floor rocks vary—the farther from the mid-ocean ridge, the older the rocks; and conversely, the youngest rocks are found near mid-ocean ridges.

In a paper titled "Further Comments on the History of Ocean Basins," which Hess delivered to the Geological Society of America in 1963, he theorized that the seafloor was spreading. He suggested that seafloor spreading was caused by magma (hot, liquid rock) erupting from the Earth's interior along mid-ocean ridges. This newly-created seafloor slowly spreads away from the ridges; it later sinks back into the Earth's interior at deep-sea trenches.

## How did **magnetic anomalies** support the **theory of seafloor spreading**?

The theory of seafloor spreading was later confirmed by the measurement of magnetic anomalies in ocean floor rocks. When molten lava is expelled from an ocean ridge, it creates new ocean floor. The magma (the hot, liquid rock) then cools and elements within the rock line up with the planet's prevailing magnetic field, retaining the orientation present at that time. Since the planet's magnetic field changes over time, evidence of this would show up in the ocean floor rocks.

In rocks from the ocean floor, scientists discovered a pattern of stripes representing these magnetic anomalies—providing a record of the changes in the orientation of the magnetic field over time. These patterns spread out symmetrically on either side of the Mid-Atlantic Ridge, a long chain of volcanic mountains cutting through the center of the Atlantic Ocean. Scientists concluded that the pattern and distribution of these rock stripes, showing the magnetic field reversals over time, could only have occurred if the seafloor had been spreading apart over millions of years.



## How many times have the **Earth's magnetic poles reversed**?

The Earth has reversed its north and south poles about 177 times over the past 85 million years. Based on the study of certain rocks and their magnetic properties, scientists estimate that a magnetic reversal took place only 2 million years ago. No one knows how or why the magnetic fields switch places—and we definitely don't know what happens when the fields *do* change.

## Where does the **crust** from spreading seafloor eventually **end up**?

The crust created by the spreading seafloor eventually subducts (moves beneath) a continental plate; from there, the crust sinks to the Earth's mantle, where it again becomes molten. Over time, this molten rock may be "recycled" into new seafloor at mid-ocean ridges—like a giant conveyor belt working in the Earth.

## Are there any **other explanations** for the **movement of the continents**?

Yes, there are other explanations for the movement of the continents—because not everyone agrees with the mechanism of plate tectonics. One reason is that, although the idea of moving plates seems sound, the mechanisms behind plate tectonics are virtually unknown. Thus some scientists believe in continental drift, but not in plate tectonics.

Another possible explanation for continental drift is that the Earth is expanding. This would give the illusion that the continents move across the surface of the planet (although no one can explain why or how the Earth is expanding). Another theory is "surge tectonics," in which there is a sudden surge of plate movement, as opposed to a constant flow. And still another suggestion is that there is no continental drift—that the continents have *always* been in the same positions they are in now.

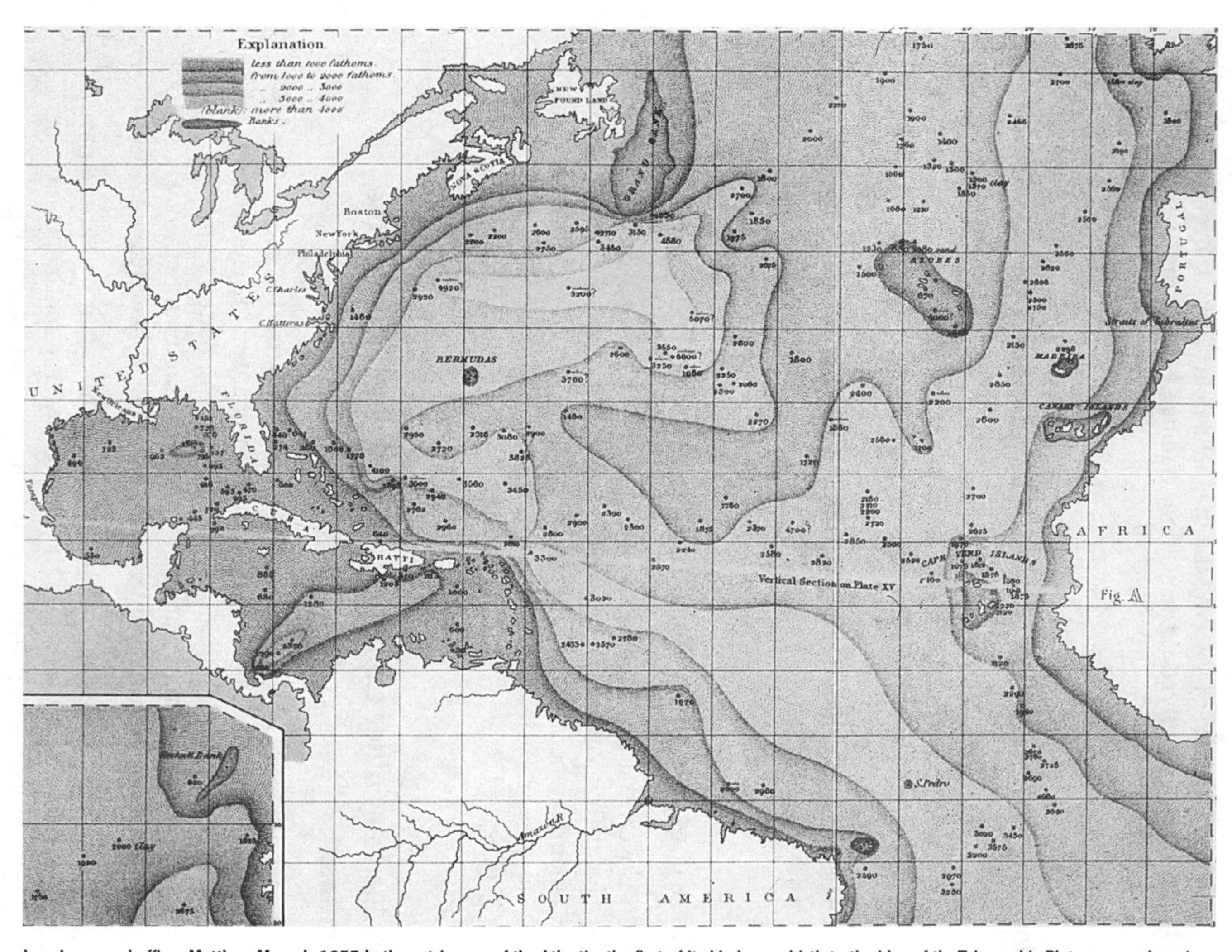


American naval officer Matthew Maury's 1855 bathymetric map of the Atlantic, the first of its kind, gave birth to the idea of the Telegraphic Plateau—a submarine land formation from Canada to the British Isles, across which the first transatlantic cable was laid. *NOAA Central Library*

# RIDGES AND RIFTS



## What is a **mid-ocean ridge**?

A mid-ocean ridge is essentially an underwater volcanic mountain range, found at the boundaries between separating plates. These ridges girdle the globe, like stitches on a baseball, and are usually sharply crested with steep sides. All together, they form the longest series of mountains on Earth, extending more than 35,000 miles (56,000 kilometers) in length.

## What is the **difference** between **mid-ocean ridges** and **rises**?

Both mid-ocean ridges and rises are used to describe underwater mountain chains. But physically, rises tend to have a gentler topography and lack the central rift valleys found in ridges. For example, the Mid-Atlantic Ridge has a deep rift valley along its crest, measuring about 6.2 miles (10 kilometers) wide, and with walls about 1.9 miles (3 kilometers) high. Such dramatic topography would not be found on a rise.

## What are some of the **major mid-ocean ridges** and **rises**?

Some of the major mid-ocean ridges and rises run in a north-to-south direction. These include the Mid-Atlantic Ridge in the Atlantic Ocean, the Reykjanes in the North Atlantic and Arctic oceans, the East Pacific Rise in the Pacific Ocean, and the Mid-Indian Ocean Rises in the Indian Ocean. Others include the Carlsberg Ridge, which runs southeast from the Arabian Plate to the Mid-Indian Rise, and the Southeast Indian Ridge, located between Antarctica and the Indian and Australian plates.

## When were **mid-ocean ridges discovered**?

The mid-ocean ridges remained unknown to humans until the nineteenth century. In the 1850s, American naval officer and oceanographer Matthew Fontaine Maury (1806–73) became the first to show that the ocean bottom isn't smooth and featureless. He discovered and named the Dolphin Rise in the Atlantic, the first indication of the Mid-Atlantic Ridge, which cuts through the Atlantic Ocean from north to south (extending from roughly 55 degrees north latitude to 55 degrees south latitude).

But it wasn't until the voyage of the H.M.S. *Challenger* (from 1872 to 1876) that the single, continuous nature of the submarine mountain range, now known as the Mid-Atlantic Ridge, was discovered. And it took until the 1960s and 1970s before the entire ridge was mapped and studied.

## What are some **common features** of the **mid-ocean ridges**?

The underwater mountain ranges that are the mid-ocean ridges have a unique characteristic: While continental mountain ranges normally have a single, prominent line of peaks, underwater ranges associated with mid-ocean ridges have two lines of peaks, separated by a rift valley that can range in width from 15 to 30 miles (24 to 48 kilometers). Some of the mid-ocean ridges are longer than any continental mountain range. If these ridges were strung together, end to end, they would form an undersea mountain range that would extend some 35,000 miles (56,000 kilometers) in length. The individual ridges' widths vary from 500 to 1,500 miles (805 to 2,414 kilometers). The peaks of most mid-ocean ridges are approximately 8,000 feet (2,438 meters) below the ocean's surface. Some can rise as high as 12,000 feet (3,658 meters) from the ocean floor, with their peaks ascending above the water level to form islands. Iceland and the Azores in the Atlantic and the Galapagos Islands in the Pacific are actually the peaks of underwater mountains.

## How do **mid-ocean ridges form**?

According to the theory of plate tectonics (the movement of the Earth's plates), mid-ocean ridges form when two continental plates pull apart, and magma (hot, liquid rock) wells up from deep in the Earth, creating new crust—and thus, new seafloor. This moving magma also pushes the plates apart (in what is called seafloor spreading), contributing to the movement of the continental plates; is responsible for the continuous build-up of the ocean's crustal plates; and in certain areas, creates a buildup of lava, which eventually forms a mid-ocean ridge. Although the number of oceanic volcanoes responsible for this upwelling of molten rock is unknown, it is thought to be quite high.

## What is one of the **most active** areas along the **Mid-Atlantic Ridge**?

One of the most active areas along the Mid-Atlantic Ridge, which extends from roughly 55 degrees north latitude to roughly 55 degrees south latitude, is near the island of Iceland.

## What is a **graben**?

A graben is a low block of rock that is bounded on two sides by parallel faults that create tall, cliff-like scarps. These fault scarps—which are slopes that occur at the boundary between two plates or at fractures in the Earth's crust—are caused by the movement of the plates.

## What is a **rift valley**?

A rift valley is an area where plates are separating; it is actually a graben—with fault scarps (slopes) on either side. Rift valleys are created by magma (hot, liquid rock) that wells up from the Earth's mantle and hardens. They can occur in the ocean, along a mid-ocean ridge (such as the Mid-Atlantic Ridge) or on land, where there are two main areas of this geological activity: around Iceland, which is a peak of the Mid-Atlantic Ridge that has grown tall enough to appear above the surface of the water; and in eastern Africa's Great Rift Valley, which stretches more than 3,000 miles (4,800 kilometers) from Syria to Mozambique.

## Is a **new ocean forming**?

Scientists believe that the Great Rift Valley in eastern Africa is where the next ocean will form. As the sides of the valley move farther apart, the floor (or graben) between them will lower. Currently, the Red Sea fills part of the Great Rift, and as the rest of the valley sinks, scientists predict that water will pour in. This will flood the valley, eventually splitting Africa in two—and creating a new ocean.

# UNDERWATER VOLCANOES AND EARTHQUAKES

## What is a **volcano**?

A volcano is a vent in the Earth's surface; it can occur either on land or under water. At these openings, eruptions occur—caused by pressures beneath the Earth's surface. Gases, ash, dust, and magma (hot, liquid rock from deep within the planet) are forced out, creating numerous geological features—the most prominent and spectacular being the volcano itself, which is an accumulation of the debris that erupted from the vent.

The word *volcano* comes from the Vulcano Island in the Tyrrhenian Sea, off the northern tip of Sicily (or off the "toe" of Italy's boot). Here there is a volcano that, though not active now (the last time it erupted was 1890), was active in ancient times, causing Romans to believe it was the entrance to the nether regions and the domain of Vulcan, the god of fire, who fashioned armor for the other gods.

## What is a **high island**?

The term "high island" describes an island of volcanic origin in the Pacific Ocean. High islands can be active or dormant. The Hawaiian, Bougainville, and Solomon islands are all high islands.

## How do **volcanoes form**?

Volcanoes can form at the boundaries of plates or in the middle of plates. At a boundary formed by separating plates, volcanoes can be created when molten rock flows to the surface through the openings. The numerous volcanic mountains associated with mid-ocean ridges were formed in this way. Where plates collide, one plate may subduct (move beneath) the other, and as the subducted plate sinks into the Earth's mantle below, it becomes molten and the resulting magma (hot, liquid rock) rises up to form volcanoes. Volcanoes that were formed in this way are normally found along the edges of continents, such as in the Andes

Mountains of South America, or rising from the ocean floor, such as the islands of Japan. Volcanoes also form in the middle of plates due to the presence of "hot spots," creating islands or chains of islands, such as the Hawaiian Islands.

The actual mechanism behind magma formation and eruption is still being debated by scientists, as are ways of predicting volcanic eruptions. Volcanoes can cause widespread destruction and loss of life, but they are also one of the few ways we have to study the interior of our planet. Our current drilling techniques can only penetrate approximately 6 miles (10 meters) into the crust, while the magma of the Hawaiian volcano Kilauea originates at depths of at least tens of miles (or at least 20 kilometers).

## Why are **volcanoes important features** of the ocean floor?

Volcanoes are important features of the ocean floor because they are newly created land, continually rising up out of the deep bottom. Whether their summits remain below the surface or rise above the waves, they provide the foundation for unique ecosystems that thrive on and around them. The active volcanoes continue to add energy, mostly in the form of heat, to the environment around them; this energy is used by many forms of life. More than 80 percent of the Earth's volcanic activity occurs in the deep-ocean environment.

## Are there different **types of volcanoes**?

Yes, there are numerous types of volcanoes, which are mainly categorized by their physical features. Some common examples of volcanoes include shield (low-sloping), caldera (having a basin or crater at the summit), fissure (having vents in the side), lava dome (resembling a raised bubble), composite (also called a stratovolcano, which has alternating layers of ash and lava), and ash-cinder (a straight-sided ash cone). Volcanoes can also be a combination of several types—such as a composite caldera. Certain types, including caldera, composite, and ash-cinder, usually experience violent eruptions (spewing pumice and ash), while the other types have less explosive eruptions, producing lava flows that cover large areas.

This black lava sand beach in Hawaii was later destroyed by lava flows. *NOAA/Commander John Bortniak, NOAA Corps (ret.)*

## Where are some of the **largest calderas** on Earth?

The ash-flow calderas (great basins at the summits of volcanoes made mostly of ash) are the largest calderas on the planet. They are characterized by their very low relief, broadness, and immense ash deposits. The volcanic calderas of Yellowstone National Park (one of the largest volcanic fields on Earth) are examples of this type of volcano.

## What **types of lava** come from **above-water** volcanoes?

There are two basic types of lava, which is ground-flowing magma (hot, liquid rock), that flows on land. Their Hawaiian names are *pahoehoe* (pronounced pa-hoy-hoy) and *aa* (ah-ah). Pahoehoe is a smooth lava that forms a ropy surface (the lava cools to form wrinkles); it comes from higher-temperature and lower-volume eruptions, and has low viscosity, allowing it to flow easily and form a skin. It usually moves at approximately a yard (or meter) per minute. But, under the right conditions, such as on a steep slope, it can move at speeds up to 14 miles (23

Hawaii's lava flows encroaching on the ocean. *CORBIS/Morton Beebe*

kilometers) per hour. The thickness of a pahoehoe deposit is typically about 12 inches (31 centimeters).

Aa lava is sharp, rough, and coarse, and it normally occurs under the opposite conditions of pahoehoe. Aa tends to flow in surges, with the front advancing slowly, building up its height as it moves along at a rate of a few yards (several meters) per hour. Then, suddenly, it will move quickly forward, returning to its original thickness while covering ground at the rate of a hundred yards (meters) in a few minutes. The normal thickness of an aa flow is 6.5 to 16.5 feet (2 to 5 meters); flows are usually large, extending more than 100 yards (300 feet, or 91 meters) in width, and are often fed by smaller streams of lava.

## What is **pumice**?

Lava pouring out from a volcano can eventually flow into the ocean and cool very rapidly. This creates a volcanic glass called pumice, a very light rock often found floating around volcanic islands. The rock has the chemical composition of a granite (quartz and feldspars), but it has a low density that allows it to float. Because the lava cools so fast, it is packed with bubbles; as it solidifies, the bubble stay, giving the rock its low density, lightness—and buoyancy.

## What happens when **lava flows underwater**?

The flow of lava underwater is different than on land. This is because the water temperatures are colder and the pressures in the deep ocean are greater than that (air pressure) on land. Because of these factors, underwater lava develops very specialized features, including pillow lava and submarine lava pillars.

## What is **pillow lava** and how does it **form**?

Pillow lava is basalt (black volcanic rock) that has cooled, forming relatively small mounds. It only forms under water. As molten lava erupts from the depths and is deposited into the ocean, the surrounding water cools the exterior of the lava, forming a flexible but glassy crust in the shape of a pillow; the interior remains molten. As the amount of molten material increases in a particular pillow, the pressure increases, expanding the pillow and eventually breaking a hole in its glassy crust. The lava slowly flows out of this opening, the water cools the exterior of the outpouring lava, forming another pillow, and the process is repeated. This sequence continues as long as the eruption goes on, many times producing a thick deposit of pillow lava over a large area. The lava itself can originate from an underwater eruption, such as from a mid-ocean ridge or underwater volcano, or from lava that flows from the land into the ocean.

## What are **lava pillars**?

Underwater lava pillars (also called basalt pillars) are gnarled columns of a black volcanic rock called basalt, some up to 50 feet (15 meters) tall.

They range from thin and spindly to as thick and stout as a giant redwood tree; they also form as lava pillar archways. The outside of the pillars are coated with a paper-thin layer of black volcanic glass; the inside is a peppery gray, lined with a network of fine cracks. They are invisible in the blackness of the deep oceans until lit by a light source. They were first discovered by the lights of a passing submersible.

Pillow lava is formed underwater by basalt (black volcanic rock) that has cooled into small mounds such as those pictured here (off the coast of Hawaii). *NOAA/OAR National Undersea Research Program*

These huge formations are created when lava erupts onto the seafloor, but the actual mechanism by which they are created is still under study. One theory is that during a submarine eruption, water trapped below the growing layer of magma (molten rock) squirts through gaps in the lava flow, allowing the pillars to grow. As more lava flows out, the lava layer on the seafloor thickens; it continues to grow upward around the jets of water, leaving the duct open. Because the water is much cooler than the lava, it "freezes" the lava into pipe-like cylinders that grow around the jets—the building of them stops only when the flow of molten rock subsides.

Scientists hope to use pieces of lava pillars gathered by submersibles to better understand mid-ocean spreading centers. The details of these areas are relatively unknown because they are located so deep on the seafloor—at depths of 8,203 to 11,483 feet (2,500 to 3,500 meters).

## Are any **volcanoes forming** in the **oceans today**?

Yes, there a number of volcanic eruptions occurring on the floor of the oceans today. In fact, scientists estimate that some 80 percent of our planet's volcanic activity happens in this deep, hidden realm. One of the most spectacular phenomena associated with volcanic activity is the growth of huge, localized volcanoes—some of which become large enough to poke above the waves, forming islands (called seamounts) and sometime island chains. Examples of currently forming volcanoes are the Loihi Seamount, off the coast of Hawaii, and the Axial Seamount, off the north coast of Oregon.

## Is there a **volcanic basis** for the **legend of Atlantis**?

Although most scientists agree that the underwater city of Atlantis did not exist, they may have discovered the origin of the myth. In ancient times, Minoans lived on the island of Thira (or Thera)—a Greek island in the Aegean Sea, where the volcano Santorini is situated. Around 1470 B.C.E., the volcano had a huge eruption and all that remains of the island today is a large, crescent-shaped caldera (volcanic crater). Centuries after the eruption, the Greek philosopher Plato (c. 428–348 or 347 B.C.E.) described the island's "disappearance." Over the years, his account may have evolved as the myth of Atlantis—the city that sunk into the sea.

## Are there **volcanoes** elsewhere **in the solar system**?

Yes, there are other volcanoes on planets and satellites in the solar system. Venus has many—seen from spacecraft such as the *Magellan,* which took radar images of the planet. Mars also has volcanoes, including one of the largest known in the solar system; it is called Olympus Mons. The *Voyager* 1 and 2 spacecraft to the outer planets discovered more volcanoes. In fact, in the late 1970s, a *Voyager* captured images of an eruption on Io, one of Jupiter's moons! Since that time, ground-based telescopes and the *Galileo* spacecraft around Jupiter have imaged many other such eruptions on the small moon.

## What is the **Ring of Fire**?

The Ring of Fire is the name given to the periphery of the Pacific Ocean—a belt of seismic activity; it is also called the Circle of Fire. Counterclockwise, it extends from the southern tip of South America, north to Alaska, then west to Asia, south through Japan, the Philippines, Indonesia, and to New Zealand. (Hawaii is in the middle of the ring.) The Ring of Fire is an area where plates are subducting (moving underneath each other), creating numerous volcanoes—more than half of the Earth's active volcanoes. For example, Japan has 77 subduction-generated volcanoes, while Chile has 75. In addition, Kamchatka has 65, Alaska and the Aleutian Islands have 68, and the western United States has 69 such volcanoes. Some famous volcanoes along this ring are Japan's Mount Fuji; Indonesia's Galunggung; Mount Katmai and Augustine in Alaska; and Mount St. Helens in Washington state.



## Which **volcano** is the **most massive**?

The most massive volcanic mountain—indeed, the most massive mountain of any type on Earth—is Mauna Loa (*see* photo next page), located on the Big Island of Hawaii. Created by a hot spot located underneath the ocean, this mountain stands 13,679 feet (4,169 meters) above sea level. It rises a total of approximately 30,000 feet (9,144 meters) from the ocean floor and occupies almost 10,000 cubic miles.

## Which **volcanic eruptions** have been the **most destructive**?

Since the nineteenth century, the most destructive, or deadliest, eruptions (listed in order of destruction, from worst to least) occurred at Mount Tambora, Indonesia, on April 5, 1815; Krakatau, Indonesia, on August 26, 1883; Mount Pelee, Martinique, on August 30, 1902. The most recent deadliest eruption occurred on November 13, 1985, in Nevada del Ruiz, Colombia; 23,000 people died.

## Are the **death tolls** from **volcanic eruptions on the rise**?

Yes, similar to most deaths from natural disasters, there is a rise in the death toll from volcanic eruptions. This is because as the global population increases, there will be more people living in the direct path of such explosive events. Scientists estimate that between 1600 and 1900, the death toll from volcanic eruptions was about 315 people annually. In the twentieth century alone, the number has increased to about 845 people annually—and the number is expected to keep rising.

The volcanic Mauna Loa, located on the Big Island of Hawaii, is the most massive mountain (of any type) on Earth—its base lies on the seafloor. *NOAA/Commander John Bortniak, NOAA Corps (ret.)*

## What are some **mid-ocean volcanic islands**?

There are numerous mid-ocean volcanic islands. One of the largest is Iceland, off the southeast coast of Greenland. This huge island was created by volcanic material from roughly 200 active volcanoes—all from eruptions along the spreading Mid-Atlantic Ridge. The Azores Islands are also situated along the Atlantic Ocean's mid-ocean ridge, just off the coast of Portugal; as are the islands in the Tristan da Cunha group, in the South Atlantic Ocean.

## What is **Surtsey**?

Surtsey is a volcanic island just off the shore of Iceland, in the North Atlantic Ocean. The island appeared only recently—on November 16, 1963, to be precise—during a spectacular display of erupting magma and steam in the ocean off Iceland's southern coast. The island took about 4 years to grow, reaching 492 feet (150 meters) above sea level by 1967, with an area of 2 square miles (5 square kilometers). Early in the island's life, ocean waves caused a good deal of erosion, but the core quickly solidified. Surtsey, like Iceland, lies over the Mid-Atlantic ridge; it is now a permanent island, colonized by plants and sea life.



## Did any **other islands** grow around **Surtsey**?

Yes, two other smaller islands formed near Surtsey: The Syrtlingur (or "Little Surtsey") in May 1965 and the Jólnir (or "Christmas Island") in December 1965. These islands were not as lucky as Surtsey. Both soon disappeared after the volcanic activity ceased in the area, the new lava worn away by the wave action of the sea. For example, Jólnir disappeared in August of the following year—its life was about 8 months.

## Why are **Iceland's volcanoes unique**?

The volcanoes on Iceland are all on the Mid-Atlantic Ridge—but they are not like the common cone-shaped volcanoes found around the world. Icelandic volcanoes are actually fissures hidden beneath a glacier that covers approximately 8 percent of the island. This means eruptions—which have occurred every five years for the past 1,100 years—can melt large parts of the overlying ice, unleashing huge floods.

## What is a **hot spot**?

A hot spot is another process created by plate tectonics—the movement of the Earth's plates. Hot spots lead to the creation of volcanoes on the ocean floor—but they are distinct from separating or colliding plates. Canadian geophysicist J. Tuzo Wilson (1908–93), one of the pioneers of the theory of plate tectonics, sought to explain the presence of volcanoes that formed in the middle of a plate instead of along a plate boundary. He postulated the presence of what he called a "hot spot," a localized, stationary upwelling of magma (molten rock) from the Earth's mantle.

Although the mechanism is still debated, many scientists believe a hot spot is created by an upwelling of magma, which flows up through weak areas in the overlying plate. This often leads to the formation of a volcano. Over millions of years, as the plate continues to move slowly over this hot spot, a chain of volcanic islands is born. Eventually, the oldest of these often inactive volcanoes erodes and "sinks" below the water's surface.

## Where are **hot spots located**?

Hot spots are located around the world, and some of them are quite well known. Scientists believe the Hawaiian Islands are a chain of volcanic islands created by the presence of a hot spot, which has generated almost 200 volcanoes over a period of 75 million years. The most recent above-surface creation from this hot spot is the Big Island of Hawaii. Other places that may be hot spots include the Galapagos and Society Islands, and Yellowstone National Park in Wyoming.

## Where does the **Hawaiian Island hot spot originate**?

Scientists may have recently pinpointed the origin of the Hawaiian hot spot at the boundary between the mantle (the part of the Earth that lies below the crust and above the core) and the metallic core of the planet. This plume of molten material is approximately 1,800 miles (2,900 kilometers) beneath the crust, where the molten outer layer of the core heats the rock at the base of the mantle. Data also suggest that the molten plume of material first flows horizontally toward the base of the Hawaiian hot spot, then rises vertically.

Using a technique called seismic tomography, scientists analyzed this hot spot by studying seismic waves generated by earthquakes in the region of Tonga and Fiji; the waves traveled through the deep mantle beneath the Hawaiian hot spot before reaching recording stations in Oregon and California. They also analyzed the waves for polarization; this showed, for the first time, variations at the base of the mantle—indicating the molten material's localized transition from a horizontal to a vertical flow.

## What is the **newest Hawaiian Island**?

Currently, the volcano called the Loihi Seamount is building just off the islands, and will eventually become the newest Hawaiian island. This mountain—about 20 miles (32 kilometers) off the coast of the Big Island of Hawaii—rises some 17,000 feet (5,182 meters) above the ocean floor and is about 3,000 feet (914 meters) below the ocean's surface. In several thousand years, Loihi will poke through the surface of the ocean water to become the newest island in the Hawaiian chain.

## Are there any **hot spots** in the **Indian Ocean**?

Yes, another well-known hot spot track extends from India to the island of Réunion, in the Indian Ocean. About 67 million years ago, present-day India was above the hot spot. A great deal of basaltic lava erupted, creating a huge volcanic field called the Deccan Trapps. As the plate moved to the northeast over the hot spot, more volcanic centers formed, creating features including the Maldives (around 57 million years ago), Chagos Ridge (48 million years ago), Mascarene Plateau (40 million years ago), Mauritius Islands (18 to 28 million years ago), and in the last 5 million years, the volcanoes of Pito des Neiges and Piton de la Fournaise, which make up the island of Réunion.



## What are **black smokers?**

Black smokers are features of the deep-ocean floor, occurring in areas where there is volcanic activity. They are created by hydrothermal vents: At these vents, super-hot water erupts in a plume that looks like a column of black smoke; but it is actually hot, salty, mineral-laden water—at temperatures upwards of 700° F (370° C)—which is released from the vent and encounters the much cooler seawater. The sulfide and sulfate minerals in the water give the plume a dark color. As these minerals are deposited around the plumes, sulfide chimneys build up, and from these chimneys the black water continually flows.

## When did scientists **discover black smokers**?

The first black smokers were discovered in 1977 off the Galapagos Islands, off the coast of Ecuador, in the Pacific Ocean. Scientists were investigating the ocean floor in the submersible *Alvin,* a mini-submarine built by the Woods Hole Oceanographic Institution. Near the Galapagos, they noticed these "smoking chimneys." Since that time, several chimneys have even been recovered from the ocean floor—some measuring 5 feet (1.5 meters) tall, and ranging in weight from 1,200 to more than 4,000 pounds (545 to more than 1,816 kilograms).

This black smoker (the super-heated plume from a hydrothermal vent) was photographed on Endeavor Ridge—an underwater land formation off the California Coast. *NOAA/OAR National Undersea Research Program; P. Rona*

## Where are **black smokers found**?

Black smokers tend to occur on sulfide mounds, which, over time, build up along mid-ocean ridges. These mounds—ranging anywhere from about the size of a pool table to the size of a tennis court—grow in fields of hydrothermal vents (a volcanic opening where super-hot water escapes from the depths of the Earth). Although black smokers have only been found on mid-ocean ridges, not every ridge has black smokers. Scientists are still trying to figure out why some ridges have these hydrothermal vents and others don't. In addition, scientists believe only a relatively few black smokers have been discovered so far, since only a small portion of the world's approximately 31,070 miles (50,000 kilometers) of ridges have been explored. More of these features will doubtless be found in future explorations.

## Why are **black smokers important**?

Black smokers are important because they form the foundation for a unique ecosystem. This ecosystem is located deep in the darkness of the ocean bottom, in the most extreme environment possible for life.

Scientists studying these chimneys, where super-heated waters escape from hydrothermal vents in the seafloor, discovered unique life forms dwelling on and around the area—life forms such as tube worms and exotic shrimp thriving without sunlight, under immense pressure (from the surrounding waters), and in high water temperatures. Also, scientists found different types of microbes living in the interiors of these chimneys, some of which are heat-loving, or thermophilic; these microbes are some of the most primitive forms of life found on Earth.

The ecosystem around the black smokers is the only one on Earth that does not rely on the rays of the Sun as its foundation. All the organisms living around a hydrothermal vent (a volcanic opening where super-hot water escapes from the depths of the Earth) are dependent on bacteria that use hydrogen sulfide from the hot water plume as their primary energy source.

The circulation process initiated by these hydrothermal vents is also important to our planet: The vents transfer large amounts of heat and chemicals from the interior of the Earth into the oceans, which in turn, greatly influences the properties of the ocean's water—and regional climate.



## Is there a connection between **black smokers** and the **origin of life**?

Yes, there may be a connection between black smokers and the origin of life on Earth. Scientists believe the presence of unique organisms around black smokers indicates that life could exist—and evolve—without the Sun's rays. And some scientists have even speculated that life on Earth originated around early black smokers. These areas were "safe havens" for these forms of life, a place to hide while other organisms went extinct due to changes in the climate or other causes. Because of these findings, many scientists now believe life may be present at deep-ocean vents on other planets or satellites in our solar system. For example, such organisms may live around vents on Jupiter's moon Europa, a satellite that may hold an immense ocean underneath its icy surface.

## What **unique phenomenon** is associated with **black smokers**?

Recently, scientists have discovered an extremely faint glow of light is given off by black smokers. This light is imperceptible to humans, and can only be recorded using cameras sensitive to low light levels. The cause of the faint glow of light emanating from black smokers is still a mystery. When this phenomenon was first discovered, scientists thought the glow was due to thermal radiation, a result of the hot temperature of the water plume. But more recent measurements have shown that thermal radiation alone does not account for the entire amount of light from the deep-sea hydrothermal vents (a volcanic opening where super-hot water escapes from the depths of the Earth)—there has to be another source of light.

Scientists have suggested four potential mechanisms for this light source:

**Crystallo-luminescence:** As the hot, saltwater plume from the vent quickly cools in the surrounding seawater, the dissolved minerals crystallize and drop out of solution, giving off light.

**Chemi-luminescence:** Another mechanism is chemi-luminescence, in which chemical reactions taking place in the vent water release energy in the form of light.

**Tribo-luminescence:** A third mechanism is tribo-luminescence, in which light is given off as mineral crystals crack from the cold, or collide in the erupting plume.

**Sono-luminescence:** The fourth possible mechanism is sono-luminescence, in which light is given off as microscopic bubbles in the hot plume collapse.

## What are **volcanogenic massive sulfide deposits**?

Volcanogenic massive sulfide (VMS) deposits are the remnants of hydrothermal vents (a volcanic opening where super-hot water escapes from the depths of the Earth) and sulfide mounds formed in the ancient oceans—some of which can now be found on land. They are large areas containing precipitated sulfide minerals (minerals that were separated from a solution), and are mined for elements such as copper, zinc, lead, silver, and gold. Equally important, certain VMS deposits contain fossils, which give scientists clues about the organisms that lived in and around these ancient vents. In particular, the Yaman Kasy VMS deposit, located in the Ural mountains of Russia, has yielded fossils from three different areas. In addition, fragments of black smoker chimneys were found there in 1995. This deposit dates from the Silurian period, approximately 430 million years ago.

## Do **volcanoes and earthquakes** have anything in **common**?

Yes, volcanoes and earthquakes have one major thing in common: They both most often occur along plate boundaries. In addition, many times a volcanic eruption will be accompanied by an earthquake—before and/or after the eruption.

## What is an **earthquake**?

An earthquake is the shaking of the earth, caused by the movement of the Earth's crust. The shaking is caused by the release of energy—in the form of seismic waves—as rock suddenly breaks or shifts under stress.

## What are **seismic waves**?

Seismic waves are produced by the energy released by earthquakes; the waves radiate from the quake's epicenter (the point on the surface of the Earth directly above the focus of an earthquake). P-waves, or primary waves, move in a back and forth direction; they are the first waves produced by a quake and travel the fastest, reaching the other side of the world in about 20 minutes. S-waves, or secondary waves (also called shear waves), move side to side; they can pass through solids, but not liquids. Love and Rayleigh waves travel at the Earth's surface, like rolling ocean waves; these surface waves are responsible for most of the structural damage caused by earthquakes.

## What is a **fault**?

Faults are fractures in the Earth's crust along which great masses of rock naturally move. They can occur between pieces of a crustal plate or as a boundary between two crustal plates. Not all faults are visible on the surface; many are found deep in the Earth's crust.

## What do **earthquakes have to do with faults**?

When there is movement along a fault, it creates an earthquake. Some movement is so gradual and subtle that only sensitive scientific instruments can detect the activity. But if the rock movement along a fault is sudden and dramatic, it can cause a correspondingly powerful earthquake.

## What is the **San Andreas Fault**?

The most famous example of a boundary between two slipping plates is the San Andreas Fault in California, in which a part of the Pacific plate slides by the North American plate. This 600-mile- (965-kilometer-) long and 20-mile- (32-kilometer-) deep fault is the "master fault" in an intricate network of faults that runs along the coastal region of California. The movement along the San Andreas measures less than an inch (some sources say up to 2 inches) a year—and scientists estimate that as a result of this movement, in 20 million years, the Pacific plate, on the western side of the fault will have slid northward so that Los Angeles and San Francisco will be neighboring cities.

## What happens when an **earthquake** occurs **in the ocean**?

Usually when an earthquake occurs below the ocean floor, nothing significant happens—just the mild shaking of nearby lands. But certain marine earthquakes can cause a shifting of the ocean floor, especially along fault lines. When this happens, the displacement of earth can generate a seismic sea wave, called a tsunami—from the Japanese words *tsu* (meaning "harbor") and *nami* (meaning "waves"). These are gigantic and potentially damaging waves.

## How are **earthquakes measured**?

Earthquakes, either in the oceans or on land, are measured using the Richter Scale, developed by American seismologist Charles Richter (1900–85) in 1935. An earthquake's intensity (magnitude) is measured on a scale of 1.0 (barely detectible) to 9.0 (extremely destructive). But the scale is not linear; instead, each unit represents an exponential increase: An earthquake that measures 7.0 is ten times more powerful than one at 6.0. Most earthquakes do not reach above 8.8. But unofficially, a 1960 earthquake in Chile reached a magnitude of 9.6—and spawned a deadly tsunami that hit Hawaii, Japan, and the Philippines.

## What **major earthquakes** has the **United States** experienced?

There have been many major earthquakes in the United States, including Alaska. Most of these quakes occurred near, but not in, the ocean. And all but one occurred along the Pacific coast.

| Richter Magnitude | Year | Location |
|---|---|---|
| 8.3–8.6 (estimate) | 1899 | Yakutat Bay, Alaska (prior to statehood) |
| 8.5 | 1964 | Prince William Sound, Alaska |
| 8.0–8.3 | 1811–12 | New Madrid, Missouri |
| 7.7–8.25 (estimate) | 1906 | San Francisco, California |
| 7.1 | 1989 | Loma Prieta, California |
| 6.8 | 1994 | Northridge, California |
| 6.5 | 1971 | San Fernando, California |

## What **major earthquakes** have occurred in **other parts of the world** in the **twentieth century**?

There have been many major earthquakes around the world in the twentieth century, most of them far from the oceans. These quakes had high magnitudes, and often caused extensive damage. But even lower magnitude earthquakes can cause great destruction and loss of life if it strikes a densely populated region where buildings are not designed or built to be earthquake-resistant. The extent of damage also depends on the number and the magnitude of aftershocks of a major quake.

| Richter Magnitude | Year | Location |
|---|---|---|
| 9.6 (not official) | 1960 | Chile |
| 8.3 | 1994 | Bolivia |
| 8.2 | 1976 | Tangshan, China |
| 8.1 | 1985 | Mexico City, Mexico |
| 7.7 | 1990 | northwest Iran |
| 6.9 | 1988 | Armenia |
| 6.8 | 1995 | Kobe, Japan |

# OTHER FEATURES OF THE OCEAN FLOOR

## What are **seamounts**?

Seamounts are isolated volcanic mountains that rise from the ocean floor; the nearest equivalent on land is the singular volcano that rises above surrounding flatlands. (Even though seamounts are geologically separate from each other, they may occur in a chain.) Data gathered from more than 50 seamounts indicate that these features are remnants of now-extinct underwater volcanoes—and have a typical volcanic cone shape (though some can eventually fill in, becoming flat-topped). If the summit of a seamount has a small depression, this feature is called a crater; a larger depression is called a caldera.

Seamounts typically rise from 3,000 to 10,000 feet (914 to 3,048 meters) above the ocean floor, but are not high enough to poke through the water's surface. (There are exceptions to this, however, such as the Loihi Seamount, off Hawaii, which is expected to surface, creating a new island.) Seamounts have been found in all the oceans of the world, but the highest concentration is in the Pacific, where more than 2,000 have been identified and some of them are still active. The Gulf of Alaska has numerous seamounts rising from its floor.

## What is a **guyot**?

A guyot is a flat-topped seamount; another name for this formation is a tablemount. American geophysicist Harry Hess (1906–69) discovered the guyot, naming it in honor of prominent geographer and geologist Arnold Guyot (1807–84). Seamounts are normally present near areas of volcanic activity, where there is a constant flow of molten rock (magma). If this magma accumulates to fill in the top of the seamount, its summit becomes level, producing a guyot.

## What is the **Cobb Seamount**?

The Cobb Seamount is one of the most thoroughly explored seamounts: Due to the relatively shallow depth of its summit (only 124 feet, or 38 meters, below the water's surface), it receives ample light, which has

Scuba diver works near the Loihi Seamount, a new volcano forming in the waters off Hawaii. *CORBIS/ Roger Ressmeyer*

allowed scientists to examine it. It was discovered in 1950, and is one of a chain of seamounts that extends from the northern Pacific into the Gulf of Alaska. The Cobb Seamount is located approximately 270 miles (434 kilometers) off the coast of Washington, and rises nearly 9,000 feet (2,743 meters) from the ocean floor. Most of the seamount's summit, covering almost 23 acres, has been mapped by divers.

## What is the **Axial Seamount**?

The Axial Seamount is located in the Pacific Ocean about 300 miles (483 kilometers) off the northern coast of Oregon. It is at the intersection of the Cobb Seamount chain and the Juan de Fuca Ridge—a seismically active mid-ocean ridge in the eastern Pacific that is spreading apart to form new oceanic crust. The Axial Seamount rises some 4,500 feet (1,372 meters) from the seafloor and its peak is still nearly 4,000 feet (1,219 meters) below the surface—but it's growing. The volcano is currently active, with more than 8,000 small eruptions detected during January 1998. In addition, glassy shards of rock have been brought up, indicating that magma is seeping or erupting through the crust.

## Why are **seamounts important**?

Studies in the South Pacific Ocean south of Tasmania (an island off Australia) have found large numbers of fish and invertebrates, some new to science, living near seamounts. Among the more interesting finds were deep-sea coral reefs that are home to unique species, such as bamboo corals. And in one sampling, scientists found 259 species of invertebrates—such as coral, crabs, and sea stars—living on the seamounts, along with 37 species of fish. Approximately one third of the invertebrates were previously unknown, and up to 40 percent of these new species were thought to only occur on seamounts in this region. These findings indicate that seamounts are important, in global terms, as habitats for many unique and previously unknown species of flora and fauna.

## What is an **ocean trench**?

An ocean trench is a very long, deep, narrow, V-shaped valley found near the edge of the continents, or close to island chains in the oceans. They form at subduction zones, as one continental plate slips under another plate, forming a deep depression on the surface of the ocean floor—the deepest points in the surface of the Earth. Ocean trenches usually occur parallel to the continents and island chains, and are areas of heavy seismic and volcanic activity. So far, explorers have identified 22 trenches—18 in the Pacific Ocean, 3 in the Atlantic Ocean, and 1 (the Java Trench) in the Indian Ocean. Among the more well-known trenches are the Tonga, Puerto Rico, and Mariana.

Major ocean trenches can exceed 18,000 feet (5,486 meters) in depth. In width, they vary from 10 to 22 miles (16 to 35 kilometers): The Tonga

Trench, located between New Zealand and Samoa, is the narrowest (and straightest); the Kurile Trench between Japan and Kamchatka is the widest. Their length can also vary widely: The Japan Trench is only 150 miles (241 kilometers) long and is the shortest, while the Peru-Chile Trench off the west coast of South America is nearly 1,100 miles (1,770 kilometers) long.



## What is the **deepest trench in the ocean**?

The deepest trench in the ocean is the Challenger Deep, in the Mariana Trench, located in the western Pacific Ocean near the Mariana Islands (east of the Philippines). It was first measured in 1899; the latest measurement places its deepest point at 38,635 feet (11,776 meters), or approximately 7.3 miles (11.8 kilometers) below sea level. To compare, the tallest point on Earth, Mount Everest, measures about 29,022 feet, 7 inches (8,846 meters) above sea level.

## What are the **major ocean trenches**?

The major trenches and their depths are:

| Name | Ocean | Approximate Depth (feet / meters below sea level) |
|---|---|---|
| Mariana Trench | Pacific | 38,635 / 11,776 |
| Tonga Trench | Pacific | 35,505 / 10,822 |
| Japan Trench | Pacific | 34,626 / 10,554 |
| Kurile Trench | Pacific | 34,587 / 10,542 |
| Mindanao Trench | Pacific | 34,439 / 10,497 |
| Kermadec Trench | Pacific | 32,963 / 10,047 |
| Puerto Rico Trench | Atlantic | 30,184 / 9,200 |
| Bougainville Deep | Pacific | 29,987 / 9,140 |
| South Sandwich Trench | Atlantic | 27,651 / 8,428 |
| Aleutian Trench | Pacific | 25,663 / 7,822 |

## What is the **abyssal plain**?

The abyssal plain, or the abyssal floor, is a flat, relatively featureless area on the deep-ocean bottom; it is found next to the continental slopes. Abyssal plains lie at depths of about 6,560 feet (2,000 meters) to more than 19,680 feet (6,000 meters); the average depth is approximately 13,000 feet (4,000 meters). Here, the water temperature is near freezing, the pressure is immense, and there is neither sunlight nor changing seasons. Only a few living organisms have adapted to life in these inhospitable regions.

There are many abyssal plains across the oceans, including the Somali Abyssal Plain off the east coast of Africa, the Hatteras Abyssal Plain off North Carolina, and the Great Bight Abyssal Plain south of the Australian continent.

## How were the **abyssal plains created**?

The abyssal plains are sometimes thought of as the true ocean floor; they have one of the smoothest surfaces on Earth, with less than 5 feet (1.5 meters) of vertical variation for every mile (1.6 kilometer). These smooth, level plains are a result of a steady and ongoing deposit of sediments, which fill in the nooks and crannies of the rough ocean floor. The sediments come from many sources; they can be washed down the continental slope by turbidity currents racing down submarine canyons, or they can drift down from the ocean waters above. And they can consist of materials from fine particles to the remains of marine life.

## What are the **abyssal hills**?

Abyssal hills are one of two major features (plateaus are the other) on the ocean's abyssal plains. They are commonly low-relief features, usually found seaward of abyssal plains, or in basins isolated by ridges, rises, or trenches. The average height of the hills is 330 to 660 feet (100 to 200 meters) and the average diameter is about 33 feet (10 meters). Approximately 80 percent of the Pacific Ocean floor, and about 50 percent of the Atlantic Ocean floor, is covered by these hills.

## What are **submarine plateaus**?



Submarine plateaus are one of two major features (abyssal hills are the other) on the ocean's abyssal plains. They are broad and more or less flat-topped features, usually more than 660 feet (200 meters) in height, though they vary in height and depth. Some plateaus are volcanic in nature; less frequently, they are the accumulation of deposited sediments. Submarine plateaus occur at various depths.

## What is the **Kerguelen Plateau**?

The Kerguelen Plateau is a huge feature in the Indian Ocean; it measures about one-third the size of the United States and lies more than a half mile (less than a kilometer) under the surface of the water. This 100-million-year-old plateau was recently studied, as it is an example of a unique geological feature—a large igneous province (LIP). It may also be the result of one of the largest and longest-lived volcanic events on the Earth.

An LIP originates where magma (hot, liquid rock) wells up from deep beneath the surface of the ocean floor, and deposits molten rock. Some scientists believe large igneous provinces, which are the result of the largest volcanic events on Earth, may have affected the planet's past environment by altering ocean circulation, climate conditions, and sea levels. Apparently, such volcanic episodes were relatively common between 50 and 150 million years ago, but have been more infrequent during the past 50 million years. The depth of these plateaus have prevented their exploration: Until recently, they were almost inaccessible. But new technology has enabled their exploration.

# LIFE IN THE OCEAN

# AN OCEAN OF LIFE

## How is **ocean life categorized**?

Marine life is classified into three general groups (this is only one of many ways to classify marine life):

**benthos:** These organisms live on the seafloor. They include corals and sponges (which are permanently fixed or immobile on the ocean bottom); snails and crabs (which scurry and creep along the ocean floor); and animals that burrow. This group also includes larger seaweeds, barnacles, and sea squirts.

**nekton:** These are swimming organisms, such as fish, that move freely and migrate from one place to another. They are able to swim in any direction, regardless of the ocean's currents. This group includes squids, herring, and many other familiar animals.

**plankton:** These are the smaller drifting and floating organisms—animals (zooplankton) and plants (phytoplankton). They usually do not have the ability to move from place to place with locomotion, but they may be transported by the action of the waves or currents.

## How are the **ocean waters categorized**?

The ocean can be divided into three major zones, which are defined by the amount of sunlight they receive.

**Sponges are categorized as benthos, which are those organisms that live on the seafloor. Here glass sponge grows atop pillow lava.** ***NOAA/OAR National Undersea Research Program***

The **euphotic (or epipelagic) zone** occurs at ocean depths to about 700 feet (roughly 200 meters), depending on the water's transparency.

Below that depth (700 feet, or 200 meters) to about 3,000 feet (roughly 900 meters) is the **disphotic (or mesopelagic) zone**, where very little light penetrates the water.

The area devoid of light, below about 3,000 feet (roughly 900 meters), which entails 90 percent of the space in the ocean, is called the **aphotic zone**. It is subdivided into the bathyal, abyssal, and hadalpelagic zones.

## Where is **most life** found in the **oceans**?

Most of the plant and animal life in the oceans is found in the relatively thin uppermost layer called the euphotic (or epipelagic) zone. This layer extends from the surface down to no more than about 700 feet (213 meters).

## Why is the ocean environment thought to be more stable than the terrestrial?

Marine organisms have the same four basic needs as do organisms on land—they require food, water, air, and a place to live. Terrestrial conditions are somewhat unstable, with relatively rapid changes in vegetation (caused, for example, by forest fires or extreme weather conditions), air quality (caused by pollution), and water (floods or droughts)—thus altering the environment surrounding terrestrial organisms. In the oceans, the environment is much more stable, continuously providing food, air, support for organisms' bodies, places to live—and especially water (with all its minerals and gases, too)—for all the organisms.

## How are **organisms distributed** in the **near-shore waters** and **open oceans**?

Along the coastlines, marine life is found in areas with the best available shelter and food—especially along the nooks and crannies of rocky shores and coral reefs. In the open oceans, marine life is present down to several miles (or kilometers). But within the water column, the populations of organisms are unevenly distributed: Most organisms live in the euphotic (or epipelagic) zone (the top layer of the ocean), or they migrate to this layer in search of food.

## How many **different types of plants and animals** live in the **oceans**?

There are estimated to be more than 250,000 different types of plants and animals living in the oceans—and there are without question many more to be discovered, especially in the deeper parts of the oceans. Some scientists estimate the number of species is closer to 400,000; and others even say there are between 1 million and 10 million benthic (deep-ocean floor) species yet to be described!

### What do the studies of **marine ecology and biology** entail?

Marine ecology and biology entail the study of the plants and animals in the ocean—and are concerned with the relationship between the plants and animals as well as their relationships with their environment. This includes the way the organisms adapt to the various chemical and physical properties of the seawater. Researchers in these fields examine such variables as pollutants, the natural movements of the ocean, light availability, and even the stability of the seafloor.

## MARINE PLANTS

### What does **flora** mean?

Flora is another word for the plant life on Earth. It is also used to describe the entire plant life of a given region or habitat, or plant fossils in a geological stratum (layer of rock). The word comes from Flora, the goddess of flowers and springtime in Roman mythology. The plural of flora is florae or floras.

### What is **photosynthesis**?

Photosynthesis is the process by which plants (and some single-celled flagellated organisms) convert inorganic carbon dioxide (found in the atmosphere), water, nitrite ions, and phosphate ions into useable sugars and amino acids by using the energy from sunlight. Plants use the process of photosynthesis almost everywhere on land. Photosynthesis also can occur in the ocean, in the euphotic zone, or in depths up to 700 feet (213 meters).

## What is a **plant**?

Contrary to popular belief, the photosynthetic process does not define an organism as a plant; in fact, it is difficult to define a plant. These organisms vary widely in shape, size, and color, and range from independent, single-celled algae to the specialized cells that make up the multicellular plants we grow in our gardens. But there is one key microscopic feature common to every plant cell: An inflexible cellulose wall on the outside of the plant cell membrane.

## What are living **cells**?

Cells are the tiny units that make up the majority of living organisms. About 60 to 65 percent of the cell is water because water is a perfect medium for biochemical reactions to take place. Cells are primarily composed of oxygen, hydrogen, carbon, and nitrogen. Within the cell, the most important organic materials are the proteins, nucleic acids, lipids, and carbohydrates (or polysaccharides). Certain cellular structures, all formed from a combination of these organic compounds, are also present in the cell.

## What are the **two types of living cells** on Earth?

The two types of living cells on Earth are: the prokaryotes and eukaryotes. One of the greatest differences between these cells lies in their DNA: The prokaryotes' DNA are single molecules in direct contact with the cell cytoplasm (the cytoplasm is the living substance of the cell, excluding the nucleus); the eukaryotes' DNA have more than one molecule and the molecules can be diverse. Plus, the DNA of eukaryotes is found within a nucleus that is separated from the cell's cytoplasm by an envelope (a membrane); certain eukaryotes also are further divided by additional internal membranes. (The prokaryotes do not have such internal membranes.) In short, prokaryotes are simple cells and the eukaryotes are complex cells.

## What are some examples of **prokaryotes** and **eukaryotes**?

The prokaryotes (simple cells) include bacteria and cyanobacteria (once referred to as blue-green algae). The eukaryotes (complex cells) include all plant (including algae) and animal cells.

## How do **plant cells differ** from **animal cells**?

Plant cells have a rigid cell wall, a large vacuole (a fluid-filled pouch), and chloroplasts that are used for the synthesis of glucose (in which light energy from the Sun is converted to chemical energy—and finally into glucose). None of these features are found in animals cells.

## Where and how did the **first plants evolve**?

Scientists believe that the first plant-like organisms evolved in the oceans about 3 billion years ago. To compare, the oldest known life is thought to be about 4 billion years old, in the form of small bacteria-like organisms. (However, to date, no such life has been discovered in the fossil record; the oldest known fossil evidence of life is about 3.75 billion years old.)

No one knows how organisms developed photosynthesis. Some scientists believe that certain early bacteria developed into chloroplasts—the special "organs" of a plant cell that carry out photosynthesis—which then released oxygen as a waste product, eventually producing the oxygen in our atmosphere.

## What **sequence of events** led to **modern flowering plants**?

It is thought that the various plants evolved in the following way (and as usual with science, the discovery of more fossil evidence may change some of these dates in the future):

**4 billion years ago:** Single-celled life developed in the oceans.

**3 billion years ago:** The process of photosynthesis, with oxygen as a byproduct, probably evolved in some microorganisms in the ocean—eventually producing an oxygen-rich atmosphere. About 1.5 to 2 billion years later, the planet's protective ozone layer formed.

**600 million years ago:** There was an explosion of animal life—which evolved and became increasingly diverse, plants also flourished.

**470 million years ago:** Certain plants made the "first steps" onto land, adapting to conditions around the edges of tidal pools, called the land-water interface. Since the first land animals had not yet evolved, these plants were the first life on land.

**430 million years ago:** Plants with roots, stems, and leaves evolved; they are called vascular plants.

**420 million years ago:** The first true plants, including ferns, flowerless mosses, and horsetails, evolved on land and rapidly filled many ecological niches (environments). About 70 million years later, ferns developed seeds.

**145 million years ago:** The first flowering plants evolved on land; about 50 million years later, they dominated the land.

## What **adaptations** did **marine plants** make in order to **move to land**?

The major change marine plants needed to make so they could move to land was to develop a new way of obtaining and keeping water, since they would no longer be submerged in the sea. Evaporation was also a problem, and plants had to develop a way to transfer necessary gases.

To solve these problems, various plants evolved many "new" mechanisms over millions of years. The one we are all familiar with is the seed—a waxy, waterproof cuticle that covers the embryo of the plant, preventing excessive loss of water. Some plants developed protective cells around the reproductive organs in order to minimize the loss of water. Still others developed spores, reproductive cells that develop directly into full-grown plants without having to be fertilized first.

## What are some examples of **modern marine plants**?

There aren't as many modern marine plants as there are marine animals. But there are several that most of us recognize—including algae, marine grasses, and mangroves.

## Do **marine plants** resemble **land plants**?

No, in the ocean, most plants do not look like the plants found on land. Most marine plants do not have stems, leaves, or roots; many cannot even be seen without the use of a high-powered microscope. But there are similarities: Marine plants make their own food through photosynthesis, just like land plants.

## What are **algae**?

The great bulk of marine plant life consists of algae (the singular is alga). They are structurally simple single- to multi-celled organisms that carry out the oxygen-producing process of photosynthesis, and are found in both the oceans and in freshwater. You can see this type of organism by filling a clear jar with pond water and placing it on a sunny windowsill; or place a pail of rainwater outside. After several days, the water will look cloudy and green. This green water is filled with thousands of single-celled algae using the sunlight to carry out photosynthesis—similar to land plants that use photosynthesis to produce food in order to survive. Thus, algae are self-supporting and can live wherever there is the right balance of light, oxygen, carbon dioxide, and water.

There are numerous types of algae, named for their color, including blue-green, green, red, and brown. They are very diverse and are nonvascular (vascular plants, mostly on land, have vessels that conduct fluids up and down the plant; algae do not have these vessels). The free-floating algae are known as phytoplankton (plant plankton), and are usually single cells. They are distinguished from other marine plants by their reproductive processes.

## How **big** are **algae**?

Contrary to popular belief, not all algae are small. Most algae are single-celled organisms, some as small as 1 to 2 micrometers in diameter (a micrometer is 1 one-millionth of a meter or 0.000001 meter—about 0.00004 inches); others are multicellular organisms, such as pond scum, seaweeds, and most green coatings on trees. On land, they can also live in symbiosis (mutually beneficial association) with certain types of fungi.

The most complicated marine algae are the brown seaweeds. Bladderwracks, tangleweeds, and oarweeds are stuck fast to rocks in the ocean, and can grow yards (meters) in length. They usually have leaf-like fronds that float in the water, and contain chlorophyll that carry out photosynthesis.

**Ribboned seaweed photographed underwater off South Island, New Zealand.** ***CORBIS/Paul A. Souders***

Like land plants, the larger marine algae use photosynthesis, but the similarity ends there: Seaweeds do not have special roots that take in water; and they lack internal tissues like a xylem (land plants use these small internal tubes to carry water to their leaves). In addition, unlike most land plants, some seaweeds are able to adapt to long periods of drying.

## **Where** are **algae** in the oceans?

The larger algae (macroscopic) are usually attached to a firm surface; they also grow on rocks in moving or stagnant water. In the oceans, they usually grow as seaweeds in the intertidal and subtidal zones as deep as 879 feet (268 meters), depending on the transparency of the water. The smaller algae (microscopic) are usually single-celled, free-floating organisms. They are a major part of the food chain in the ocean's upper layers.

## What is **phycology**?

Phycology, also called algology, is the study of algae. Phycology is from the Greek *phykos* or "seaweed"; algology is from the Latin *alga,* or "sea wrack."

## Why is the word **algae** often **misunderstood**?

The term is often misunderstood because there are so many types of algae. "Algae" merely refers to any aquatic organism capable of photosynthesis.

## Why are algae important to the study of ancient life?

Fossilized single-celled algae have been found in rock that is more than a billion years old. It is thought that green and blue-green algae were some of the most successful plants that arose on the Earth. For such small single-celled plants to survive so well—and for more than a billion years so far, is amazing.

## How do scientists **classify algae**?

Like all organism classifications, there is no one way in which scientists classify algae. Traditionally, the non-motile (mostly attached) forms of algae were thought of as plants, and the motile (mostly free-swimming) forms—even if they used photosynthesis—were considered to be both plants and animals. Now some scientists classify algae in multiple kingdoms; another classification puts most algae under the plant kingdom; and yet another classification puts algae in the kingdom Protista—except for cyanobacteria, which it places in the kingdom Prokaryotae.

The debate won't be over for a while. Research has suggested at least 16 lines in which groups of algae have a common ancestry, including the cyanobacteria, diatoms, and brown, green, and red algae. There will have to be more research into algae before any agreement will be reached as to classification.

## What are the various **types of algae**?

There are numerous types of algae, including the following:

**red algae (Rhodophyta):** These algae are eukaryotic (made up of complex cells) and are mostly found in marine environments. They lack the chlorophyll-b that most other algae possess (they have chlorophyll-a instead), and there are special blue and red pigments within the algae. Because they have incomplete cell division when going through the reproductive process, they have "pit connections" between each cell; their actual reproductive history is very complex. The cell walls of certain red algae are also responsible for the production of two important polysaccharides—agar and carrageenan. Both have suspending, emulsifying, stabilizing, and gelling properties and are used in the production of some commercial foods.

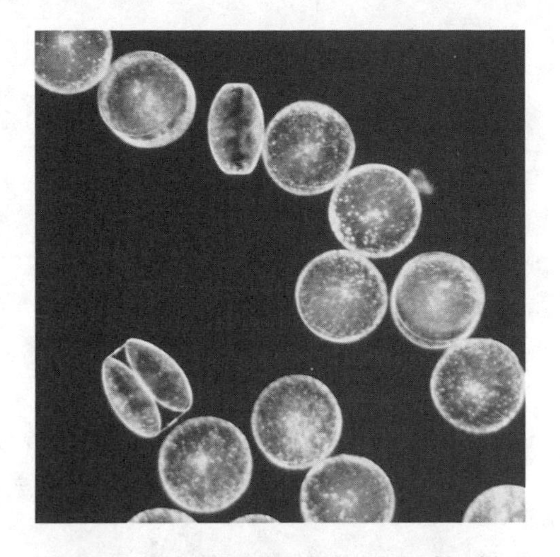

**Diatoms are single-celled phytoplankton incapable of moving under their own power. They are extremely common and can be found in both salt- and freshwater.** *CORBIS/Douglas P. Wilson; Frank Lane Picture Agency*

**brown algae (Phaeophyta):** These algae are found mostly in marine environments, but they lack the chlorophyll-b that most other algae possess. Instead, they have another type called chlorophyll-c, along with special photosynthetic yellow to deep-red pigments. Many of the brown algae grow to great sizes, and we know them as kelp. Kelp can reach up to 100 feet (30 meters) in length. Commercially, alginate is obtained from brown algae kelp, a polysaccharide used in the same way as agar and carrageenan (as suspending, emulsifying, stabilizing, and gelling agents in foods); other brown algae are used as sources of vitamins and fertilizers.

**green algae:** These algae are single-celled, or form cells in a group (colonial). They are similar to most plants, as they have both chlorophyll-a and -b, and also store food as starch. There are several green algae that live in the ocean. For example, one type has its cell walls infused with aragonite (a form of calcium carbonate); this algae makes an important contribution to a coral reef's formation and survival.

**A small school of sturgeon feed on filamentous algae, which grew around a manmade reef of submerged pipes.** ***NOAA Sea Grant Program; Dr. James P. McVey***

**dinoflagellates:** These small organisms are an important part of the marine food chain; they are phytoplankton (small plant organisms) and have long flagella, or whip-like tails, that allow them to move up and down in the water column. They often have a multi-layered covering of cell material.

**diatoms:** These single-celled phytoplankton (small plant organisms) are found in both salt- and freshwater; they can even be found in moist soil. They are one of the most common types of phytoplankton. Diatoms do not have flagella (whip-like tails), and thus cannot move under their own power.

## What are **cyanobacteria**?

Cyanobacteria are microscopic organisms; they are similar to bacteria because they lack a nuclear membrane (they are simple cells, or prokaryotes). They are also photosynthetic, and thus can manufacture their own food—but are classified by most scientists as bacteria, not as algae.

The cyanobacteria reproduce by binary fission (in which the cell splits into two identical cells), production and germination of spores, or by the breaking of multicellular filaments. Cyanobacteria are important to the growth and health of many plant species—especially those on land. This is because they are only one of a few organisms on Earth that can take inert atmospheric nitrogen and turn it into an organic form (as nitrate or ammonia). These organic forms are called "fixed" forms, which plants need in order to grow. In fact, fertilizers work in part by adding fixed nitrogen to the soil, where the plant roots absorb it. Certain types of cyanobacteria can also cause problems. For example, a species of *Lyngbya* can cause "swimmer's itch," a form of skin irritation that can develop from exposure to water infested with this cyanobacterium.

## Why were **cyanobacteria once thought** to be **blue-green algae**?

Because they are photosynthetic and aquatic, cyanobacteria were once referred to as blue-green algae; but they are not related to any of the numerous eukaryotic algae. However, there really is such a thing as blue-green algae, and it is eukaryotic (has complex cells). Cyanobacteria are simple-celled relatives of bacteria.

## Are all **cyanobacteria blue-green**?

No, not all cyanobacteria are blue-green in color. They can range from blue-green to purple. They get their name and part of their color from a bluish pigment called c-phycocyanin that is used to capture light for photosynthesis; they also carry a red pigment, c-phycoerythrin. Amazingly, cyanobacteria also carry chlorophyll-a, the identical photosynthetic pigment used by plants—but this green pigment is masked by the other two pigments.

There are also types of cyanobacteria that are red or pink—due to the pigment c-phycoerythrin—often found growing around sinks, drains, and even lurking on greenhouse glass. Another species is the *Spirulina,* a cyanobacterium that gives African flamingos their pink color.

## Why were **cyanobacteria** important to the **evolution of life** on Earth?

Cyanobacteria were important to the evolution of life on Earth because of their relation to eukaryotes (complex cells): Sometime around the early Cambrian period (505 to 544 million years ago), cyanobacteria began to live within certain eukaryote cells, making food for their host in return for a place to call home. This is called endosymbiosis—and was the origin of mitochondrion structures (specialized cells containing energy-producing enzymes) that eventually evolved within eukaryote cells.

The other way in which cyanobacteria were important to the evolution of life on Earth dealt with the origin of plants: The chloroplasts—the place where the plants make food for themselves—are actually cyanobacteria living within plants' cells.

Seaweed can take many forms: Here it carpets the rocks along the shores of Cashel Bay near Connemara, County Galway, Ireland. *CORBIS/Macduff Everton*

## What are **stromatolites**?

Stromatolites are mostly ancient blue-green cyanobacteria; they were probably one of the earliest forms of life in the oceans. These organisms secreted lime, creating hard, large, layered, mushroom-shaped structures. Fossilized stromatolites have been found in many places; the oldest specimens were found near an area called North Pole, Australia. These ancient stromatolites, which released oxygen into the air, are believed to have played an important role in the build-up of oxygen in the Earth's early atmosphere. Stromatolites still live on Earth; one of the places they are found is in Western Australia, particularly in Shark Bay (in the Indian Ocean).

## What are the most **common types of large algae** found along the **ocean shore**?

The most common types of large algae found in saltwater are mainly in the intertidal zones and clear waters less than 300 feet (100 meters) deep: they are green (*chlorophyceae*), blue-green (*cyanophyceae*), brown (*phaeophyceae*), and red (*rhodophyceae*) algae.

## What is **seaweed**?

Seaweed is actually a larger form of algae. The most common are the brown and green algae that form brown- and green-looking seaweeds.

## Is **seaweed** really **edible**?

Yes, one of the most popular seaweeds (actually red algae) is *nori,* which is an important part of many peoples' diet, especially in Japan. Various brown algae species include the popular *wakame, kombu,* and *hijiki*—all part of many people's diets, especially in Japan.

## What is **sargassum**?

It is a seaweed that, unlike its algae relatives (which grow near the ocean shore), thrives in the open ocean. The sargassum weed—a brown algae—grows on the surface of the Sargasso Sea, a large tract of relatively calm water in the western Atlantic Ocean, near Bermuda. Here, sargasso forms a floating "island" of seaweed that covers an area measuring about two-thirds the size of the United States. This seaweed has air bladders located on its short stalk; they are also pelagic, or free-floating, and reproduce through a form of asexual reproduction (by breaking off pieces of the plants).

A remarkable community of animals lives in the floating mass of sargassum weed, including worms, bryozoans, shrimp, crabs, fish, and hydroids. Many creatures have adapted to the environment by imitating the color and sometimes texture of the seaweed in order to hide from predators. Other organisms use the seaweed for transportation: baby turtles, for example, often "ride" the algae out into the ocean.

Although 7 million tons of the seaweed lives in the Sargasso Sea, it is not thick enough to hinder navigation by humans. Occasionally, large chunks of the seaweed mass do break off and float away, washing up along the East Coast of the United States.

## Are there any **other types** of **sargassum** seaweeds?

Yes, other sargassums are also found in waters along the East Coast of the United States; in all, there are about 15 different species stretching from

**Kelp, a form of brown algae, can reach up to 100 feet (30 meters) in length and can form underwater forests or beds, such as the one the diver swims through here.** ***NOAA/OAR National Undersea Research Program; W. Busch***

Maine to Florida. One sargassum, called *Sargassum filipendula,* is benthic (or attached to the bottom), and is found from Cape Cod southward.

## What is **kelp**?

Kelp is a large group of brown algae (or brown seaweeds). They can grow into sizeable structures, sometimes reaching up to 200 feet (60 meters) in length; or they can form as branching mats. The largest are found off the western coast of North America where they grow on the rocky bottom, beyond where the waves break. Kelps along the Atlantic coast are smaller, reaching only about 10 feet (3 meters) in length and grow mainly below the tide line. They are often collected and used as a food source or for industrial raw materials. Bull kelp and feather-boa kelp are common types of the seaweed.

## What are **kelp forests**?

Kelp forests are huge growths of the brown algae (or brown seaweeds). They can cover an ocean area of several miles with towering vegetation and thick canopies. These forests contain abundant marine life, including the popular sea otters, which dwell in the kelp beds along the central California coast.

Humans have also harvested the rapidly growing kelp beds, using large barges with paddle-wheel "mowers" to collect the seaweed (the blades of kelp rapidly grow back). Most of this kelp is processed into algin, a stabilizer and emulsifier used in many products, including paint, cosmetics, and ice cream.



## What have scientists discovered after studying one of the world's **largest kelp forests**?

Just off the coast of San Diego is the Point Lomo kelp forest community. Researchers have studied the ecology of these kelp beds for years, trying to determine the various processes that affect the forest—including seasonal climate variations. They've determined that the effects of global climate episodes such as El Niño (a periodic warming of the waters off the west coast of South America) and La Niña (the following periodic cooling of the same waters) affect the kelp, while smaller, localized climate shifts do not. For example, during the 9-year study, *Macrocystis* species were not affected by the competition from other kelp species. But *Pterygophora californica,* an important understory (under the top canopy) species showed a reduction in growth and reproduction due to light-limited conditions and competition with *Macrocystis*; this occurred during La Niña periods when *Macrocystis* thrived. Conversely, when El Niño conditions led to poor *Macrocystis* growth, the understory kelps did much better.

# PLANKTON

## What are **plankton**?

Plankton are mostly tiny animal and plant organisms that float or weakly swim in the ocean's surface waters. Some plankton are single-celled while others are multi-celled—with some forming colonies. The word plankton comes from the Greek *planktos,* meaning drifting or wandering.

The organisms that make up plankton are very numerous and diverse; they can be grouped in many different ways. One division is based on the ability for photosynthesis; those plankton (mostly algae) that are capable of this process are called phytoplankton. Non-photosynthesizing (often termed animal) plankton are called zoo-plankton. The divisions of plankton are not made based on plant or animal characteristics—because some plankton can exhibit traits of both plants and animals.

## Which **plankton** exhibit **characteristics of animals *and* plants**?

One example is the euglenoid, which is usually thought of as phytoplankton (plant plankton) and contains chlorophyll (for photosynthesis). But if the water changes and the light levels drop, these creatures will be forced to hunt for their food rather than manufacture it through photosynthesis. Thus, they bear characteristics of both animals and plants.

## How are **plankton divided**?

One division (or classification) of plankton is by their size—from invisible to visible with the naked eye. The following table lists the names given to the various sizes of plankton and some common examples.

| Name | Size (in microns*) | Example(s) |
|---|---|---|
| Ultraplankton | Less than 5 | |
| Nanoplankton | 5–50 | protozoans (single-celled zooplankton) |
| Microplankton | 50-500 | invertebrate eggs and larvae |
| Mesoplankton | 500 to 5,000 | |
| Macroplankton | 5,000–50,000 | copepods |
| Megaloplankton | larger than 50,000 | large jellyfish |

*one micron is one micrometer, or one millionth of a meter (0.000001 meter)

## Which **marine animals eat plankton**?

In healthy oceans, where there are few or no harsh pollutants or natural disasters, plankton are eaten almost as fast as they are produced. The common animals that feed on plankton include jellyfish, comb jellies, certain shrimp, herrings, anchovies, and even huge blue and gray whales.

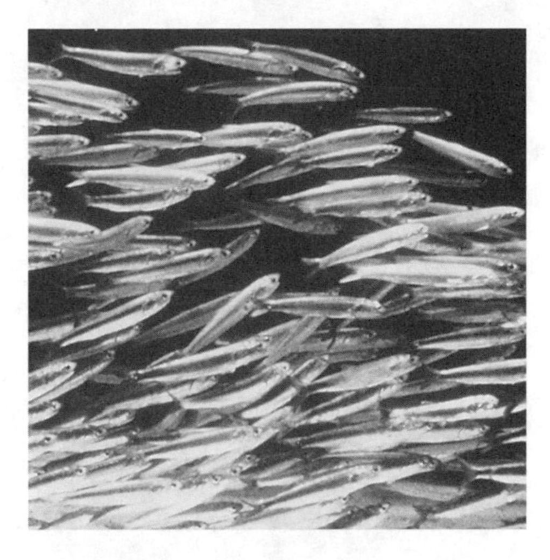

Tiny anchovies, among the fish that eat plankton, are near the bottom of the food chain. *NOAA/OAR National Undersea Research Program*

## Why are **plankton important** to ocean life?

Plankton are the foundation of the ocean food chain. All plankton are eaten by larger predators, such as fish or whales, or they die and sink to the ocean floor.

## Who was **Viktor Hensen**?

German scientist Viktor Hensen (1835–1924) was the director of what is popularly known as the Plankton Expedition, a scientific expedition in 1889. The goal of this venture was to systematically categorize all the organisms in the sea. Hensen was responsible for naming the smallest organisms found by the expedition—the plankton.

## What are **phytoplankton**?

Phytoplankton are the plant-type plankton living in the oceans and most other surface waters, such as freshwater lakes, rivers, and ponds. These small plants are found in water to about 100 to 130 feet (30 to 40 meters) deep. Similar to plants on land, the phytoplankton use photosynthesis—the process of using sunlight to make the food that gives them energy.

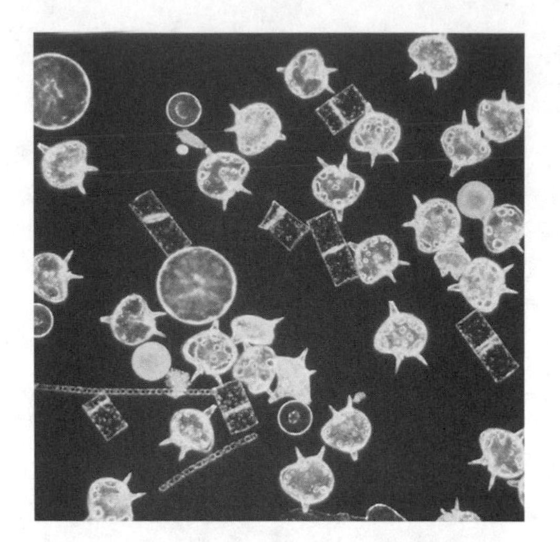

Dinoflagellates are phytoplankton with whip-like tails that allow them to move up and down in the water column. *CORBIS/Douglas P. Wilson; Frank Lane Picture Agency*

## What **percent of marine plants** are **phytoplankton**?

Phytoplankton make up approximately 90 percent of the ocean's plants—and are the most prolific types of plankton (in other words, they far outnumber animal plankton, also called zooplankton). Individually, they are extremely small, but they are great in number!

## What is a **dinoflagellate**?

A dinoflagellate is a phytoplankton that has a long flagella, or whip-like tail, that allows the organism to move up and down in the water column. Dinoflagellates often have a multi-layered covering of cell material; certain types are even armored with overlapping layers of cellulose plates. Some dinoflagellates are even bioluminescent, or light-producing; this group also includes those organisms that cause the red tides, when the sea water turns "red" from a profusion of dinoflagellates.

## What is a **diatom**?

A diatom (Bacillariophyta) is a single-celled phytoplankton that is found in all types of water (ocean and freshwater) and even in moist soil. In the oceans, they are divided into two types: elongated or round (wheel-shaped) forms. They also have a type of pigment similar to brown algae, and their cell walls are made of silica.

Diatoms are considered one of the most common types of phytoplankton; along with dinoflagellates, diatoms are among the most abundant phytoplankton found in temperate coastal regions. Unlike dinoflagellates, diatoms do not have flagella (whip-like tails), and thus cannot move through the water under their own power. Instead, they have evolved certain ways to keep themselves afloat in the surface waters of the ocean, including specialized spines; certain species will also connect themselves to other diatoms to form long chains and increase their buoyancy.

Krill, a large zooplankton, are the foundation of the Antarctic marine food chain. *CORBIS/Peter Johnson*

## What are **fossilized diatoms** called?

Accumulations of fossilized diatom shells are called "diatomaceous earth," a chalk-like deposit composed of silica. It is often used in filtration and as an abrasive.

## What is a **common type** of **zooplankton**?

Copepods are the most abundant zooplankton in the surface ocean waters. They are the most numerous crustaceans in the world, with more than 6,000 species found in fresh and ocean water. They are about a fraction of an inch (just over 1 millimeter) in size and have varied body shapes.

Still other zooplankton are the larvae of much larger marine animals, such as corals and fish. Released in great numbers during the mating season, few of these tiny animals ever reach adulthood—as predators feed on the massive numbers of these tiny animals.

## What are some **plant-eating zooplankton**?

Several types of zooplankton, such as copepods and krill, eat phytoplankton (plants). But this makes sense: The small size of these animals makes chewing on larger plants—or animals—impossible!

## What is one of the **largest zooplankton**?

One of the largest zooplankton is the krill, a small, shrimp-like crustacean of the genus *Euphausia*; krill range in size from about 0.5 to 3 inches (1 to 7 centimeters). Many larger animals feed on these small zooplankton, including the filter-feeder blue whale, as do other whale species.

## Why are **copepod plankton** so **important** to ocean life?

Copepod plankton are important links in many marine food chains. Not only do they feed on smaller microorganisms and phytoplankton, but they are also important food for larger animals.

## What is a **carnivorous zooplankton**?

A carnivorous zooplankton called the arrow worm (or chaetognath) is known as the "tiger of the zooplankton." This organism feeds on animals, attacking and devouring prey as large or larger than itself. Most arrow worms range from about 0.4 to 1.2 inches (1 to 3 centimeters) long, and are shaped like darts. They are also free-swimming and have numerous chitinous teeth—and are themselves a prime target of larger species, which see them as a food source.

## What happens to **plankton** that **die**?

While plankton are an important food source in the ocean, they are also prolific, which means there can be great quantities of them that are never consumed by predators. One place where plankton are plentiful is the Indian Ocean, where about 90 percent of the silt on the seafloor consists of the remains of plankton. Plankton blooms during the course of this ocean's summer monsoon, a time when the rains are plentiful, the weather is warm, and there are more nutrients for the organisms to feed on. Only the toughest parts (the resistant organic material or skeletons) of plankton are preserved when they die; this material is often broken up and slowly settles from the ocean surface to the ocean floor. One reason scientists are interested in this silt is because it is a record—a natural "climatological archive"—of the past several thousands of years, and therefore, worth studying.

## Is there a connection between **plankton** and the **global climate**?

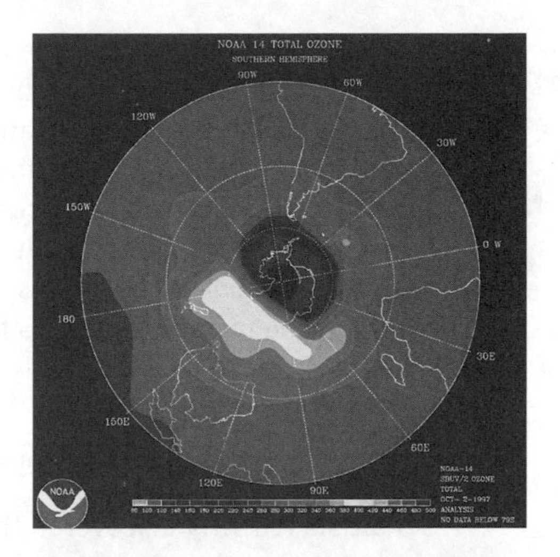


In October 1997 the National Oceanic and Atmospheric Administration (NOAA) released this graphic showing the "hole" in the ozone layer over the South Pole. (Also *see* the color graphic of this image, in the color insert.) Many scientists believe the enlargement of the ozone hole severely impacts the marine food chain. *AP Photo/NOAA*

Some scientists believe there is an important connection between plankton—particularly phytoplankton—and the global climate. These plant-plankton cycle (use) carbon and other elements in the atmosphere. If carbon dioxide increases in the atmosphere, the Earth's average temperatures could rise. Plankton remove carbon from the carbon dioxide in the air and use it in their respiration; they also store some of the carbon when they die and sink to the bottom of the oceans. But if certain nutrients, such as iron, nitrogen, or phosphorus are limited, the phytoplankton may not grow to their full potential—meaning less carbon would be absorbed from the atmosphere. Whether a reduction in the number of matured plankton would be enough to cause a major climate change is still unknown, but scientists are studying the possibility.

## Does the **ozone hole** affect **plankton**?

Yes, some scientists believe that the ozone hole—the area in our atmosphere where the Earth's protective blanket has been destroyed—may have an important impact on plankton. This may especially be true in the Antarctic and Arctic oceans (around the South and North Poles, respectively). For several months each southern spring, two-thirds of the ozone shield protecting Antarctica and the surrounding ocean from the Sun's ultraviolet radiation is destroyed; up to half of the shield over the Arctic region is also lost for a short time each year. (These changes in the ozone layer are thought to be caused by chemicals from industrial and commercial emissions.)

Since plankton are susceptible to an increase in solar radiation, reductions in the ozone layer (or, put another way, enlargements in the hole in the ozone) affect these ocean organisms. When the ozone levels are low, scientists have noted a plankton loss of between 6 and 12 percent in the Antarctic Ocean. And since plankton are the bottom of the marine food chain, their reduction could affect all life in the oceans. Plankton can be thought of as the oceans' meadows: For example, krill feed off the plankton and marine mammals feed off the krill.

### Why does **radiation from the Sun reduce** the number of **plankton**?

Scientists recently discovered that the reproductive cells of planktonic algae are several times as sensitive to the Sun's ultraviolet radiation as the organisms' mature cells. This means that the impact of solar radiation on plankton could be greatest in the spring—during so-called "blooms," when plankton reproduction is at its peak. If this proves to be true, marine plankton may be experiencing stunted growth. And because these animals are at the base of the food chain, an increase in the ozone hole could have serious effects on marine life.

Researchers found that the plankton's asexual spores (one way of reproduction among certain algae) are 6 times more sensitive to ultraviolet-B radiation from the Sun than are mature algae. This was determined by measuring the slowing rate of photosynthesis when the organisms are exposed to increasing radiation. Free-swimming gametes, which is the sexual means of reproduction of some types of plankton, were found to be even more susceptible: their photosynthesis fell by 65 percent after a 1-hour exposure to ultraviolet-B radiation—the equivalent of an ozone layer reduced by 30 percent—causing a reduction in plankton growth rate of about 17 percent.

## MARINE ANIMALS

### What does **fauna** mean?

Fauna is another term for the animal life on Earth. The word is also used to describe the entire animal life of a given region, habitat, or geological stratum (layer of rock). The word is derived from the Latin Fauna, the Roman goddess of nature. The plural of fauna is faunas or faunae.

## How can **ocean animals** be described?

There is a wide range of animal life in the oceans, and many ocean-dwellers are unique and quite different from their land-dwelling cousins. Some are quite bizarre—having no legs, eyes, or ears. Other marine animals look and behave like plants, permanently attaching themselves to rock or the seafloor, where they siphon off oxygen and food from the water that passes by them.

Whether static or mobile, all marine animals obtain their food from their environment; they cannot produce their own food. They must gather and consume organic material in order to live. As on land, marine animals can be herbivorous (they eat plants); carnivorous (they eat meat, or other animals); or omnivorous (they eat both plants and animals).

## What **animals** inhabit the **oceans**?

There are hundreds of animals inhabiting the oceans. The following lists only the major groups. (The following two chapters—Marine Mammals, Birds & Reptiles, which begins on page 261, and Fish & Other Life, which begins on page 303—describe most of these groups in detail.)

Sponges (phylum [order] Porifera)

Coelenterata
- hydroids (phylum Hydrozoa)
- jellyfish (phylum Scyphozoa)
- sea anemones (phylum Anthozoa)
- coral

Ctenophora
- comb jellies

Marine Worms

Bryozoans (phylum Bryozoa)

Mollusca (100,000 species; 7 classes; these are found close to shore)
- snails and other single-shelled (univalved) animals (Gastropoda)
- chitons (Polyplacophora)
- two-shelled (bivalved) mollusks (Bivalvia)
- squids and octopuses (Cephalopoda)

Arthropoda (75 percent of all animals—land and ocean)
- chelicerates (horseshoe crabs, spiders, and mites)
- insects
- crustaceans (crabs, shrimps, lobsters; nearly all marine arthropods are crustaceans)

Echinoderms
- sea stars
- brittle stars
- sea urchins
- sea cucumbers

Tunicates (phylum Chordata)
- sea squirts or ascidians

Fish (vertebrates; nearly 50 percent of the 40,000 known species of vertebrates are fish)
- Cartilaginous fish (phylum Chondrichthyes—sharks, rays, and skates; 10 percent of all fish)
- Bony fish (phylum Osteichthyes—tuna, cod, salmon, etc.; 90 percent of all fish)

Marine Reptiles (14 percent of the 40,000 known species of vertebrates are reptiles)
- crocodiles
- sea turtles
- sea snakes

Birds (sea and shore)
- puffins
- shorebirds (shallow water waders)
- egrets, herons, and ibis
- sea ducks
- gulls, terns, and skimmers
- cormorants
- pelicans
- gannets

### What is the fastest marine animal?

The fastest known marine animal is the sailfish, *Istiophorus platypterus.* It has been clocked at speeds of 68 miles (109 kilometers) per hour.

frigatebirds
pelagic birds (such as storm petrels)
kingfishers
ospreys

Marine Mammals
whales (cetaceans)— baleen and toothed (toothed includes dolphins and porpoises)
seals, walruses, sea lions (Pinnipedia)
manatees or sea cows (Sirenia)
sea otters (Mustelidae)

## What is **extinction**?

Extinction occurs when a species completely disappears from our planet. Extinctions of organisms have happened throughout the history of the Earth, often by natural occurrences, such as the impact of a space object (a comet, for example). Such an impact would cause material to be thrown high into the atmosphere, dimming the Sun and changing the local or global climate—depending on the size of the impacting body.

In modern times, the rate of extinction has dramatically increased—and most scientists believe there is a link between the increased extinctions and the growth of the human population. In other words, as the human population increases, more species become extinct—and this could happen at an increasing rate. By the year 2000, the world population is estimated to be approximately 6 billion. The rate of species extinction is currently at one species a day.

## What is an **endangered organism**?

An endangered organism—plant or animal—is in immediate danger of becoming extinct. In most cases, the populations of these organisms are very low in number, and they need active protection (usually by a government) in order to survive. Unfortunately, even though animals are protected while on the endangered lists, their populations often decline. For example, the southern sea otter was driven to the brink of extinction due to hunting (for its fur) early in the twentieth century; it still numbers only in the few thousands even though it is protected. This is because these otters are very vulnerable to the effects of pollution, including oil spills. Still other endangered organisms make a comeback, such as the gray whale population, which was recently removed from the endangered species list.

## Are there any **endangered marine animals**?

There are many endangered marine animals. The survival of these animals is at risk for many reasons, including the disappearance (or severe reduction) of their habitats, pollution, and even human activities (for example, marine mammals can get tangled in fishing nets or collide with boats). Marine mammals that are on the endangered list include southern sea otters; manatees; monk seals; and humpback, blue, fin, sei, right, and bowhead whales.

## What is a **threatened organism**?

A threatened organism is one that is at risk and, if its population dwindles (or rather, continues to dwindle), it will be reclassified as endangered.

## Are there any **threatened marine animals**?

Yes, there are several threatened marine animals. For example, the Stellar sea lion is now on the list of threatened marine animals because of its tremendous drop in population. Researchers and conservationists constantly monitor the populations of marine animals to determine if they are threatened; thus, the list changes often (some fall off of it while others are moved to the endangered list).

## What are some of the **smallest animals** in the **oceans**?

Some of the smallest animals in the oceans are the marine protozoa—complete one-celled animals. These microscopic animals obtain food, breathe (respire), and eliminate waste products just like multi-celled organisms; they are found with plankton in the upper to deep-ocean layers. Protozoa can climb, crawl, swim, and scuttle, feeding on bacteria, other protozoans, diatoms (minute plants), and tiny organic debris. Three groups of protozoa are common in the ocean:

Sarcondinians: Sarcondina means "creeping flesh," which describes how these animals move. They have jellylike bodies, and almost ooze to form a "foot" to pull themselves along. The most common marine sarcondinians are the forams (found mostly in the deep oceans) and radiolarians (mostly in the upper layers of the ocean). Amoebas are also sarcondinians, but there are few marine species of amoebas.

Ciliates: These organisms—numbering more than 8,000 species—are covered with cilia, or hairlike structures. They use the cilia to move, eat, and breathe. Most of these organisms are solitary and free-swimming, but can be attached to colonies. They are often found between sand grains, where they eat bacteria and plant cells—and are food for other creatures that inhabit the same area.

Flagellates: Flagellates have whip-like tails called flagella. They move around using this projection; or they live in colonies attached to rock with a kind of flagella. Most of these animals feed on fine organic microparticles or bacteria.

# THE FOOD CHAIN

## What is a **food chain**?

A food chain is a complex, interlocking series of dependencies that life forms have on each other and on their environment. It is also the passing of nutrients from one organism, plant or animal, to another. No life form remains isolated and independent of its environment or of other life forms—all living things in the ocean, as on land, are bound together in an unending cycle of life and death. Numerous, small organisms form the base of the food chain; at the top are the largest organisms.

## How does a **food chain work**?

A food chain normally begins with phytoplankton, which are the simple plant organisms that have the ability to manufacture food out of inorganic substances—they use the energy of the Sun's rays in the process of photosynthesis. These small organisms are eaten by herbivorous zooplankton (small, plant-eating animal organisms), who are themselves eaten by carnivorous zooplankton (small, animal-eating animal organisms). In turn, these zooplankton can be eaten by larger animals, such as fish, who are themselves eaten by larger fish. Organisms in the food chain that avoid being eaten eventually die, sink to the bottom, and sustain bacteria, which turn their complex, organic substances into simple inorganic nutrients; bacteria are also, in turn, eaten by larger organisms.

The nutrients produced by the bacteria eventually work their way back up toward the ocean surface, mostly through currents, to the phytoplankton; these nutrients are then used by the microscopic plants (phytoplankton) to begin the cycle again. In fact, the food potential of an area is often dependent on the rate at which these nutrients are recycled back into the top layers of the ocean.

## Is there **only one food chain** in the ocean?

No, there are many individual food chains in the ocean. These chains can overlap and intersect to form complex food webs.

## What are the **producers, consumers,** and **decomposers**?

In the oceans, the producers are the plants, as they make their own food through photosynthesis. The consumers are the animals, as they consume organic materials rather than making it themselves. And finally, the decomposers are bacteria, as they break down organic substances (dead organisms) into inorganic nutrients.

Sharks, at the top of the marine food chain, are predators and scavengers. *NOAA/OAR National Undersea Research Program*

## What happens to the **top predators**—the animals at the top of a food chain?

Animals that nothing else prey on are known as the top predators; eventually, even they die and sink to the bottom of the ocean. Their bodies are eaten by animals known as scavengers, such as crabs, lobsters, and sharks. Bacteria also attack the remains, decomposing them. This process breaks down their organic material into simple, inorganic nutrients, which are recycled to the ocean surface where they are eaten.

## What can affect a **marine food chain**?

A marine food chain is directly affected by the populations (organisms) within its food chain and it can also be greatly affected by major external changes (since these have an effect on the organisms within the chain itself). Populations may be greatly affected by such activities as natural disasters (for example, earthquakes, floods, or volcanic explosions); weather and climate changes (for example, El Niño and La Niña ocean patterns); and the result of human activity (for example, chemical dumping, oil spills, or over-fishing).

## Which **organisms** are the **basis of the ocean food chains**?

At the bottom of the food chain are billions of microscopic plants—mostly diatoms and other phytoplankton (tiny plant organisms)—the first links in most marine food chains.

## Are there any **food chains** that **do not depend on sunlight**?

Yes, there are food chains that do not depend on sunlight—or on photosynthesis. The hydrothermal vents found in the ocean floor in volcanically active areas have organisms at the base of their food chain that depend on the warm, mineral-rich waters around the vents—not on sunlight—to make their own food.

## What is a **food web**?

A food web is the complex intersections of food chains in a specific place and at a certain time. Scientists use the food chains to examine the relationships between plants and animals within the chain, between the plants, and between the animals themselves. The food webs are much harder to determine, as the relationships between species in various places (and in different food chains) is not always easy to interpret. Scientific interpretations of food webs are often educated guesses.

# SMALL LIFE IN THE OCEAN

## How are **ocean bacteria** described?

Bacteria (single-celled or noncellular tiny organisms) living in the oceans play a substantial role in what are called bio-geochemical processes—especially in the process of breaking down organic substances into inorganic nutrients. Bacteria are found from the coastal regions to the deep oceans.

There are many types of marine bacteria. For example, in the deep oceans, bioluminescent bacteria, or microbes that glow, are often found. And these bacteria may be indicators of pollution: Scientists have used these bacteria to determine the toxicity of water, because when these creatures are exposed to certain compounds, they darken.

## Are **microorganisms** found in the **deepest part of the ocean**?

Yes, microorganisms (organisms, or bacteria, that can only be seen with a microscope) have been found in the deepest part of the ocean—on the floor of the Challenger Deep in the Mariana Trench (in the Pacific Ocean). In 1996, the submersible *Kaiko* dove to the Challenger Deep, collecting the first sample ever from such a depth. After growing organisms from the mud in a lab, researchers identified hundreds of species of bacteria, archaea, and fungi; most of the creatures were similar to those found living at other deep-sea sites. They also found bacteria that love the cold—a chilly 39° F (4° C)—surviving as resistant spores. The researchers also determined that these bacteria can handle pressures 1,000 times that at sea level. Scientists are working to identify the genes that help these microorganisms resist enormous pressures.

## Is there a **connection** between **ocean bacteria** and **iron**?

Yes, scientists have recently discovered that oceanic bacteria are ruled by the iron content of the oceans. Some of these ocean bacteria need a diet of both iron and carbon to grow: Without iron, the microorganisms release most of the carbon into the atmosphere instead of using it to build new cells. In addition, scientists now understand that these bacteria compete with phytoplankton (plant organisms) for iron; the bacteria, gram for gram, contain twice as much iron as phytoplankton. But scientists have yet to determine the importance of any of these connections.

# MARINE MAMMALS, BIRDS & REPTILES

## DEFINING MARINE MAMMALS

### What are **marine mammals**?

Marine mammals are air-breathing, endothermic (warm-blooded), vertebrate (having a bony or cartilaginous skeleton) animals. Their young develop within the mother; after babies are born, the mother cares for and feeds them.

### How did **marine mammals evolve**?

About 350 million years ago (during the Paleozoic era), some amphibious creatures left the sea for the land. Some of these amphibians evolved into reptiles; some of the reptiles evolved into mammals. And some of these mammals eventually returned to the sea, evolving into marine mammals such as today's whales, seals, and manatees.

Marine mammals are relatively recent life forms, only appearing in the Cenozoic era, evolving approximately 65 million years ago. Seals and walruses appear to have developed approximately 54 to 65 million years ago, during the Paleocene epoch of the Tertiary period. Sea cows probably originated in the mid-Eocene epoch approximately 38 to 54 million years ago.

Whales are thought to have evolved during the Upper Eocene epoch, but this is highly debated. In fact, recently scientists digging in the Himalayan foothills found a fossilized jawbone from a new whale species named *Himalayecetus subathuensis*. This fossil indicates that whales may be older than previously thought—approximately 53.5 million years old.

## What characteristics do **marine mammals** share with **land-dwelling mammals**?

Marine mammals have retained the main characteristics of land-dwelling mammals, such as a four-chambered heart, biconcave red blood cells, a diaphragm breathing muscle, and both the hard and soft palates of the mouth (oral) cavity. They also, like most of their land-dwelling cousins, bear their young alive, and feed them milk from mammary glands.

## What **adaptations** did **marine mammals** develop to live in the oceans?

Over millions of years of living in the ocean, marine mammals adapted to their environment by developing special features that enable them to survive and flourish. These adaptations include powerful tails for propulsion in the water; lowered metabolic rate, which enables them to use less oxygen; a thick layer of fat, which insulates these organisms from the cold temperatures of the ocean; and milk that has a high fat and protein content, used to nourish their young.

## What are the **main groups** of **marine mammals**?

The main groups (or orders) of marine mammals are the Cetacea, or whales, dolphins, and porpoises; the Pinnipedia, or seals, walruses, and sea lions; the Sirenia, or manatees (sea cows); and the Mustelidae, or sea otters, which are members of the weasel family (along with badgers, wolverines, and minks).

In early March 1998 this 40-ton, 60-foot (16-meter) blue whale was found floating in Rhode Island's Narragansett Bay. The rare mammal, one of the world's largest animals, was towed by the U.S. Coast Guard to a nearby beach where scientists (shown) could examine it. *AP Photo/Robert Button*

## What is the **fastest marine mammal**?

The fastest known marine mammal is probably the sei whale, *Balaenoptera borealis,* reaching speeds of 35 miles (60 kilometers) per hour over short distances. Some scientists believe that the orca (*Ocrcinus orca,* or so-called "killer whale") may be a close contender; orcas may reach speeds of 42 miles (70 kilometers) per hour as they chase their prey.

## What is the **largest marine animal and mammal**—and coincidentally the **largest animal** known to have lived on **Earth**?

The largest animal and mammal in the ocean—and thought to be the largest animal that ever lived on Earth—is the blue whale. This whale can reach a length of almost 100 feet (30 meters). It is estimated that approximately 6,000 to 10,000 blue whales remain in the world; they are found in the Atlantic, Pacific, and Indian oceans, and near the North and South Poles.

## How **intelligent** are **marine mammals**?

According to behavioral biologist Edward O. Wilson, the smaller toothed whales (a group that includes several species, notably the orca) and dolphins (many of the approximately 80 species) are two of the most intelligent marine mammals. In fact, on a list of the ten most intelligent animals on Earth (excluding humans), the smaller-toothed whales come in seventh, while the dolphins rank eighth (two species of chimpanzees come in first).

## How does the **breath-holding** capability of **humans** compare with that of **marine mammals**?

There is definitely a difference between the breath-holding capability of humans and certain marine mammals. While humans can and do swim, we are land-dwellers, and this is the environment to which we are best accustomed. Marine mammals, on the other hand, are accustomed to their watery environment, having made certain adaptations over time. The following list gives the average time certain mammals can hold their breath.

| Mammal | Average Time (minutes) |
|---|---|
| human | 1 |
| pearl diver (human) | 2.5 |
| sea otter | 5 |
| porpoise | 15 |
| sea cow | 16 |
| seal | 15 to 28 |
| Greenland whale | 60 |
| sperm whale | 90 |
| bottlenose whale | 120 |

## How **deep** and **for how long** do some **marine mammals dive**?



The following table lists the maximum depth certain marine mammals can reach, and the duration they can stay underwater.

| Mammal | Maximum Depth (feet / meters) | Maximum Time Underwater (minutes) |
|---|---|---|
| bottlenose whale | 1,476 / 450 | 120 |
| fin whale | 1,148 / 350 | 20 |
| porpoise | 984 / 300 | 6 |
| sperm whale | more than 6,562 / more than 2,000 | 75–90 |
| Weddell seal | 1,968 / 600 | 70 |

# WHALES AND DOLPHINS

## What are **cetaceans**?

Cetaceans are marine mammals of the order Cetacea, which includes all whales, dolphins, and porpoises. The order is divided into two suborders: toothed whales (Odontoceti) and baleen whales (Mysticeti). Dolphins and porpoises are toothed whales, as is the famous orca whale (or "killer whale"). The baleen whales (also called toothless or whalebone whales) include the humpback and gray whales.

## What are the **characteristics** of the **cetaceans**?

The cetaceans have broad tails, tiny ears, and are hairless. They are streamlined to make swimming easier, and are propelled by the up and down movements of their horizontal tail flukes (the two lobes that protrude from their tails). They steer and stabilize themselves by using paddle-like flippers. These animals also have layers of blubber under their skin that are used to store energy, and for insulation from the cold.

## What is the **origin** of the term **Mysticeti**, used to describe baleen whales?

The term Mysticeti is from the Greek word for "mustache." This is because the baleen on these whales somewhat resembles a hairy upper lip.

## What is a baleen?

The baleen whales have a structure in their mouths called a baleen, a series of flexible keratin plates with hairy-like fringes that hang down from the whale's upper jaw. These long plates can be 2 to 12 feet (0.6 to 4 meters) long; the right whale is among the species having longer baleen. To feed, these animals merely open their mouths, letting water that is full of krill and other tiny plankton pour in. They then partially close their mouths, which forces the water out through the baleen. As the water passes through, the krill and other plankton are trapped on the inside of the baleen—which essentially acts as a filter. This food is scraped off the baleen by the whale's tongue, and is swallowed whole.

## What are some **examples** of **baleen whales**?

The species that make up the baleen whales include the blue, finback, right, minke, gray, and humpback whales. The blue whale, *Balaenoptera musculus,* grows up to 98 feet (30 meters) in length; the finback whale, *Balaenoptera physalus,* can reach up to 82 feet (25 meters) long; the right whale, *Eubalaena gracialis,* reaches up to 56 feet (17 meters) in length; the minke, *Balaenoptera acutorostrata,* ranges between 26 and 33 feet (8 and10 meters) long; the gray, *Eschrichtius rubustus,* from 40 to 50 feet (12 to15 meters); and the humpback whale, *Megaptera novaeangliai,* grows up to 49 feet (15 meters).

## What are some examples of **toothed whales**?

The toothed whales, or suborder Odontoceti (derived from the Greek for tooth, *odont*), includes more than 65 species, including dolphins and porpoises. Some of the better known examples of these whales are the sperm whale, *Physeter catodon,* 36–66 feet (11–20 meters) long; the pilot whale, *Globicephala melaena,* which can grow to 20 feet (6 meters) long; the har-

**Bowhead whales (viewed from above), from the baleen suborder, are Arctic-dwellers.** ***NOAA***

bor porpoise, *Phocoena phocoena,* usually about 4–5 feet (1.5–2 meters) long; the bottlenose dolphin, *Tursiops truncatus,* ranging between 6 and 12 feet (2 and 4 meters); and the orca, *Orcinus orca* (also known as the "killer whale"), which can grow to 23 feet (7 meters) in length.

## What do **toothed whales eat**?

The toothed whales are predators, grabbing their prey with their teeth, then swallowing them whole. The primary items on the toothed whales' menu are squid and fish, although some species will eat other marine animals.

## How do **toothed whales locate** their **underwater prey**?

All toothed whales locate underwater prey by using their highly developed sense of hearing. Toothed whales have, over millions of years of evolution, lost their sense of smell but have retained fairly good vision.

But vision cannot always be used, especially when the water is murky, or when there is little to no light. To overcome these problems, toothed whales use their own form of sonar in a process known as echolocation.

Though they live in the water, whales are mammals and must breath air. Here, a whale is seen (white puff near center of picture) exhaling through its blowhole (its nose), situated on top of its head. *NOAA/Commander John Bortniak, NOAA Corps (ret.)*

## How do **whales breathe**?

Whales are mammals and must breathe air; unlike fish, they cannot extract oxygen from the water around them. In order to breathe, whales have a nose (called a blowhole) on top of their heads. This means the nose is exposed to the air as soon as the animal surfaces. The whale first exhales old air (the familiar "blowing"), then inhales. Its nose is then pinched closed, and the animal dives under the water, essentially holding its breath until it comes up again for air.

Baleen whales have two blowholes; toothed whales have one. The blowhole of the sperm whale is located on the left side of its forehead, so the spouting comes from the whale at a 45-degree angle, unlike the 90-degree top-spouting of other whales.

When a whale dives deeply, its bodily processes slow, cutting off the flow of oxygen to non-essential areas of the body. This conserves the amount of oxygen it has inhaled and increases the time needed between breaths.

## Do **whales migrate**?

Yes, whales do migrate. For example, Korean gray whales journey from the Okhotsk Sea off the Siberian coast to the islands of South Korea. The California gray whale has the longest migration of any mammal, traveling along the West Coast of North America from the Bering Sea to Baja California in a yearly round trip of approximately 12,700 miles (20,434 kilometers). These whales journey south in the fall from their summer feeding grounds in the Bering Sea to shallow lagoons off Baja

A finback whale and her calf. At maturity, the finback, which is among the baleen (or toothless) whales, can reach up to 82 feet (25 meters) in length. *CORBIS/Judy Griesedieck*

California, where they give birth to their young during the winter season. In the spring, they make the return trip north to the Bering Sea.

## Why do **whales migrate**?

Most whales seem to migrate for two main reasons: 1) the availability of food and 2) the mating and birthing cycles—both dependent on the season. For example, humpback whales migrate in the western Atlantic Ocean from around the Bahamas, north to Georges Bank, Cape Cod, and even Iceland during the summer to feed; they migrate back southward during the winter to breed. In the eastern Pacific Ocean, some humpbacks migrate up and down the North American coast and others migrate between the Hawaiian Islands (where they spend winters to breed) and the Gulf of Alaska (where they summer and feed).

Just how the whales navigate through the open ocean is unknown. Some scientists believe that the whales use the Earth's magnetic field as a guide, allowing the animals to orient themselves.

## How do the **larger whales** compare in **weight and length**?

Here is how several of the larger whales, which come from both suborders (baleen and toothed) stack up in weight and length:

| Whale | Average Weight (tons) | Greatest Length (feet / meters) |
|---|---|---|
| blue | 84 | 98 / 30 |
| finback | 50 | 82 / 25 |
| sperm | 35 | 59 / 18 |
| bowhead | 50 | 59 / 18 |
| right | 50 (estimated) | 56 / 17 |
| humpback | 33 | 49 / 15 |
| sei | 17 | 49 / 15 |
| Bryde's | 17 | 49 / 15 |
| gray | 20 | 39 / 12 |
| minke | 10 | 30 / 9 |

## What is **unique** about the **gray whale**?

The uniqueness of the gray whale (a toothless whale) comes from its method of feeding. Like other baleen whales, the gray whale uses its baleen to trap food in its mouth as it expels seawater.

But this animal—an active, medium-sized whale of about 39 feet (12 meters), which is found in the North Pacific—goes one step further than just filling up its mouth with water and forcing it out. In captivity this whale has been observed swimming just off the bottom, then turning on its side and sweeping its head back and forth. This behavior disturbs any crustaceans resting on the bottom, causing them to rise. The gray whale then creates a suction by pushing its tongue against the bottom of its mouth while expanding its throat grooves. This pulls the prey into its mouth under the short baleen. Think of this as the gray-whale version of the vacuum cleaner, removing crustaceans from the ocean bottom!

Although this behavior has only been observed in captive whales, there are several factors that suggest this is also a typical mode of feeding for these animals in the wild. The gray whale's baleen is usually shorter on one side of its mouth—and that side is often relatively free of barnacles,

**The gray whale is a North Pacific–dweller: Here one of three that got caught in ice in the Beaufort Sea and became the subject of an American-Russian rescue effort.** ***NOAA/Office of NOAA Corps Operations***

suggesting that this feeding behavior is not only normal for these animals, but that they adapted physically to make this suction trick possible.

## What has **happened** to the **right whale**?

The right whale (a toothless, or baleen, whale) is the most endangered of the big whales—and may be at the brink of extinction. Worldwide, their numbers have dwindled to between 350 to 400. Many right whales, indigenous to the North Atlantic Ocean, migrate as far south as Florida during the Northern Hemisphere's winter, and back toward Canada in the summer. But the whale counts during these migrations indicated that only a few whales have borne calves in recent years.

There have been attempts to protect the critical habitats of the right whales. For example, in 1994 the New England Aquarium began operating an early warning system in the southeastern United States to let ships know the relative positions of calving right whales. Every winter, the aquarium researchers conduct daily flyovers of the right whale's calving grounds; they then relay the information to the Coast Guard,

**Beluga whales are among those species that swim in pods. (Here at least ten are seen from above.)** ***NOAA/Captain Budd Christman, NOAA Corps***

Navy, and local harbor pilots—who in turn, relay the information to military and commercial vessels.

## Do **whales** swim in **groups**?

Yes, most whales travel in groups called pods, with each species varying in its mode of travel within the pods. For example, some whales, such as the orcas (so-called "killer whales"), remain in tight family units; others swim in large pods, such as the bottlenose dolphin, which lives in groups of up to 12; while still others, such as gray whales, are often seen swimming alone or a mother is seen swimming with her calf or calves.

## Do **humpback whales sing**?

Yes, humpback whales (which are toothless) do produce "whale songs," but scientists don't know how the sounds are generated. Humpbacks have no vocal cords, yet their sounds (classified as true songs) cover the widest frequency range of all the whales. And as strange as it may seem, when the whale sings, no air is released into the water—unlike humans who have to force air through their vocal cords in order to sing.

Most whale "singers" are males, with most of the songs sung at the breeding grounds of the animals. Many researchers believe the songs are linked to the courtship of the animals—either to attract a mate, advertise the mammal's availability, or establish a territory. The songs travel fast and great distances, too, since water conducts sound 3.5 times faster and 4.5 times farther than in air: Some humpback songs can be heard miles away—even as far as 100 miles (161 kilometers) if the frequency of the song is low enough.



## Are the **humpback whales in danger**?

Yes, humpback whales are in danger. They are listed as an endangered species, and are thus protected worldwide. Less than a century ago, there were about 120,000 of this species of baleen whales; because of commercial whaling, their numbers eventually dwindled to about 10,000. Today, the numbers are not even close to previous numbers, and only a few scattered breeding sites remain. Not only is there the threat of whaling (not all countries have stopped), but there are also problems with pollution and human interference (mainly in the form of the humpbacks having mishaps with marine vessels).

## What are **sperm whales** like?

Almost everyone's vision of a sperm whale comes from the famous tale of the sea told by American writer Herman Melville (1819–91) in his classic novel *Moby Dick*. (In this story, Moby Dick is described as a "great white whale," which is actually a white sperm whale, and he was the nemesis of Captain Ahab.) Sperm whales are toothed whales that have large, blunt noses, and squared-off heads. In fact, the head of this whale is about one-third the animal's entire length. They grow to weigh as much as 35 tons and can be as long as 59 feet (18 meters). They also have the largest brains of any mammal on Earth (maybe this is why Melville chose this particular whale for his story). They live in all the oceans and are the most numerous of the great whales.

A dolphin escort for the National Oceanic and Atmospheric Administration ship *Peirce* (1985). *NOAA; Personnel of NOAA Ship* Peirce

## Why have **sperm whales** been **hunted**?

Sperm whales have been hunted extensively—in the past, and to a modest extent today—for a variety of reasons. They are a source of spermaceti, a white, waxy substance obtained from the oil in the animal's head; and ambergris, a material from the animals' digestive tracts. Both these substances are used in the manufacture of cosmetics and the spermaceti is also used for candle wax or long-lasting lamp oil. Other parts of the animal, such as the blubber, have also been valued—thus the sperm whale's decrease in numbers.

## What is the **difference** between **dolphins** and **porpoises**?

Though perhaps indistinguishable in pictures, if you saw a dolphin and a porpoise side by side, you would notice the differences between these two toothed whales. These marine mammals—with about 40 species between them—differ in the structures of their snout and teeth. True dolphins have a beak-like snout and cone-shaped teeth; true porpoises have a rounded snout and flat or spade-like teeth.

## How are **dolphins similar to humans**?

They may not seem like us, but researchers recently discovered that the dolphin genome (set of chromosomes) and the human genome are basically homologous—the same. There are just a few chromosomal rearrangements that changed the way the genetic material was put together. Because of this, scientists are now studying diseased dolphins and sporadic die-offs closely, trying to determine how these animals react to disease and to unfortunate exposures to pollutants. Some researchers wonder if dolphins will turn out to be the proverbial "canary

Killer whales (Orcinus orcas) come up for air in the icy waters of the Antarctic. *NOAA*

in the coal mine," signaling to humans when there are too many toxins in our environment.

## What are **"killer whales"**?

Killer whale is another name for the orcas. These carnivorous mammals are the largest of the dolphins; the males may be as long as 30 feet (9 meters); females are smaller. Most orcas are black and white, although some have been reported to be all black or white; their young are black and orange. This toothed whale hunts in packs (pods) and is found worldwide.

Like many larger creatures of the sea, orcas have no natural enemies. But they do hunt other creatures—and only for food, not sport (the name "killer whale" is highly exaggerated, and they rarely attack humans). While they are known to eat birds, fish, and squid, they are also the only cetaceans to feed on mammals, including smaller dolphins, seals, and porpoises. And while in a pod, they may even attack a baleen whale.

## What is **spyhopping**?

Orcas, or killer whales, exhibit a behavior called spyhopping. This is when each orca "stands" vertically—up to halfway out of the water. They do this to look around and see above the water.

## What is **echolocation**?

Echolocation is the process used by certain animals (including bats) to determine the shape, size, and distance of an object; it is based on the sound waves generated and received by the animal. Scientists don't know how echolocation works in most marine mammals, such as the sperm whale. The most complete studies to date have been done on the smaller toothed whales, such as the dolphin. (It's much easier to study an animal that is not large enough to swallow you whole!)

Dolphins can generate a wide range of sounds underwater. Some sounds, such as chirps, moans, and squeals, are thought to be used in communication with other dolphins. Dolphins can also emit series of very short clicks, which are used in echolocation. These clicks can be generated at a rate of up to 800 times per second, and are thought to be emitted from the animal's blowhole—as an outgoing beam of sound pulses from an organ in the dolphin's forehead. (This organ is the noticeable "melon" on the head of the dolphin—a large, lens-shaped organ made of a waxy tissue and located between the top of the head and the upper jaw.)

When these out-going sound waves strike a target, such as a fish, some of the waves reflect back in an echo toward the dolphin. These reflected sound waves are received by the bony lower jaw of the animal, transmitted to the bone-enclosed inner ear, and converted to nerve impulses that are sent to the brain. There, the amount of time that elapsed between the out-going clicks and the returning echoes are used to determine the distance to the target.

The dolphin varies the rate of click generation, so that incoming echoes can be received in between out-going sounds. The animal uses low-frequency clicks to scan and higher frequency clicks to make more precise determinations. Using this sophisticated process, a dolphin can continuously determine the shape, size, and distance to an object underwater—as well as the direction of that object's movement.

## Do any other **marine mammals** use **echolocation**?

Yes, in addition to the toothed whales, some baleen whales—such as the blue and gray minkes—also use echolocation. And this process isn't just limited to the cetaceans: Some pinnipeds, such as the California sea lion, Weddell seal, and walrus, also use echolocation.

## How have **recent fossil finds** provided clues about **whale evolution**?

Scientists recently discovered a previously unknown genus and species of whales—a whale that has so far, not been named. The almost complete fossil was found by amateur fossil-hunters in a quarry on Washington's Olympic Peninsula, between Port Angeles and Clallam Bay. The bones were widely scattered and many of them were broken; they were found in an area that was a seafloor 28 million years ago, during the Oligocene period. This was also a period when whales were going through major evolutionary changes.

Scientists painstakingly reconstructed the animal—and discovered the uniqueness of the specimen. Based on this rebuilt skeleton, the whale was 15 to 18 feet (4.6 to 5.5 meters) long—about the size of a modern pygmy right whale, which weighs 3.25 to 4.6 tons and is among the smallest of the baleen whales. It was toothless, having baleen (plates along the upper jaw) to trap food while expelling ocean water from its mouth. Its blowhole was on its snout instead of the top of the head. In addition, the length of the arm bones and the way the rib bones attached to the backbone vertebrae more closely resembled land mammals. Since whales from more recent periods evolved bones that don't as closely resemble land animals, scientists are looking into the possibility that this whale represents a step in the evolution of the species.

# SEALS, SEA LIONS, AND WALRUSES

## What are **pinnipeds**?

Pinnipeds are a group of carnivorous, marine mammals. The word Pinnipedia, which means fin-footed (or those with "winged feet"), is

On land, sea lions move around on all fours. They make themselves at home along the Pacific coastlines of North and South America as well as New Zealand. *NOAA/Captain Budd Christman, NOAA Corps*

used to describe this group, since all four limbs of the pinniped were, through evolution, modified into flippers. Pinnipeds are carnivorous (meat-eaters). This group includes seals, sea lions, and walruses, all of which are very awkward on land, but are swift and agile in the water.

## How can **seals, sea lions, and walruses** be **identified**?

Members of this group (Pinnipedia) can be confusingly similar in their appearance, but there are important differences that distinguish each. True seals (phocids) have no external ears and use only their forelimbs for motion on land. Sea lions, on the other hand, have ears and when on land these animals move around on all fours. Fur seals are an intermediate group between the true seals and the sea lions—they have ears, move on all four limbs, and have thick underfur. Finally, the walrus is really a close relative of the seal; it has no external ears, but it moves on all fours. These large animals are also known for their ivory tusks.

**Walruses sunbathing on smooth rocks. Other than polar bears—and humans, who have hunted the animal for its ivory tusks and blubber—the walrus has no other predators.** ***NOAA/Captain Budd Christman, NOAA Corps***

## What are **walruses**?

Walruses are giant marine mammals related to the seals, but constitute a distinct family of their own. They are large, tusked pinnipeds that inhabit the cold, northern waters of the Atlantic and Pacific oceans, near the edge of the Arctic ice pack. Males are larger than the females, with an average length of more than 10 feet (3 meters); these marine mammals frequently weigh more than 2,500 pounds (1,135 kilograms). A layer of thick blubber insulates the walrus from its harsh, cold climate. The most prominent feature on walruses, both male and female, are their two prominent tusks—really enlarged canine teeth—which are used for territorial defense and courtship displays.

Although walruses are much larger, they do share many of the characteristics and body shapes of the seals. Like true seals (phocids), they have no external ears. These animals have whiskers and brushlike mustaches on their faces, and short reddish hairs on their hides. They feed on shellfish; when not searching the bottom for food, they spend most of their time hauled out on ice floes, resting and sleeping. Walruses are social animals, living in big groups dominated by the largest males, who collect harems of females. The only predators of this large pinniped are polar bears, who mainly prey on the young animals; humans also constitute a danger to walruses, which, for their ivory tusks and blubber oil, have been the object of hunters.

## What are **sea lions** and where do they **live**?

Sea lions can be thought of as eared seals—they are related to the fur seal, but do not have the coat that has made their cousins the object of hunts. Sea lions are Pacific Ocean–dwellers. Among the varieties are the Stellar sea lion, which lives in Alaska and the Pribilof Islands (in the southeast Bering Sea); the California, which inhabits western North America and the Galapagos Islands; the South American, which can be found along the western coast of South America; and Hooker's sea lion, which lives in New Zealand.

## What are **seals**?

The term seal is often used very loosely, to refer to any marine mammal in the order Pinnipedia, including walruses and sea lions. But scientists narrowly define a seal as only the phocid (from the family Phocidae), which is the "earless" seal. These seals do have ears, but they are internal; in other words, they lack the visible external ear of other animals. Further the "true seal" uses only its forelimbs for motion on land.

Of the roughly 17 species of seals, only 1 lives in warm water; the others live in cold and temperate waters. True seals are awkward on land, but agile in the water. Carnivores, they like to eat squid, fish, and assorted invertebrates.

## How do **seals hunt**?

Recently, scientists attached a tiny video camera to a Weddell seal. The seal took a deep breath and dove 330 feet (100 meters) below the Antarctic ice sheet, stalking its dinner for about 20 minutes before resurfacing for air. The video of the Weddell seal stalking its prey revealed several interesting things: The seal has sharp eyesight—and hunts by sight

**A Weddell seal pup; though the Weddell uses echolocation (sonar), scientists recently found that this animal has keen eyesight, which it uses to hunt for food.** ***NOAA/Commander John Bortniak, NOAA Corps (ret.)***

rather than sound as was once thought; the seal blows a blast of air from its nostrils to "encourage" prey hiding in an ice crevice to come out into the open; it can come within inches of its prey without being detected; and finally, even after traveling more than 1 or 2 miles underwater, seals can find their way back to a small air hole in the ice's surface.

## What are some **true seals** and where do they live?

The following table lists some of the true seals—or phocids—and their habitats.

| Common Name | Habitat |
|---|---|
| ribbon | Siberian coast |
| ringed | Arctic |
| larga | North Pacific |
| northern elephant | western North America |
| southern elephant | South America |
| Weddell | Antarctic |

| Common Name | Habitat |
|---|---|
| Ross | Antarctic |
| leopard | Antarctic |
| crabeater | Antarctic |
| harbor | Gulf of Maine |

## What are **fur seals**?

Fur seals, often called sea bears, belong to the same family as sea lions, both of which have external ears. They have a double coat of fur, with the underfur (or undercoat) being very dense and soft, and unfortunately, prized by hunters. Varieties include the northern, Alaska, Galapagos, Australian, and Antarctic fur seals.

## What is the **largest pinniped**?

The largest pinniped is the elephant seal. The males can be more than 20 feet (6 meters) long; whereas the female measures about half that size.

## What is an **elephant seal**?

An elephant seal is the largest pinniped—growing to 16 to 20 feet (5 to 6 meters) in length; some males weigh up to 7,500 pounds (3,405 kilograms). They come ashore only twice a year—once to molt (they shed their skin in a month) and once to mate. At other times, they are offshore, searching for food. They live an average of 20 years.

## How are the **male elephant seals** distinguishable?

One of the main features of a male elephant seal is its large nose, or proboscis. This large structure on the front of the animal's face is actually a secondary sex characteristic: As the male grows and matures, the nose grows, too—becoming fully developed at about 8 to 10 years of age. The males use the nose for display purposes during mating seasons—the males holding their noses high in the air while bellowing challenges to each other.

**The elephant seal is the largest pinniped; males can grow to more than 20 feet (6 meters) in length.** ***NOAA/Commander Richard Behn, NOAA Corps***

## Are **elephant seals** near **extinction**?

Elephant seals are not quite near extinction, but they are struggling. During the 1800s when they were valued for their oil (as a light source for humans), elephant seals were hunted to near extinction. They recovered during the first part of twentieth century, but now they are threatened once again. The demand for the animals' oil is not the problem this time, but rather a danger is posed by the fishermen that ply the same waters where the elephant seals live and eat. The problem stems from the size of these animals: Males are up to 7,500 pounds (3,405 kilograms), and even the pups can weigh 300 pounds (136 kilograms) by the time they are weaned and on their own. To maintain this huge size, the animals must eat great quantities of fish and squid—the same food that the commercial fishermen collect. In other words, the elephant seal now vies with the fisherman for food.

To learn more about this marine mammal, scientists are conducting a decades-long study of the elephant seal population at the Peninsula Valdes (part of Argentina's Patagonian reserve), the fourth-largest elephant seal population in the world—and the only growing population.

# SEA COWS, DUGONGS, AND MANATEES

## What are **sirenians**?

The marine mammals that belong to the order Sirenia are the sea cow, dugong, and manatee. These aquatic creatures are herbivorous (plant-eaters).

## What are **sea cows**?

Sea cows are marine mammals that grow up to 10 feet (3 meters) in length, have little hair, and rely on a layer of blubber for insulation. Their scientific name, Sirenia, is derived from the comely Sirens who, in the Greek poet Homer's *Odyssey,* lured Odysseus' crew into the rocks—a reference to ancient sailors who thought the sea cows were actually mermaids.

## What is a **dugong**?

A dugong is a type of sea cow found off Africa and in the South Pacific—but mostly near Australia, where it is endangered. The dugong's closest terrestrial (land) relative is the elephant; it is also the only herbivorous mammal living totally in the ocean. Dugongs eat mostly sea grass, and use their short-bristled muzzle to dig and root in the shallows of bays and estuaries. They have two flipper-like forelimbs, but no hind limbs. The tail is broad—allowing the animals to propel themselves through water at speeds reaching about 13 miles (21 kilometers) per hour. Although they may seem slow, they are active—and some scientists believe the animal's intelligence is comparable to a deer.

## What is a **manatee**?

A manatee—a thick-skinned and wrinkly marine mammal—is a close relative to the dugong, and is also a sea cow. In fact, the only major difference between the two is their habitat: The manatee is usually found along Florida shores, although their range once spread from North Carolina to Florida. Manatees perform a service to marine craft: They use

**The lure of the sirens? The wrinkly and thick-skinned manatee belongs to the order Sirenia.** ***CORBIS/Brandon D. Cole***

their bristly muzzles to dig and eat the bothersome water hyacinth that can clog slow-flowing river channels. Like the dugong, the manatee is endangered.

## Why are **manatees** and **dugongs** endangered?

One of the main reasons has to do with their appetites: Because they root for the sea grasses in estuaries and shallow bays, they encounter more recreational craft than do most ocean animals. A manatee comes up for air every 10 to 15 minutes, but the rest of the time, it is often lying placidly in the shallow water, just under the surface—where they are often hit by boats. In addition, they are very slow, and are not able to get out of the way of such craft. Thus, many of the animals are injured by propellers or fishing nets; they are also extremely susceptible to silt filling their shallow water habitats—not to mention pollutants and petroleum spills.

There are major efforts underway to protect the manatees and dugongs—especially by studying their populations and by not disturbing the animals' habitats. For example, manatees off the coast of Florida are closely watched. Many animals, usually captured as abandoned young or hurt animals, are tagged, released into the wild, and monitored by satellite—in hopes of keeping an eye on the number of calves born, so that population shrinkage and growth can be gauged. And in western Australia's shallow waters, there are laws that limit how close vessels and people are allowed to come to the dugongs: no closer than 330 feet (100 meters) for vessels and 100 feet (30 meters) for people.

# SEA OTTERS

## What are **sea otters**?

Sea otters are the smallest marine mammals; they are also the least changed in form from their terrestrial (land) ancestors. Sea otters are relatively large animals—the males often grow to 4.5 feet (1.35 meters) in length, including the tail, and weigh about 45 to 100 pounds (20 to 45 kilograms). They are closely related to the smaller river otters found in freshwater streams, and are members of the weasel family (the order Mustelidae).

## Are there **different** types of **sea otters**?

Yes, there are three subspecies of the sea otter: the southern sea otter (found off the California shore), Alaskan sea otter, and Asian sea otter. The differences between these animals includes the shape of the animals' skulls and the size of the otters (the Alaskan sea otters usually grow the largest).

## What do **sea otters eat**?

Sea otters are carnivorous (meat eaters) and they consume all different kinds of ocean creatures including sea urchins, shellfish, fish, and various marine invertebrates. Each individual otter has its own favorite foods, too. One may prefer sea urchins; while another prefers abalone.

A familiar position for the sea otter, which eats, rests, and sleeps while lying on its back. *NOAA/Commander John Bortniak, NOAA Corps (ret.)*

These animals spend a considerable amount of time looking and diving for food because they must eat constantly to survive—they consume the equivalent of 20 to 25 percent of their body weight each day. They often use their front paws (forepaws) to dig into the seafloor to find burrowing animals such as clams. They are also one of the few animals to use "tools" to get to their food. For example, when an otter dives for abalone, it may use a stone to "hammer" the shellfish if it is stubbornly stuck to a rock. It may also use a stone to dig for clams. The sea otter has a small pouch under its left armpit, where it can store a rock or some food while swimming.

Once the sea otter resurfaces with its catch, it eats—or prepares to eat. These animals have strong teeth that can bite through crab shells and a few other types of shellfish. But hard-shelled fish such as abalones must be cracked open; some sea otters accomplish this by hitting the shell with a rock, as the shell sits on the animal's chest, while others will put the rock on their chests, then smash the shell against the rock. Even if a sea otter doesn't use rocks as tools, as is the case with the Alaskan sea otter, which has plentiful soft- and thin-shelled animals to prey on, they still have the ability to do so.

## How does a **sea otter rest**?

Sea otters seem to do everything on their backs—and they do. They eat, rest, and sleep while lying on their backs. When an otter sleeps, it will generally wrap itself in kelp to keep from drifting away. These animals lie in the waters with all four limbs—which are not protected by fur—projecting out of the water. They may do this to conserve heat; unlike other marine mammals, the sea otter does not have a layer of insulating blubber. Instead, their fine, dense fur—the densest of all the animals, with up to 1 million hairs per square inch (15,000 per square centimeter)— traps air and helps the animals maintain warmth.

## Have any **sea otters** become **extinct**?

At one time, scientists thought the southern sea otter, found off the shore of California had become extinct—mainly because of the fur trade during the nineteenth century. But now we know a small group survived around Big Sur, a group responsible for eventually repopulating the otters along the California coast. Today, there are about 2,200 sea otters along a 250-mile (402-kilometer) range along the state's central coast; the original population was about 20,000. And there are still some concerns: A survey during the late-1990s indicated there was a drop in the southern sea otter population over the previous few years. Scientists are not sure if the decrease is due to disease, contaminants, or to accidental drowning in fish traps.

## What happened to **sea otters** along **Alaska's Aleutian Islands**?

The population of sea otters in certain habitat "pockets" along Alaska's Aleutian Islands has plummeted by 90 percent in less than a decade. Scientists were not sure why until just recently: They believe that hungry orcas have been eating the sea otters like popcorn. The problem extends even further: Without the sea otters, the population of sea urchins (which the otters normally eat) is booming, stripping the undersea kelp forest.

Why have the orcas (killer whales) developed an appetite for sea otters—which they usually ignore while hunting sea lions and harbor seals? Scientists believe that over-fishing and rising water temperatures may be the problems, as both have caused a decline in the fish populations. Commercial fleets over-fished the area; then, in the late 1980s, the populations of sea lions and harbor seals—animals that depend on the fish for survival—declined to 10 percent of what they once were. With fewer sea lions and harbor seals to prey on, the orcas turned to the sea otters as a food source; the whales moved from the deeper, open ocean to the coastlines in search of them. The problem in the Aleutian Islands is a clear example of what happens when just one link in the food chain is disturbed.



## MARINE BIRDS

### What is a **bird**?

A bird is a warm-blooded vertebrate that reproduces by laying eggs; they are thought to have evolved from reptiles. Birds are members of the animal kingdom, and have their own class, the Aves. They have four limbs, with the front two limbs modified into wings.

### How many **species of birds** are found along the **coasts** and **oceans**?

There are only about 300 species of birds that live on the coasts and oceans, which is equal to about 3 percent of the total number of birds in the world (on land and water). What they lack in diversity, they make up in their huge numbers. Most of the bird concentrations occur where sea life is prolific, such as in the Antarctic Ocean (also called the Southern Ocean), off the coasts of Peru and Chile. Birds that live off the ocean are usually divided into two groups, seabirds and shorebirds. The diet of these birds is varied and can include worms, crustaceans, bivalves, fish, snakes, and mice.

### What are **seabirds**?

The name seabird, also sometimes called oceanic bird, is used by some ornithologists (zoologists who study birds) to describe any bird that spends much of its life on or over the ocean waters. These birds include some familiar water bird species, such as boobies, gannets, frigate birds, loons, cormorants, puffins, pelicans, certain ducks, gulls, and terns.

"Seabird" especially applies to albatrosses, shearwaters, petrels, and storm-petrels—those birds that are seen most often over the ocean waters. Except during breeding season—or when they are seen following ships for wastes thrown overboard—these birds live far from land. Amazingly, the seabirds are some of the most abundant birds that fly over the oceans. In fact, one of them, the Wilson's storm-petrel, was once referred to as the most abundant bird in the world.

Red-crested cormorants and distinctive horned puffins (with their black topcoats and compressed beaks) share a perch atop a rocky precipice. *NOAA/Captain Budd Christman, NOAA Corps*

## Does any **bird** spend its **entire life at sea**?

Yes, the albatross spends almost its entire life on the ocean, only coming to land every other year to nest.

## What are **shorebirds**?

Shorebirds are the many species of birds that wade into shallow waters—of estuaries, rocky shores, beaches, tidal flats, jetties, and marshes—to find their prey, but do not swim. These birds seem quite similar to each other, especially the sandpipers. Shorebirds are usually "counter-shaded," or dark on top where they receive the most sunlight, and light on the bottom. This makes the birds hard to see on a beach—and even more difficult to determine their species. Thus, bird watchers and ornithologists identify the animals by their beaks, legs, or behavior.

## Do **shorebird and seabird bills** differ from one another?

Yes, just as with land birds, they differ greatly, and there is even a variety of bills within each group. Most birds have bills that are specially adapted so they can hunt and eat food in whatever environment they live in—on land or in the ocean. For example, the red-breasted merganser, a shorebird, has a spike-like bill with saw-tooth margins, which it uses to grab and tear open crustaceans. The greater yellowlegs, another shorebird, has a long, thin beak it uses to stab or grab its fish or crustacean prey, or to probe in the mud. The bill of the albatross, a seabird, is specialized for catching fish in the open ocean.

**An osprey guards its nest high atop U.S. Coast Guard navigation equipment in Indian River inlet, Delaware. This shore-dweller feeds only on fish, and is often called fish hawk. It can be found along seacoasts as well as along inland lakes and rivers.** ***NOAA/Personnel of NOAA Ship PEIRCE***

## What are some **common shorebirds**?

There are many shorebirds around the world—too many to mention in these pages. But some of the more familiar ones are the sandpipers (wading birds having long, slender bills, which they use to probe the shallow waters or muddy areas for crustaceans or worms); turnstones (short and squat birds that turn over rocks and seaweed looking for food); oystercatchers (a chicken-sized bird with a long, chisel-like bill to open up bivalves); and yellowlegs (sandpiper-like birds that prey on fish and crustaceans).

## What are **wading birds**?

Wading birds are those shorebirds that walk along and in shallow-water areas; they have extremely long legs and widespread toes that keep them from sinking in the mud or sand. Most also have very long necks, which the birds fold into an S or hold straight out when they fly. Their bills can easily probe the mud or snap up (or stab) fish in shallow water. They usually nest with other shorebirds in large colonies called rookeries, making their twig-lined nests in low shrubs or empty trees in marshes and mangrove swamps.

The red-backed sandpiper, with its long beak, is a shorebird. *Field Mark*

## What are some **common wading birds**?

There are many common wading birds, including egrets, herons, and ibises. Some of the more familiar ones are the snowy egret (a white-feathered bird, having a thin, black bill, black legs, and bright yellow feet); great blue heron (a 4-foot, or about 1-meter, grayish-blue bird that eats a varied diet, from mice and snakes to fish); and a white ibis (a white, heron-looking bird, but with a curved bill).

## What is a **sea duck**?

Although ducks are not usually associated with the oceans, there are many species that live in bays and estuaries and along open shorelines. Most of these ducks are divers—they dive under the water to search for food, rather than feeding at the surface. Sea ducks include the red-breasted merganser (it lives mainly on fish, which it chases underwater); common eider (the largest species of duck; blue mussels are their favorite food, which they swallow whole using their stomach muscles to grind up the shells); and the oldsquaw (it feeds on fish, shrimp, and mollusks down to 200 feet [61 meters], using its wings to propel itself underwater).

## How are **gulls** defined?

Gulls are very common along most any shoreline—and even inland in larger country fields. What is commonly called a seagull is actually a herring gull, seen along the East Coast and interior (near waterways) of the United States. They have long wings and slightly hooked bills and usually have a mix of white, black, and gray feathers; very rarely do they have any color. Gulls are omnivores, meaning they eat both meat and plants, and many scavenge for food; they feed on nesting sites (eating the eggs), crabs, shellfish, and almost anything that is considered food—even the sandwich of a picnicker at the beach.

## What are **terns**?

Terns are shorebirds that are similar to gulls, but are usually smaller and more streamlined; they have forked tails, whereas gulls have a splay of tail feathers. Terns also hover in the air, then dive into the water to catch fish. The most prevalent is the common tern, a pigeon-sized bird with a severely forked tail. Terns are sensitive to human intervention during their mating season—thus many nesting sites have to be protected while these birds are breeding.

## What are **skimmers**?

Skimmers are crow-sized shorebirds, with large, knife-like bills. One of the most common is the black skimmer—black on top, white on the bottom, and a red beak that the bird uses to skim across the water in order to catch fish and crustaceans.

## What are **cormorants**?

Cormorants are black, goose-sized shorebirds that fly with their necks outstretched. They are often seen diving underwater from a swimming position, not from the air, in search of fish to eat. A cormorant uses its webbed feet and sometimes its wings, to propel itself through water, employing its tail as a rudder. Varieties include the double-crested cormorant (with an orange throat-pouch and a barely visible double-crest on its head) and the great cormorant (larger with a white patch on its throat). These birds can be an important part of their environment: They rapidly process fish into nutrient-rich droppings, or guano; algae grow in these areas, supplying food to large populations of invertebrates and fish.

## What are **pelicans**?

Pelicans are large birds that feed on fish and crustaceans. This bird is best known for the pouch under its beak—a fleshy throat pouch that inflates when the pelican is underwater catching prey. This pouch is not to store the fish, but is used as a scoop to separate the fish from the water. The most common are the brown pelican (with a wing span of about 7 feet [just over 2 meters]; they also dive from the sky to catch fish such as herring, mullet, and menhaden near the surface) and the white pelican (with an amazing 9-foot [almost 3 meter] wing span; they nab fish while swimming rather than diving from the sky).

## Why are **brown pelicans** a **protected** species?

Brown pelicans were close to extinction in the 1970s. Like many susceptible birds, they were decreasing in population because of exposure to DDT, a now-banned pesticide. Similar to the California condor and other bird species, their egg shells became thin, and they experienced poor reproduction rates. Thanks to the protection they received as an endangered species, they are now common along both the East and West coasts of North America, from Virginia to Florida, and from San Francisco to Mexico—and have been upgraded to a protected species.

Brown pelicans are still protected because of humans: Much of what were once prime habitats for the pelicans are now the sites of condominiums and sea walls. There is another problem: Scientists have recently discovered that near places like Miami, Florida, people who feed the birds chunks of fish left over from a day of fishing are unwittingly causing the birds a painful death. Many of the fishermen clean their catch, then feed the pelicans the leftover carcasses. But this doesn't help the birds: Much of the fish that is being fed to the birds is grouper or even dolphin—and pelicans cannot digest the bones of these larger fish. After consuming these handouts, the pelicans fly away, and the bones either get caught in their throats or press against their stomach lining, puncturing the stomach or other organs. In Florida, marine agents are now posting educational signs at marinas statewide, alerting fishermen to the dangers of feeding the pelicans. If people still want to feed the birds, marine agents recommend giving them boneless chunks or the smaller fish that are part of their natural diet—such as grunts, mullet, or pinfish.

## Do any birds fly **over the oceans** when they **migrate**?

Yes, many of the migration paths of various birds—seabirds and land birds (mostly the songbirds)—pass over the oceans. For example, a seabird called the Leach's storm petrel breeds on rocky islands and deserted coasts from Canada to Massachusetts in the spring, then migrates to the open ocean during the winter. The inland-dwelling ruby-throated hummingbirds migrate south from the United States every fall, covering 600 miles (965 kilometers) including the Gulf of Mexico, to follow the seasonal growth of flowers (besides beetles, bugs, flies, gnats, and other insects, hummingbirds also eat the nectar of flowers). In the spring, they migrate in the opposite direction.

**The penguin may not be able to fly but it sure can swim. Here, a pair of Gentoo penguins ably navigate the waters.** *CORBIS/George Lepp*

## How did certain **ocean birds** become **flightless**?

Some ocean birds became flightless because in their environments, mostly oceanic islands, they had no natural predators—so they no longer needed their wings for escape. This group may include the flightless penguins in the Antarctic, as well as the flightless grebes, rails, and cormorants that live, or have lived, on oceanic islands. Many of these flightless birds have also become extinct. One of the most famous was the dodo, an ocean island bird hunted to extinction by humans.

## Which **birds** are the **best swimmers**?

Among birds, penguins get the award for best swimmers. They are flightless marine birds but their streamlined shape makes them excellent swimmers. They can stay in the water for a long time without being affected by the cold: They carry about a 1-inch- (2- to 3-centimeter-) layer of fat and have waterproof feathers, with a layer of air trapped underneath for added protection from the cold. Penguins can move swiftly through the water, up to 22 miles (35 kilometers) per hour, but only stay underwater for about 2 to 3 minutes at a time. They are so fast that they can shoot out of the waves and land securely on the ice—upright on both feet. And they are agile and quick on land, too: They walk upright, and in snow, they often move by sliding on their bellies.

### How many **types of penguins** are there?

There are many types of penguins around the world, mostly found along the coasts of the Antarctic, in the southern temperate cool zone, on the Galapagos Islands, and on the subtropical coasts of South America, South Africa, and Australia. Among the most well known penguins are the emperor and king (these two penguins are the largest and they are members of a group called the Giant penguins); other groups are the Adelie, crested, black-footed, yellow-eyed (or yellow-crowned), and dwarf—all of which include many species. They feed on planktonic animals, especially krill, and small fish, crabs, and squids.

## MARINE REPTILES

### What are **marine reptiles**?

Marine reptiles are very similar to terrestrial reptiles: They are cold-blooded, air-breathing animals covered with many scales. And although 14 percent of the 40,000 known species of vertebrates are reptiles, there are very few marine reptiles. They include the crocodile, sea turtle, and sea snakes. Most of these animals are still tied to the land, as they have to lay their eggs on the shore; when the eggs hatch, the hatchlings resemble miniature versions of the adults.

### What is a **crocodile**?

Crocodiles are found in warmer climates worldwide. They have long bodies and powerful tails used for swimming or defense. In water, they use their limbs to maneuver and steer; on land, they use them to walk with a slow gait, with their bellies held high off the ground. Crocodiles are usually fairly aggressive and may bite or swing their tail in defense. These animals eat a wide variety of food, especially fish, large vertebrates, and carrion.

A yellow-bellied sea snake, a marine reptile, slithers along a South African beach. *CORBIS/Anthony Bannister; ABPL*

One of the more well-known marine crocodiles is the American crocodile, which lives exclusively in the saltwater areas of southern Florida and the Florida Keys. Because of years of over-hunting (their hides were highly valued), they are now an endangered species. But this designation has not protected them enough; their habitats are still being encroached upon (due to development) and they are the targets of poaching (illegal hunting).

## What is the **difference** between a **crocodile** and an **alligator**?

There are two ways to tell these animals apart: by their size and their head. Crocodiles (*see* photo, next page) are slightly smaller and less bulky than alligators, but the crocodile's head is larger and it has a narrower snout and a pair of enlarged teeth in the lower jaw that fit into a notch on each side of the snout; the upper and lower teeth are visible when the jaw is closed. Alligators have broader bodies and snouts; when this animal's jaw is closed relatively few upper teeth are visible.

The American alligator and the American crocodile are also differentiated by their habitats: the alligators live in fresh to brackish (slightly salty) water; the crocodiles live in salty water.

**It would be a menace at half the size: This huge crocodile was caught in 1925 off Borneo, but the marine reptile can be found in warm climates worldwide.** ***NOAA/Family of Captain Jack Sammons, C&GS***

## What is a **sea turtle**?

A sea turtle is a reptile with a lightweight, streamlined shell, which protects the turtle's vital organs. It has a heavy neck that, unlike many land turtles, it cannot pull all the way into its shell. Its legs are muscular and powerful, allowing some species to swim at speeds of up to 35 miles (56 kilometers) per hour. There are 8 species of sea turtle, including the green, loggerhead, hawksbill, and leatherback. Sea turtles range widely in size: the olive ridley grows to less than 100 pounds (45 kilograms), whereas the leatherback can be anywhere from 650 to 1,300 pounds (295 to 590 kilograms) at maturity! Certain species can also live very long—more than 100 years.

## What **problems** have plagued **sea turtles in Florida**?

Most sea turtle species in Florida are either endangered or threatened. One of the more recent problems they face is papillomas, a potentially fatal disease that causes tumor-like growth on the soft tissue of sea turtles in this region. The growths can cover the eyes of the turtles, causing blindness and eventual death, as the animal cannot see to find food. The actual cause of the papillomas is unknown, although several researchers believe it may be caused by waters that are polluted by runoff.

Another problem endangering the sea turtle is the rise in boat-related injuries—something that may increase during high nesting years (when there are a number of nests on the beaches). During this time, there are more turtles—and, usually, more people on the water. Though the turtles aren't usually at the surface, where they might be in danger of being hit by a boat, they do surface for fresh air; the more often they do so, the greater the likelihood for a mishap.

**Sea turtles have muscular and powerful legs, allowing some species to swim at speeds of up to 35 miles (56 kilometers) per hour.** ***NOAA/OAR National Undersea Research Program; G. McFall***

**Loggerhead sea turtles, hatched on a Virginia beach, head for the ocean.** ***CORBIS/Lynda Richardson***

## How are **endangered sea turtles** being protected?

Many people are trying to find ways to protect endangered sea turtles. For example, in Indian River County, Florida, officials have developed a habitat conservation plan as part of a legal settlement. When the county wanted to put up sea walls, the Sea Turtle Survival League brought a lawsuit, citing that such walls would interfere with nesting habits: They argued that the proposed walls would not allow the turtles to dig nests in the area, and would also lead to the erosion of turtle-nesting sites farther down the beach. The main reason for the concern was that this area, along Florida's Atlantic coast (about two-thirds of the way down the Florida peninsula) hosts the world's second-largest population of threatened loggerhead turtles—and almost all the nesting sites of endangered green and leatherback turtles. This ranks Indian River County as a globally important sea turtle nesting area.

## What is a **sea snake**?

A sea snake is similar to a land snake, except it spends its time in the water. These reptiles are found in shallow tropical and subtropical waters (but not in the Atlantic Ocean)—sometimes numbering in the hundreds. The sea snakes are all related to the cobra family, and have a very potent venom (it can cause severe injury to humans). They use their flattened tails as a paddle as they swim, and they usually feed on fish. Between breaths at the surface, they can stay under water for more than 30 minutes. An example of this animal is the yellow-bellied sea snake.

# FISH & OTHER OCEAN LIFE

## DEFINING FISH

### What is a **fish**?

A fish is an aquatic animal—a vertebrate (skeletal) creature that lives the majority of its life underwater. But there is such a wide variety of fish—from the flying fish that live just above the water's surface to those that are found deep in the ocean trenches—that neatly defining them is difficult. Still, some generalities can be drawn: Most fish are cold-blooded, breathe by means of gills, and have a heart with only two chambers (as compared with the four-chamber heart of humans); others are not cold-blooded and do not have gills. Some have scales, others have rough skin composed of tiny "teeth" called denticles. Most have fins for swimming, but they come in various combinations, sizes, and shapes.

The reasons for such diversity are these: In adapting to their diverse environments, fish have made many evolutionary changes over hundreds of millions of years. For example, there are blind fish, fish that can crawl about on land, and electric fish; others build nests, change color in a few seconds, or even puff up to ward off predators.

### When did **fish evolve**?

It is difficult to say when fish actually evolved, but scientists do know they are the descendants of jawless fish. The following lists the possible evolution of the fish over time:

About 460 to 480 million years ago: The first jawless fish evolve. (It's worth noting here that this starting point is hotly debated in the scientific community.)

About 450 million years ago: The first jawed fish emerges.

About 390 million years ago: The ancestors of the bony fish evolve.

About 380 million years ago: The first shark-like fish evolves.

About 360 million years ago: The early offshoots of the bony fish (Osteichthyes) evolve into the first amphibians (intermediates between fish and reptiles).

About 175 million years ago: The first true bony fish emerge.

Between 190 and 135 million years ago: The first modern sharks evolve.

## Why were **fish important** to the **evolution of life** on Earth?

Fish were very important to the evolution of life on Earth because they were the precursors to amphibians, the first animals to venture onto land.

## How did **amphibians evolve** from **fish**?

Amphibians—an offshoot of fish and the first creatures to crawl on land—adapted to their new terrestrial environment by developing primitive lungs and using their fins for crawling. Still, the earliest amphibians spent most of their time in the water. They were very fish-like, and laid fish-like eggs in a moist place, usually in the water.

## What is **ichthyology**?

Ichthyology is the study of fish. Swedish naturalist Peter Artedi (1705–35) is considered the father of ichthyology; he made several investigative writings about fish. His knowledge of the relationships

Salamanders, such as the Cheat Mountain salamander shown here, are among the vanishing species of amphibians—living links in the evolutionary chain. This species survives only in West Virginia. *Associated Press/Marshall University File*

among the various species and his concepts of fish groups have become classics in the field.

## What does it mean to be **cold-blooded**?

When it comes to ocean fish, cold-blooded means that the body temperature is just about the same as the surrounding water. This means that fish are particular about the temperature of the water in which they live. In fact, most species are sensitive to temperature changes, and usually cannot stand rapid fluctuations of more than 12° to 15° F (7° to 8° C).

But not all marine fish are cold-blooded. As far as we know, certain tunas and bonitos are exceptions to the rule. For example, the blue-fin tuna can regulate its own body temperature, similar to a mammal.

## What is the **smallest known fish**?

So far, the *Pandaka pygmaea* of the Philippines is the smallest known fish. It measures just under a half inch (1 centimeter) at maturity.

## What is the **largest known fish**?

The whale shark is the largest known fish, measuring up to 60 feet (18 meters) in length—and amazingly, this giant only feeds on plankton. The whale shark can weigh up to 15 tons—or 5 billion times as much as the common goby.

## Where do most **ocean fish live**?

Most ocean fish live along coastal areas, especially on the continental shelves to depths of 400 to 600 feet (122 to 183 meters). These coastal fish occupy almost every niche (habitat in this area), and represent two-thirds of all known living fish species, with most of the fish living within 20 miles (32 kilometers) of the shore. There are fish (such as flounder, sole, and cod) that blend in with the bottom sands in the shallow coastal waters; others spend their lives in the ocean, but travel upstream through inland rivers to spawn, such as salmon and shad; still others occasionally walk on land, such as the walking catfish.

Fish are also prolific in the open ocean waters, at depths from the surface to 6,000 feet (1,828 meters). Fish such as tuna, mackerel, and marlin prefer waters no deeper than 3,000 feet (914 meters)—a layer of water where temperatures are most comfortable for them.

The fewest number of fish live at the deepest ocean depths, past about 18,000 feet (5,486 meters). We may be surprised in the future as more ocean depths are explored: Some deep-water areas may have more fish than we realize.

## Have any **new ocean fish** been **discovered**?

Many times when scientists explore new parts of the oceans, they discover new types of fish. For example, in 1978 a research group lowered baited traps and a camera through the Ross Ice Shelf and into the Antarctic waters, to a depth of 1,960 feet (597 meters) below sea level. There, they found many new types of fish—adding to the already huge list of fish species.

## What is a **living fossil**?

A living fossil is any organism that was once thought to be extinct, but was actually found, very surprisingly, to still be alive. Because the discovery of such creatures provides scientists with new information about evolution, the organisms become part of the fossil record. Perhaps the most famous example of the discovery of a living fossil occurred in 1938: A small trawler (fishing boat) brought up an amazing catch from deep in the Indian Ocean, near the Comoros Islands off Madagascar—a coelacanth (SEE-la-kanth), a fish that was thought to have gone extinct millions of years ago. Other coelacanths have been found since. The surprising discovery caused scientists to revise the thinking about the life that exists in the deep (as of yet, mostly unexplored) ocean: It's possible that there are all kinds of ancient-looking fish living in the depths.

The coelacanth first appeared during the Devonian period of the Paleozoic era, almost 400 million years ago. Some scientists believe this species of fish was the first to haul itself out of ancient oceans and onto land—and are probably the ancestors of many land animals. But based on the fossil record, it appeared that the fish went extinct toward the end of the Cretaceous period, about 60 million years ago. But evidently, enough coelacanths made it through millions of years to maintain a population—and now live deep in the oceans.

## What **missing fish** was recently rediscovered?

A giant roughy, also known as a giant sawbelly (*Hoplostethus gigas*), was first recorded and last seen in 1914. Now this fish has been found again after 85 years—it was discovered during a survey of commercial catches from the Great Australian Bight.

The fish was caught in water between 590 and 1,148 feet (180 and 350 meters) deep, in the same general location as the one found in 1914. Its decades-long absence is attributed to its habitat: The giant roughy apparently lives in the deep, central part of the Great Australian Bight, which is lightly fished, or not fished at all. In addition, it probably lives on or close to the rough bottom, avoiding the nets of trawlers.

The impetus for this discovery was the creation of the *Australian Seafood Handbook*—an identification guide for edible Australian

## How does a fish breathe?

A fish breathes oxygen present in water: The water enters the fish's mouth and passes through a chamber over the gills; as the oxygen-rich water flows over the gills, a series of membranes allow air, but not water, to pass through; the oxygen then passes into the bloodstream; finally, carbon dioxide is released into the water as a waste product.

This system is more efficient than the one used by those of us on land: Most land mammals absorb only about 20 percent (or less) of the oxygen from the air into their bloodstreams, while fish absorb up to 80 percent of the oxygen available in the water.

fish. Scientists had been working closely with commercial fishermen to sample and record catches from the deep oceans and coasts. While recording and photographing commercial species that are caught off the Great Australian Bight, a scientist was shown a selection of fish caught by a trawler. One of the fish was the long-missing giant roughy.

The discovery of this long-lost fish provides yet another example of how elusive the life forms in the deep ocean can be.

## How **long** can a **fish live**?

Although not all fish species have been observed over time, it is thought that most fish live an average of about 25 years. Scientists do know the average age (or life span) of those fish that have been extensively studied or monitored; for example, Atlantic cod and herring live about 22 years, haddock and barracuda live about 15 years, and a bluefin tuna can live about 13 years. But some fish remain to be studied, so the average life span estimate may change as more research becomes available.

## How do **ocean fish stay afloat**?

Most ocean fish stay afloat with the use of a gas, or air, bladder. This gland-lined airtight sac is located in the gut of the fish; the size dependent on the size of the fish. The glands extract gases from the fish's bloodstream, and direct these to the bladder. The fish has the ability to regulate the amount of gas in the bladder, allowing the fish to travel up and down very rapidly in the water column—but the bladder usually only works down to about 1,200 feet (360 meters). In fact, hooking a fish and bringing it to the surface too fast can cause the fish's bladder to burst—causing its stomach to be forced out of its mouth.

Some fish, such as mackerels, do not have gas bladders. These fish have to swim and move constantly to maintain a certain level in the water—if they don't, they will sink.

## How do **fish swim**?

Most fish swim by pushing water aside—by wiggling the head and tail and sometimes the fins, depending on the fish. The fins are used for steering and stabilization: The pectoral (side) fins are used for balance or turning, and for pitch (tendency to ascend or descend); the dorsal (top) and pelvic (bottom) fins are used as stabilizers, controlling the body's rolling motion; and tail fins are multi-functional, and are used for propulsion, steering, and stabilizing. Most fish also reduce their resistance to water with a coat of slimy mucus, permitting water to slip freely over their bodies.

There are exceptions, of course. For example, flatfish, because their bodies are flattened sideways, undulate their bodies from side to side to move. Flat rays and skates also undulate their bodies, and move their pectoral fins up and down instead of side to side.

## How **fast can fish swim**?

Amazingly, there are some fish that swim extremely fast—moving faster through the water than many of the speediest terrestrial animals can move across the land. Some of the fastest swimmers are the sailfish and swordfish (one unconfirmed report clocked a swordfish at 150 miles

### What is a school of fish?

A school is a group in which all the individual fish are headed in the same direction. Like a flock of starlings in the air, the fish are uniformly spaced and swimming at the same speeds. It is no doubt an evolutionary feature in many species of fish: As the group moves in unison, it is more difficult for a predator to catch its prey. Plus, because there is such a large group, it is easier to confuse a predator—or even fool the predator into thinking the "creature" (the school) in its path is huge. Changes in the behavior of a school of fish depend on the surroundings, with the school responding to all sorts of variations in the environment, including noise, sudden motion, or a strange presence (such as a human swimmer or wader).

[241 kilometers] per hour). The following list gives the approximate recorded speeds of certain ocean and, for the sake of comparison, freshwater fish (indicated by an asterisk, *).

| Fish | Speed distance per hour |
|---|---|
| sailfish | 68 miles (109 kilometers) |
| swordfish | 60 miles (97 kilometers) |
| blue-fin tuna | 50 miles (80 kilometers) |
| shark | 22.4 miles (36 kilometers) |
| trout* | 21.7 miles (35 kilometers) |
| pike* | 15.5 miles (25 kilometers) |
| carp* | 7.5 miles (12 kilometers) |

## Can **fish fly**?

Not really, although there is something called a flying fish. But they do not fly—they glide. Propelled by the movement of its strong tail, this fish leaps into the air at speeds of up to 20 miles (32 kilometers) per hour. It uses its wide pectoral (side) fins as wings, gliding close to the ocean surface. By flicking its tail against the water, it is often able to get an additional push for a longer "flight." Some flying fish have been known to soar as high as 20 feet (6 meters) and glide for 1,300 feet (396 meters) along the surface.

**The duckbill eel, like other eels, is a bony fish (order Osteichthyes) that, surprisingly, has an acute sense of smell. This one was found off the coast of Hawaii, at a depth of about 2,600 feet (780 meters).** ***NOAA/OAR National Undersea Research Program***

## Can **fish walk on land**?

Yes, there are some fish that walk on land. Walking catfish can survive long periods on land and are capable of breathing air, using a lung-like organ as a supplement to their gills. The lungfish can also walk on land, using the swim (or gas) bladder as an air reservoir, and a nose-to-mouth connection that allows it to breathe the air.

## Can **fish smell**?

Yes, fish do have a sense of smell. The nostrils of most fish are located in the front, just behind the mouth opening. These organs are so powerful that some fish can detect the secretions from a predator's skin long before the animal gets close.

In fact, some fish don't depend on their sight as much as they do their sense of smell. Sharks and rays depend on smell to detect prey; eels have the most acute sense of smell, using long nasal sacs to detect a scent. Certain species migrate across the oceans to spawn by detecting very minute changes in seawater chemical variations—which they pick up by smell.

## How do **fish see** underwater?

In most cases, because fish have to see in low light, they have no eyelids and no irises; (in the human eye, the pupil responds to the amount of light available, opening or closing as needed). These characteristics allow the fish's eyes to gather the most light—which also gives the animals what is often described as a "bug-eyed" appearance. Fish have monocular vision—the eye on one side of the head is sensed by the opposite side of the brain. A fish's eyes are located on each side of the head, allowing the fish to see in any one direction at a time. Many also have upward-directed eyes; during daylight, they detect prey by spotting its silhouette against the sunlight from above. And of course, there are exceptions: a few surface-dwelling fish have eyes adapted to see in both water and air.

Although scientists believe that most fish have some color vision (except sharks and rays, which may only see in black and white), no one really knows if the animals are affected by color. Are they actually coming after the pattern of a fisherman's hook or the color(s)?

Some believe a fish's pattern may be most important to other fish. For example, adult and young angelfish are brilliantly colored and patterned—but they look completely different from each other. One theory is that the adult recognizes the young's patterns among the huge variety of fish found in a reef—and thus they will not threaten the young of their species.

## Can **fish hear**?

Yes, fish can hear. Sound is as important to underwater communication as it is to communication on land. Fish, like other organisms in the sea, use sounds to attract mates and to detect prey or predators.

Fish hear much the same way as humans, although they lack the air-filled middle ear—in other words, there are no ossicles like those found in a mammal's ear. Fish also do not have a cochlea, but do have hair-bearing cells in a chamber-like organ called the lagena. Most fish have three semicircular canals—although lampreys have two and hagfish only one.

## Are some **fish "electric"**?

Yes, some fish use electric fields to detect prey, navigate, or even to defend themselves or attack other fish. These fish, including skates and rays, have organic "batteries" that evolved from muscles or nerves. Electric rays found in sandy or muddy shallows or in moderately deep seas can generate more than 200 volts of energy. After a number of shocks, mostly to immobilize prey, its "batteries" are completely depleted, and the animal takes several days to recharge. Electric eels, which are not true eels, can generate up to 600-volt pulses from "batteries" that extend almost half the length of their bodies—which can reach up to 8 feet (2.5 meters) long.

### Why are **fish important** to **humans**?

Fish have been, and continue to be, one of the most important commercial products of many countries. For years, various species of fish (such as tuna, herring, and sardines) and certain "industrial fish" (monkfish, sea robin, squirrel hake, shark, and ray) have been used in livestock feeding (the fish are an excellent source of nutrients and oils). Others have been used as major food sources for humans, especially for populations living on islands or along the coasts, where fish are, in most cases, more cost-effective to harvest and eat than are other types of animals.

### Are **fish indicators** of **environmental change**?

Yes, fish—especially their demise—can be indicators of environmental change. Plankton, the tiny mobile organisms in the upper levels of the sea, are at the bottom of a huge food chain (or food pyramid), supplying a major part of the dietary requirements of many fish. The progression of eating in the oceanic food chain is from tiny to huge (fish are in the middle of this chain) and keeps the ocean ecosystem in balance. If the plankton population is affected by such problems as environmental pollution, ozone hole depletion, or temperature changes, fish will suffer, too. The local food chain can then break down—and in turn, may eventually affect the global food chain or food web.

## CLASSIFYING FISH

### How are **fish classified**?

Fish are classified in the kingdom Animalia, phylum Chordata, and class Pisces (although in some classifications Pisces is listed as a subclass). In the Pisces class (or subclass), 90 percent of the fish are Osteichthyes (bony fish), which accounts for some 30,000 different species. The remaining 10 percent are Chondrichthyes (cartilaginous fish), with 625 species, and Agnatha (jawless fish), with 50 species. Nearly half of the 40,000 or so known species of vertebrates ("segmented backbone" animals) are fish. (All together, there are seven classes of vertebrates: Agnatha, Chondrichthyes, Osteichthyes, amphibians, reptiles, birds, and mammals.)

But this classification system, though widely accepted, is not standard; there is still disagreement over the classification of fish, which means there are alternate systems of organization. For example, one classification breaks down classes of fish into Agnatha, Placodermi (known only as fossils), Chondrichthyes, and Osteichthyes. Further breakdown of the Osteichthyes includes the subclass Actinopterygii (ray-finned fish), superorder Teleostei (spiny-finned fish, with 95 percent of all living fish in this group), and subclass Choanichthyes (lobe-finned fish).

## Do any fish have a **combination** of **bony and cartilaginous** characteristics?

Yes, even though these two major classes dominate, there are some fish that carry both characteristics. One is the coelacanth, a fish that was at one time thought to be extinct.

## What are some common **characteristics** of **Osteichthyes**?

The Osteichthyes (bony fish)—including tuna, sailfish, eels, cod, and salmon—are all bony and have jaws. An early offshoot of this group may have eventually led to amphibians, the first animals to venture on to land millions of years ago (although some scientists believe the first amphibians were offshoots of a creature resembling or related to the coelacanth).

Most fish are bony fish (Osteichthyes), with skeletons made mostly or solely of bones. On each side of the bony fish, a cover protects the gills; many also have an internal organ called a swim bladder (also called a gas bladder) that helps the fish remain afloat. These fish also generally spawn rather than mate, with the males fertilizing the eggs after the females lay them. They eat a wide variety of foods, including other animals (insects, larvae, and other, smaller fish) and some plants (including phytoplankton).

Salmon, seen here in Alaska's Naknek River Estuary (at Bristol Bay) are one variety of bony fish (order Osteichthyes). *CORBIS/Natalie Fobes*

## What is the **oldest bony fish** fossil found?

The oldest bony fish fossil found so far is the *Psarolepis romeri,* a fish that lived more than 400 million years ago. The fossil was uncovered in southwestern China, and has features from both the primitive and more advanced branches of the fish family. But the true ancestry of this predator is still debated—at least until scientists can find more such fish in the fossil record.

## How do **flatfish differ** from most **bony fish**?

A flatfish, such as flounder or halibut, begins its life as a normally-shaped fish—complete with one eye on each side of its head. But as it matures, the fish changes: One eye moves and the mouth twists so that as an adult, it can lie on one side—making it truly a flat fish. These fish can also change color and patterns, allowing them to match the shading and texture of the seabed (this is a means of self-defense). They can also bury themselves in the sands of shallow waters to protect themselves from predators.

## Is a **seahorse** a fish?

Yes, a seahorse is one of the strangest fish in the ocean. This fish actually has a head that resembles a horse; the rest of its body is long, with a curled tail at the end. It spends most of its time in a vertical position, and moves up and down in the water using its swim bladder. But it is a poor

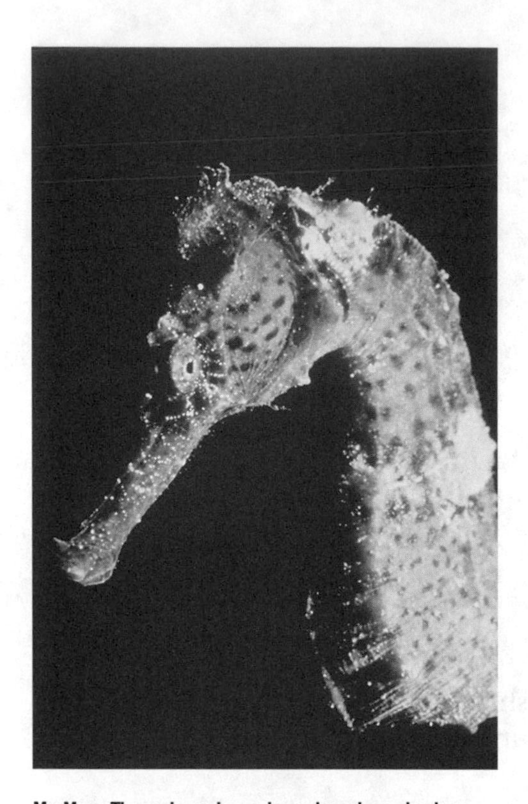

Mr. Mom: The male seahorse has a brood pouch where the female places her eggs. When they reach maturity, the offspring swim out of the pouch, looking like miniature versions of the adult. *NOAA/OAR National Undersea Research Program*

swimmer; and often uses what is called a prehensile tail—a tail that can wrap around things, such as seaweeds or sea fans—to stay in one place.

Unlike most fish, the male seahorse watches the young: the female places her eggs in the male's abdomen, in an opening called the brood pouch. When the eggs hatch, they emerge from the male as miniature versions of the adult.

## What **fish** lives in **floating sargassum**?

A frogfish called a sargassumfish lives in floating sargassum algae. It rarely moves away from the sargassum, climbing over the seaweed, stalking smaller fish and other creatures that live in the algae community. Its spiny-like protrusions give the frogfish a very strange look—and its colors help it remain camouflaged in the algae. The sargassumfish can also inflate its body by swallowing air or water, preventing a predator from dragging it away as its spine catches on the sargassum.

## Which **fish** is called the "**hitchhiker of the ocean**"?

Long-bodied remoras are often called the "hitchhikers of the ocean," as they take rides on sharks and other marine animals. This is called a symbiotic relationship, as the remoras eat parasites and any leftover food that surround the host. The remora clings to its host using a suction disk on the top if its head, and to release the hold, it slides forward. Certain species are highly selective about their hosts. For example, the marlinsucker and whalesuckers are so named because they prefer marlins and whales, respectively.

**Looks can be deceiving: The spiny pufferfish can inflate itself to as much as three times its normal size to ward off would-be predators. This one was caught on film doing just that in the waters off Malaysia.** *CORBIS/Jeffrey L. Rotman*

## Is there a **fish** called a **dolphin**?

Yes, there is a group of surface-feeding fish called dolphins—but they don't look anything like the marine mammals of the same name. The dolphin fish has a long back fin that extends from the animal's head to its tail, and its tail fin is deeply forked. They can measure from 1 to 6 feet (0.3 to 2 meters) in length, weigh up to 75 pounds (34 kilograms), and are brilliantly colored. But they don't live long—only about 2 to 3 years.

## What is a **pufferfish**?

A pufferfish (or puffer) does what its name implies: It inflates itself with water or air, enlarging itself to about two or three times it normal size. As it puffs up, it floats upside down and its spine sticks out, deterring a predator from taking a bite. When the predator leaves, the puffer deflates and flips upright again. The jaws of a puffer look like a beak and can crush hard sea urchins, mollusks, and crabs. Certain puffers are among the most poisonous marine animals known. Eating them is not advised, but some countries believe they are delicacies—and consequently, each year many people die from consuming puffers that were not prepared correctly.

## What is a **goby**?

Gobies are the largest group of predominately ocean fish, with at least 800 known species—and no doubt more will be found. Most of them are very colorful, and average 1 to 3 inches (2.5 to 7.6 centimeters) in length—although some can grow up to 20 inches (51 centimeters) long. They live in many different marine niches, including mud (into which they burrow), wave-pounded shores, in sponges, or in the burrows made by crabs and shrimp. This group also includes the smallest known fish: the dwarf pygmy goby.

## Is there a **fish** with **four eyes**?

Not really—but the so-called four-eyed fish does have two eyes that are essentially split in two. As the fish swims at the surface looking for insects, its eyes are divided into two by a horizontal bar. The upper part is for seeing in the air; the lower part for seeing under water. It also has the ability to see both places at the same time—or four images—as the eyes have two distinct retinal areas.

## What **fish** is often called one of the **"ugliest sea creatures"**?

Some people consider the anglerfish one of the ugliest sea creatures. They are flabby and lumpy, with huge mouths filled with sharp, long, irregular teeth. They usually eat anything—and have very elastic stomachs that stretch to accommodate their prey. The anglerfish has a "lure" on the top of its head (hence its name): a long stalk with a fleshy flap at the end. This "pole and lure" wiggles around in the water right in front of the anglerfish's mouth, attracting small fish. As the small fish gets close, the angler then sucks or gulps the prey into its mouth.

The fast-swimming swordfish is considered a trophy fish. This specimen was caught by Doris Lovett Dennis, using the pole she holds in her left hand, near Santa Catalina, California, in 1936. *CORBIS/Bettmann*

Even deeper in the oceans are the deep-sea anglerfish—also considered to be very ugly. Because they live in such dark waters, the "lure" of this fish is luminous. This light is created by bioluminescence—a combination of chemicals and bacteria that glow, giving the stalk a natural light, which attracts prey.

## What is a **swordfish**?

A swordfish is a fish with an elongated snout that comes to a point that measures about a third of its body length; it uses the "sword" to impale

They may not pose a serious danger to ocean swimmers and divers, but they sure look the part: Barracudas, as evidenced by the one shown (investigating a shipwreck), can look menacing indeed. *CORBIS/Stephen Frink*

or stun its prey. Swordfish average about 4 to 10 feet (1.2 to 3 meters) in length, with some reaching 15 feet (4.6 meters); although they average less than 400 pounds (182 kilograms), they can weigh up to 1,000 pounds (454 kilograms). They are among the fastest fish in the ocean, reaching close to 60 miles (97 kilometers) per hour.

## Are **barracudas dangerous**?

Barracudas are perceived to be dangerous because they have attacked swimmers and divers. But in reality, they usually only follow the diver, but don't attack. When they do attack, it is usually in sediment-filled waters: if they see a hand or foot coming their way, they think it is prey and they bite. There are about 20 species of this fish, and they average 4 feet (1.2 meters) in length. Some, such as the great barracuda (Sphyraena barracuda), can grow as long as 8 feet (2.4 meters). Barracuda are found in the Atlantic and Caribbean oceans, around the West Indies and Brazil, and north to Florida; in the Pacific, they are found from Indonesia to Hawaii.

## What are some common **characteristics** of **Chondrichthyes**?

The Chondrichthyes (cartilaginous fish)—including sharks, skates, rays, and chimaeras—all have cartilaginous (fibrous versus hard) skeletons, paired fins, and jaws that evolved from gill arches. The living forms of these fish all have teeth, and are either covered with small scales or rough skin. These animals have multiple (usually five) gill slits and have essentially the same fin patterns as do bony fish (Osteichthyes). But they have different tail fins, and unlike bony fish, they have no swim bladders (or gas bladders) to keep them afloat and therefore, they must keep moving or else they will sink. They reproduce by mating rather than spawning, with most females producing live young, not eggs. Because of these characteristics, these fish can only live in certain regions of the world—primarily in saltwater.

## What is **cartilage**?

The Chondrichthyes fish have cartilaginous skeletons, or skeletons made of cartilage, which is softer and more flexible than bone, but strong enough to support the animal's body. Humans also have cartilage—the "bones" in the end of your nose and in your ears are made of cartilage.

## Have **shark fossils** really been found **in Montana**?

Yes, the rocks at Bear Gulch, Montana, have yielded one of the world's richest collections of fossil fish—some 113 species, including 70 species of prehistoric sharks, some of which had never been seen before. About 320 million years ago, this land was situated close to the equator and was covered by a warm, shallow sea. Here, about 60 percent of all the fish were sharks—measuring in length from a few inches (several centimeters) to about 97 feet (30 meters). Scientists believe there was a cataclysmic die-off and sudden burial of the creatures. Thus the Bear Gulch sharks are beautifully preserved, many with heart, veins, stomach contents, and skin still intact.

## What is the **skin** of a **shark** like?

A shark's skin may look smooth and slippery, but it is covered with small scales that look—and feel—like tiny, sharp teeth.

## Are all **sharks alike**?

No, not all sharks are alike. There are at least 370 different types of sharks, including the megamouth, sixgill, dogfish, cookie-cutter, nurse, thresher, and dwarf sharks.

## What is the **largest shark**?

The largest shark—and the biggest fish in the world—is the whale shark. It can grow to more than 50 feet (15 meters) in length and weigh more than 15 tons.

## What is the **smallest shark**?

The smallest shark is the *tsuranagakobitozame*—a Japanese name that means, "the dwarf shark with a long face," also called dwarf dog-shark This shark can fit in the palm of your hand; the adults measure about 5 inches (13 centimeters) in length.

## What is the **least dangerous shark**?

Given their reputation (reinforced in popular media), the list of harmless sharks is surprisingly long. Most scientists agree that the least dangerous is also the largest—the whale shark. Divers have even been known to grab the back fin of this fish and take long rides through the ocean waters. These sharks feed on small fish and plankton.

Another huge but harmless shark is the basking shark, with teeth less than a half inch (1.3 centimeters) long. This shark swims slowly through a school of plankton with its mouth open, using gill rakers (a kind of strainer) to capture its food.

Even though there are numerous sharks that are thought to be relatively harmless, most scientists agree that any shark, if provoked, can use its powerful jaws and teeth to pose a real threat to whatever has provoked it.

**Jaws: The great white shark, which has been noted for occasional attacks on humans, can reach more than 36 feet (11 meters) in length; its teeth can grow to an impressive 2 inches (5 centimeters).** ***CORBIS***

## What is the **most dangerous shark**?

Although some scientists will disagree, most point to the great white shark as the most dangerous. This fish can reach more than 36 feet (11 meters) in length, and its pointed teeth can be up to 2 inches (5 centimeters) long. But this shark is rarely seen along coastal areas, as it rarely strays into shallow waters.

## Do **sharks really attack humans**?

Yes, sharks have been known to attack human swimmers and divers, but humans are normally not on the shark's menu. In most cases, a shark will try to avoid humans. Many divers note that most sharks they encounter swim away—and may be afraid of people.

But each year, about 100 people are attacked by sharks around the world, mostly by the bite-and-let-go method. It is thought that the bite-and-let-go tactic is a way for the shark to communicate to the person to leave its territory—or perhaps we humans leave a bad taste in the shark's mouth. In one study, scientists showed that many surfboards (and even people) can look like a seal to a shark swimming deep below the water's surface. And because seals are one of the favorite foods of these fish, in a case of mistaken identity, the shark will attack the human or the surfboard. Sharks have also been known to be attracted to blood in the water; and they will attack if they are bothered.

The great white shark has been noted for some attacks on humans. American author Peter Benchley wrote the famous novel *Jaws,* the story of a great white shark that repeatedly attacks swimmers off an Atlantic resort town. But other sharks, including the mako, tiger, and some hammerhead sharks, have also been cited in attacks on humans. Carpet (or wobbegong) sharks in Australia lie on the bottom of the sea during the day, their colors blending in with the surroundings. Unlucky swimmers have been known to accidentally step right on the shark, causing the fish to attack.

But there really isn't any cause for concern: The odds of being attacked by a shark are 1 in 100 million. To put this in perspective, Americans have a 1 in 66 chance of being audited by the IRS; and yet, because of the popular image of the shark, more people probably fear an attack

**Though it does not prey on humans, the hammerhead shark has been cited in attacks on people. This was one of hundreds seen on a July day in 1982 off the ship *Albatross IV. NOAA/Commander John Bortniak, NOAA Corps (ret.)***

than they do an audit. And it is estimated that in any given year more people in the United States are killed by pigs than by sharks.

## Are **great white sharks** really **dangerous** to humans?

New studies have shown that the danger great white sharks pose to humans is greatly exaggerated. In fact, these sharks may actually dislike the taste of humans. The great white shark, which can grow up to 20 feet (6 meters) in length with a weight of more than 5,000 pounds (2,270 kilograms), has been portrayed as a ruthless, eating machine from which no one is safe. However, although there have been 78 reported attacks along the California coast since 1926—there have been only eight deaths.

To gain a better understanding of these creatures, scientists have spent four years studying their behavior at Año Nuevo, some 23 miles (37 kilometers) off the coast of central California. These rocky outcrops are a rookery for elephant seals, the favorite prey of great white sharks.

Tagged with ultrasonic transmitters linked to a computer system, sharks were tracked night and day by a series of sonobuoys placed 1,650 feet (503 meters) apart off the western shore of Año Nuevo.

The results from this study give a much different picture of this ocean predator than the one commonly held. For example, great white sharks were found to hunt around the clock, not just during the day. These sharks were not solitary, but had social connections between pairs—and were even seen to slap their tails against the water to ward off others after a kill. Also, great white sharks were found to have picky eating habits, gently mouthing items to determine whether or not they're edible. While the soft-blubbered elephant seals feel and taste edible to the great white shark, surfboards, sea otters, buoys, and humans are not soft enough or tasty enough to eat.

## What are **sixgill sharks**?

Scientists often call the sixgill sharks the "dinosaurs of the deep." There are some located off central Vancouver Island, British Columbia, Canada—in the rich waters of the Strait of Georgia. They can be up to 25 feet (8 meters) long and are often seen in water about 400 to 600 feet (122 to 183 meters) deep. They have six gill slits (thus the name), wide bodies, only one dorsal fin—and have remained virtually unchanged for millions of years.

## Why is the **shark's immune system** considered **unique**?

One of the most remarkable findings about sharks concerns their immune system. In their natural habitat, sharks can ward off almost any disease, including cancer.

## What are **manta rays**?

Because of the points on their fins, manta rays are sometimes called devilfish. They are the largest of the skates and rays (cartilaginous fish)—and the "flattened" relative of shark. The eyes are on the upper surface and rays have a large, wing-like pectoral fin on each side of the

**The "wings" of the stingray are actually large pectoral fins that allow this fish to glide through water. This one was caught—and returned live to the waters—off the Carolina coast.** ***NOAA/Commander John Bortniak, NOAA Corps (ret.)***

head. The "wingspan" (distance from one end of the pectoral fin to the end of the other) of the Pacific manta can reach 25 feet (7.6 meters); the ray can weigh up to 3,500 pounds (1,600 kilograms). The Atlantic manta is slightly smaller, growing to about 22 feet (6.7 meters) and weighing about 3,000 pounds (1,300 kilograms). These graceful creatures glide through the surface waters of tropical oceans by flapping their "wings." A manta ray's mouth is small; the animal feeds on tiny fish and invertebrates it filters out of the water, often jumping completely out of the sea and diving back in to catch its prey.

## What are **stingrays**?

Stingrays are often confused with manta rays, as both animals have "wings" (actually large pectoral fins on each side of its flattened body) and glide through the water. Stingrays also use their wings to excavate the seafloor, exposing shellfish, which they crush with flat, strong tooth plates. The stingray measures about 1 to 5 feet (0.3 to 1.5 meters) in diameter, its body forming a disk that includes the head and pectoral (side) fins. As the name indicates, certain types of stingrays are also known for their stinging, poisonous tail—a whip-like feature that often sports one or more sharp, barbed spines.

## What is the **Agnatha**?

The Agnathans, including lampreys and hagfish, are cartilaginous fish that are jawless and have disc-shaped mouths that either grasp or suck. Unlike most fish, they have no scales and lack paired (symmetrical) fins. These vertebrates were originally filter-feeders, straining mud and water through their mouths and out their gills. Fossil remains of agnathans are evidence of the earliest vertebrates, appearing in the Ordovician period about 500 million years ago. They had bony skeletons and bony armor plates covering their bodies. Modern agnathans are greatly changed from their ancestors: over time, they have lost their bone and replaced it with cartilage.

## Are **hagfish** and **lampreys parasitic**?

Yes, many of these animals are parasitic. For example, most hagfish bore into dead or dying fish, feeding until only the skin and bones remain.

Some, but not all, lampreys are parasitic, preying on other fish by sucking out their blood. For example, sea lampreys found in the Great Lakes are not native to this freshwater area; they reached the lakes after man-made canals were opened—and wreaked havoc on the fish populations until they were eventually controlled.

## Do we know **all of the fish** that inhabit the **oceans**?

No, we do not know of all the fish that inhabit the oceans—far from it! In fact, scientists recently discovered four more previously unknown fish—two from the group plunderfish and two similar to jellyfish—in the frigid waters off Antarctica, near Franklin Island in the Ross Sea. This area is not known for its diversity of fish; there are only about 130 fish species there, so adding to this number was surprising.



In more detail, the newly discovered fish were:

The Antarctic gravelbeard plunderfish (*Artedidraco glareobarbatus*), which lives near sponge beds and uses a chin extension called a barbel to attract its prey (the other fish may think it's a worm). The plunderfish have an easy life—they just wait around on the seafloor, waiting for smaller fish or crustaceans to wander by, then gulp them down. They are 15 inches (38 centimeters) long; these tan-and-brown bottom-dwellers have broad, flat heads and big mouths and eyes.

The Antarctic brainbeard plunderfish (*Pogonophryne cerebropogon*), which was collected at about 1,000 feet (305 meters) below the sea surface.

Two teardrop-shaped fish were also discovered, but have not yet been named. They have jelly inside that makes them buoyant, similar to jellyfish.



## How can there be **fish** in the **cold waters off Antarctica**?

Compared with the warmer waters of the world ocean, there are few fish species off Antarctica, although there are many more than you would expect. Some scientists believe that up until about 35 million years ago, a wide range of fish occupied these areas, probably because the water was much warmer then. Over time the currents and landmasses changed so that today cold waters are brought to the area—but the fish remained and adapted to the changes in the local water.

Over the past 15 million years, many of these fish have flourished. They also have some specific, distinguishing traits when compared with other types of fish in warmer waters. For example, certain Antarctic fish, including the plunderfish, don't have swim bladders—so they don't rely on air for buoyancy; some have increased their body fat and decreased the density of their skeletons, which increases their ability to float. Many Antarctic fish are also thought to produce an "antifreeze" protein, which protects them from the cold waters. Another reason for their success is the lack of competition: Relatively few fish species venture into such frigid waters. This is referred to as adaptive radiation, in which a species evolves to fill niches (habitats) where other species do not venture—thus there is no competition. The lack of competition can also jump-start the evolutionary process, which may be happening to the fish in the chilly waters off Antarctica. In fact, some researchers estimate that the new species recently found off Antarctica have only been there for a few million years.

## OTHER MARINE CREATURES

### What **other marine creatures are there**—besides mammals, fish, birds, and reptiles?

The phyla (groups) of marine animals that have not been mentioned so far are:

Marine Worms

Porifera (sponges)—some 10,000 species

Coelenterata or Cnidaria (polyps, jellyfish, corals, sea anemones)

Ctentophora (comb jellies)

Ectoprocta (bryozoans)

Brachiopoda (lampshells)

Mollusca (this phylum is being revised, but it traditionally includes snails, conches, abalones, clams, mussels, scallops, oysters, conches, nautiluses, squid, and octopuses)—some 80,000-100,000 species

Arthropoda (horseshoe crabs, sea spiders, lobsters, crabs)—some 30,000 marine species

Echinodermata (sea lilies, starfish, brittle stars, sea urchins, sea cucumbers, sand dollars)

Tunicata (sea squirts, salps)

Arcania (lancelets)

### What are **marine worms**?

Marine worm is a loose term to describe invertebrates that are either independent or parasitic. There are about 9 phyla (groups) that contain marine worms:

**Acanthocephala:** These animals are spiny-headed, sausage-shaped, parasitic worms. When they are eaten by a fish, they attach themselves to the fish's gut, and can often impair the digestion of the host. If the fish is eaten by another animal, such as a bird or seal, these worms can become a parasite in the new host.

**Echiuroidea:** These are spoon worms found around the world, and range from 0.4 to 24 inches (1 to 60 centimeters) in length. They have a strange folded organ (internal), which is used for food gathering: the proboscis may be 30 to 40 inches (90 to 100 centimeters) long—although the worm's body averages only about 3 inches (8 centimeters).

**Platyhelminthes:** This phylum includes 25,000 species, mostly parasitic flatworms such as flukes and tapeworms. Tapeworms may have fish as hosts; they are also known to infest humans.

**Nemertea:** The nemerteans are called ribbon worms, with about 800 species worldwide. The ribbon worm can grow to as much as 100 feet (30 meters) in length, and is found in both the open ocean waters and on the seafloor.

**Aschelminthes:** These are round worms, although this group has been questioned: All the classes within the phylum may be actual phyla themselves. Most of these worms are small scavengers that feed on dead organisms.

**Phoronida:** These are the horseshoe worms, and are usually less than 0.4 inches (1 centimeter) in length. They live in a membrane tube on the bottom of shallow-water areas. They can bore into corals, rocks, or shells, using an acid secretion that breaks down and dissolves calcium carbonate.

**Sipuncula:** These are peanut worms, and include 325 species.

**Chaetognatha:** This phylum is the arrow worms, and include 150 species. They are usually found in shallow water, and, when small, have also been classed as plankton—making them an important link in the continental shelf food chain. The worms average about 0.4 to 1.2 inches (1 to 3 centimeters), but can grow to more than 4 inches (10 centimeters) long. They have several fish-like characteristics, including numerous chitinous (fibrous) teeth, external swim fins, and a distinct tail.

Marine life includes tiny worms, such as the Sabellid shown here, climbing out of a tube at the University of Hawaii Institute of Marine Biology. *NOAA*

Annelida: This phylum is the segmented worms, such as marine segmented worms, leeches, and earthworms. The list of annelidans is long, and the phylum is divided into two large classes, the Polychaete and Clitellata (or Oligochaeta and Hirudinea). The Polychaete are mostly marine organisms and are highly diversified. Many move around the ocean using parapodia (appendages resembling feet), and are filter feeders, swimmers, or burrowers.

## What are the **longest known worms**?

The ribbon worms are among the longest such creatures in the world. For example, a form of ribbon worm called the bootlace worm, living off the coast of Great Britain, measures an average of 15 feet (4.6 meters)—but some as long as 100 feet (30 meters) have been reported.

## What is a **sponge**?

Sponges are not the synthetic kind you find in the kitchen. Filter-feeding sea sponges are members of the Porifera phylum, with more than 10,000 species in existence—all aquatic and mostly ocean-dwellers. Sponges live from the intertidal zone to depths of 26,000 feet (8,500 meters) below sea level. These simple life forms have no organs, and are not mobile. They have outer and inner layers of cells, with the inner layers grabbing microscopic food particles from the passing water currents. The currents also carry oxygen to the animal and take away carbon dioxide wastes. Certain species have chemical defenses, producing offensive tastes and smells to deter predators. Still other sponge species are parasitic, burrowing into corals and shelled animals, killing the host in order to survive.

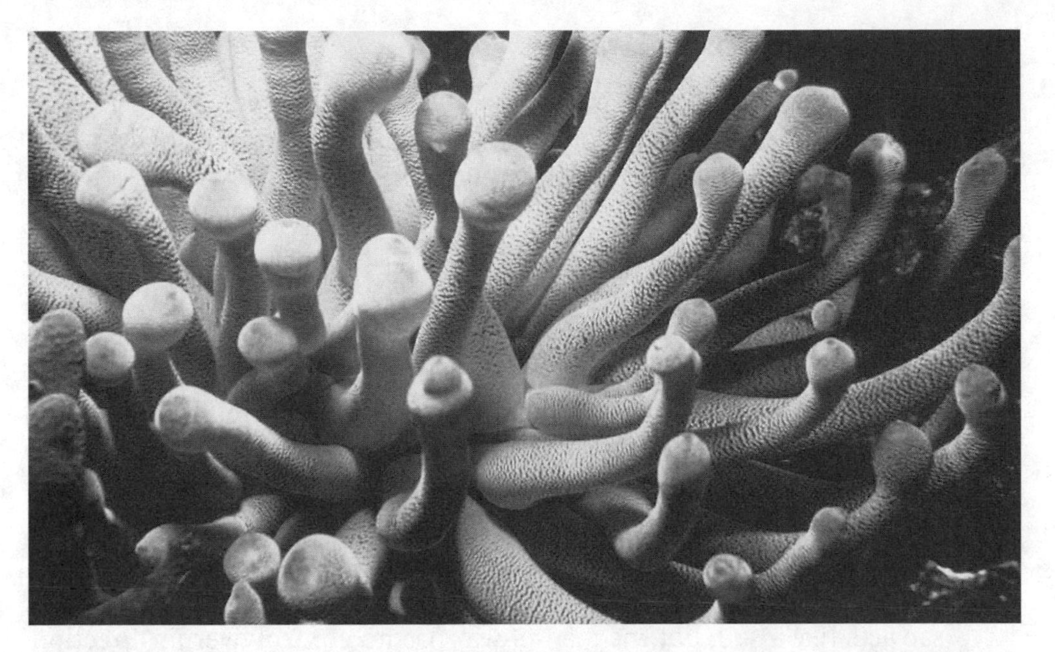

The sea anemone is a relative of the jellyfish, but unlike its cousin, it spends its entire life attached to rocks. This one was found on the ocean floor, off the Cozumel (Mexico) coast. *AP Photo/University of Wisconsin—Superior, Margie Mullet*

## Are all **sponges passive**?

No, while most sponges have cells that move about, grabbing and digesting bacteria or debris that float by in the ocean currents, some sponges don't just sit around waiting for microscopic food particles to float by. In the shallow, underwater caves of the Mediterranean Sea, scientists have discovered a new—carnivorous species—of sponge with tentacles, covered with tiny velcro-like hooks. These tentacles snag prey swimming by, usually shrimp-like crustaceans; more tentacles then surround the victim as the sponge begins to digest its catch.

## Are the **sea anemone** and **jellyfish related**?

Yes, the sea anemone and jellyfish are very closely related. A sea anemone resembles an upside-down jellyfish, but instead of swimming freely, like its cousin, it spends its entire life attached to rocks.

## How much of a **jellyfish is water**?

Even the firmest jellyfish is mostly water—about 94 percent. The jellyfish catches its food using long tentacles that hang from its body; many of these tentacles are armed with stinging cells.

## What is a **Portuguese man-of-war**?

One of the most well-known "jellyfish" is the Portuguese man-of-war (*Physalia physalia*), but it is not a true jellyfish, although it is in the same phylum. A man-of-war is actually a colony of similar animals attached to a gas-filled sac at the ocean surface, trailing a mass of tentacles that average 50 feet (15 meters), with the longest recorded being 160 feet (50 meters). The tentacles can give a powerful, painful sting, and have been known to cause a fatal shock to humans. The gas-filled sac acts like a sail, but the man-of-war has no control over its direction of travel—thus you cannot scare the organism away. The overall organism has specific places for reproduction, capturing prey, and digestion. Its predators include sea turtles and janthina snails.

## Does the **jellyfish** really have the **longest known reach of any animal**?

Yes, in particular, the Arctic jellyfish has the longest known reach of any animal. These creatures can grow to immense sizes. One picked up on a Massachusetts beach in 1865 measured 7.5 feet (2.3 meters), with 120-foot- (37-meter-) tentacles—giving it a total span, from tentacle to tentacle, of more than 240 feet (74 meters).

## How do **jellyfish swim**?

Most jellyfish are shaped like a mushroom, but the weak body wall does not really produce powerful contractions. The animal's rounded body (sometimes called a bell) forces out jets of water diagonally and downward—keeping the animal up and propelling it mostly in a vertical direction.

## Is a **comb jelly** a jellyfish?

Bryozoans often live in colonies, such as these that formed around a submerged pipe in Pokai Bay, Oahu, Hawaii; a goatfish is among the visitors. *NOAA*

No, comb jellies are not jellyfish, but they are often mistaken for the other animal. Both comb jellies and jellyfish are gelatinous (thus the word "jelly"), and have a large "head" and tentacles. But there are important differences: Most comb jellies cannot sting, whereas most jellyfish can. Comb jellies are divided into eight sections by bands of hair-like cilia that allow them to move forward and backward through the water; jellyfish move vertically by pulsations of their bell, but they can only move horizontally by the movement of waves and currents.

## What is a **bryozoan**?

Bryozoans are mainly ocean invertebrates that gather in colonies and are abundant along coasts, where they might be seen on pilings, shells, rocks, and algae. They are often mistaken for seaweed, moss—or water-logged spaghetti! Most of these organisms measure less than 1/32 of an inch (less than 4/5 of a centimeter) long—their main bodies attached to oval-shaped, box-shaped, or tubular calcareous shells with muscles. The organisms start out as free-swimming larvae that eventually land on a solid structure or rock. A bryozoan's tentacles move in and out of its shell, capturing minute food particles; the organism excretes wastes from an anus tube right next to the tentacles.

## What are **brachiopods**?

Brachiopods are similar to bivalve mollusks—they are ocean invertebrates with bivalve shells (in other words, they are soft-bodied but are enclosed within two shells). They are divided into those that have hinged shells and those that have non-hinged shells. The brachiopod has a pair of "arms" inside its shell; these are attached with tentacles, which the animal uses to grab microscopic food particles from the passing water. They were much more prolific in the past; today, they include the lampshells.

## What **animals** represent the **Mollusca phylum**?

Although this phylum is under revision (there are disagreements as to divisions within the group), most of the members have familiar names: snails, oysters, clams, mussels, squids, octopuses, and nautiluses. These animals are highly-developed, unsegmented invertebrates. With more than 80,000 different species, the mollusks are a diverse group.

## What are **bivalves**, **univalves**, and **cephalopods**?

These terms all describe members of the Mollusca phylum: The bivalves (which are protected by two similar shells, or valves) include clams, mussels, scallops, oysters, and cockles; the univalves (having one shell, or valve, and also called gastropods) include snails, conches, and abalones; and the cephalopods include nautiluses, squids, octopuses, and cuttlefish. Since most people probably equate mollusks with shellfish, this last group requires some explanation: The cephalopod has a group of muscular arms around the front of the head and these arms usually are equipped with series of suckers; these creatures also have highly developed eyes and they usually have bags of inky fluid, which can be squirted as a defense mechanism.

## Were **bivalves** really used to make **gloves**?

Yes, as strange as it sounds, these ocean creatures once supplied Italian glove-makers with a thread-like material. The pen shell bivalve, the largest in the Mediterranean Sea, attaches its shell to the seafloor using thick, strong thread. Fishermen would harvest the shell—and the strong thread would be turned over to the leather craftsmen to be woven into gloves.

## What are some **characteristics** of **clams**?

Clams are one of the more peculiar species of animals. Because they lack a recognizable head, no one can tell the difference between the front and the back of this bivalve mollusk. They are able to move freely, but they rarely change their locality. The clam secretes enough calcium carbonate to create the two plates—one a bit larger than the other—that surround its main body. Along the hinge line between the two shells, there are usually internal locking teeth. When the shell is closed, there is a slight gap, allowing the passage of its one foot.

## How can you tell the **age of a clam**?

Like rings telling the age of the tree, the clam shell has rough striations on its shells, each division representing yearly growth. Some of the rings differ in size depending on the surrounding environmental conditions or seasons—again, similar to trees.

## What is a **mussel**?

Mussels are bivalve mollusks—they have two elongated shells, live in intertidal and shallow subtidal zones, and are often found in large groups called mussel clumps. They attach themselves to rock using threads secreted as a liquid from a gland near the foot; the thread hardens on contact with water. They are extremely adaptable and have a worldwide distribution.

The principle predator of the mussel is the starfish; flounder and seabirds also consume mussels. Humans harvest the mussel as a food source, as the animals have a high concentration of vitamins, protein, and minerals.

## How does a **scallop move** across the ocean bottom?

Unlike most bivalves, scallops move across the ocean floor very quickly. They don't use a foot, but rapidly open and close their hinged shells, squirting out water that "hops" them across the seabed. They also can see where they are going as they clap their valves—they use the eyes around the edge of their shells.

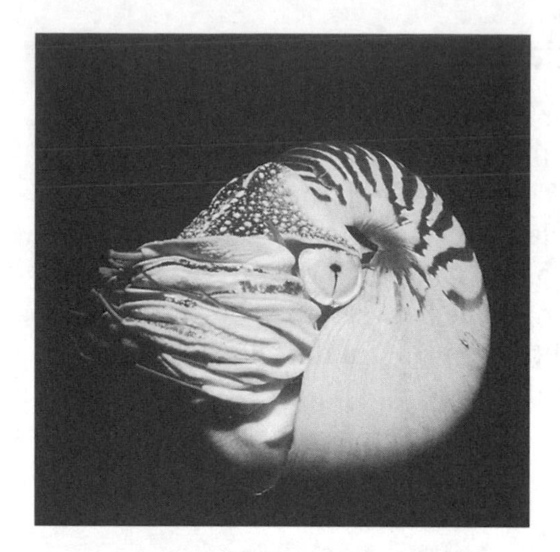

As the nautilus grows, it creates new chambers to produce the spiraling shell for which this primitive-looking mollusk is known. *CORBIS/Amos Nachoum*

## What are **gastropods**?

The gastropods (univalves or one-shelled) are the largest class of mollusks—about 15,000 fossil forms have been identified, and there are well over 35,000 species living today. Most gastropod shells are corkscrew-shaped; the spirals are called whorls. The tip of the shell, or the apex, is the smallest whorl, laid down in the animal's early life. As the gastropod grows, it lays down more whorls, with the final spiral called the body whorl, ending at the aperture or opening. The gastropods include the tulip snail, abalone, conch, and cowry.

## What are the "**butterflies of the sea**"?

Shallow water, shell-less gastropods (univalves) are also known as sea slugs or "the butterflies of the sea." Instead of being protected by shells, these animals have developed biological or chemical methods of defense against predators. But it is their beautiful and vividly colorful warning patterns that give them their nickname. These animals include the sea hare and the nudibranch.

## What is a **nautilus**?

The nautilus is a cephalopod (a member of the class Cephalopoda) of mollusks; because of its shell, the chambered nautilus is considered one of the most primitive known living cephalopods. The shell is sectioned into chambers by septa, all made of the same material. As the animal grows, it creates new, larger chambers—thus the increasing spiral of the shell. The nautilus has more than 90 tentacles topped with a hood. It

The squid (seen here in a close-up of the tail end) is the fastest-moving invertebrate: In a form of jet propulsion, it squeezes water out of its body to propel itself through the water. *NOAA/OAR National Undersea Research Program*

also has well-developed eyes, but they are less complex than those of other cephalopods.

There are only 6 known species of nautilus, most living in deep tropical waters—thus scientists know very little about the creatures. What is known is that by day, this animal remains on the ocean floor, either resting or holding onto the bottom with its tentacles. When it swims, it propels itself by forcing water out of its front cavity (similar to all cephalopods) moving the nautilus backward. At night, the nautilus rises into shallower water: By secreting gas into the shell chambers and removing the water, the nautilus floats upward. Once in shallower water, it feeds on reefs or in rocky areas, using its tentacles to capture small, slow moving fish or invertebrates.

## What are some **common characteristics of giant squids**?

The giant squid (*Architeuthis dux*) is a huge cephalopod, with 8 arms and 2 long tentacles that end in sucker-lined "clubs," which it uses to grasp prey. No one has ever seen a giant squid alive. The largest to wash up on shore measured 59.5 feet (18 meters) in length and weighed half a ton. They apparently range in size from about 20 to 43 feet (6 to 13 meters) long and weigh 100 to 660 pounds (45 to 300 kilograms). The giant squid is thought to have the largest eyeball of any animal: It measures almost the size of an adult human head! These creatures are hunted by sperm whales; the squid, in turn, eats mainly fish and other squids.

## Do **squid** really shoot **ink** to protect themselves?

Actually, both the squid and octopus produce a dark brown ink known as sepia. They squirt the liquid not only when they're scared—but also to confuse predators.

## How does a **squid swim** through the open ocean?

A squid—the fastest swimmer among the invertebrates—propels itself by forcefully squeezing seawater out of its body in one direction. This kind of "jet propulsion" allows the animal to quickly catch its primary prey—fish—or even propel itself out of the water, occasionally landing on the decks of ships! Squids have 10 extensions (8 arms and 2 tentacles) with suckers attached, that surround the head; they also have a streamlined body that helps minimize drag through the ocean waters. Even with all this speed and agility, it is thought that a squid must keep swimming or it will sink.

## How **fast** can large **squids** swim?

Larger squids, especially over shorter distances, can reach speeds up to 20 miles (32 kilometers) per hour. They are among the fastest of all marine organisms.

## Does an **octopus** like to **swim**?

No, considering the octopus is an ocean animal, it doesn't do much swimming. Only when threatened does the octopus swim about. This is because the cephalopod lacks the streamlining that makes other ocean-dwellers good swimmers. The octopus prefers to remain in contact with solid structures, pulling itself along with its sucker-lined arms. An octopus is also a solitary animal, seeking shelter in a cave, den, or under rocks—and because it does not have a shell, it can squeeze into some very minute holes and cracks. These animals usually only come out of hiding to find food or ward off predators.

Most octopus species have about 240 suckers on each of its eight arms—or nearly 2,000 suckers per octopus! *NOAA/OAR National Undersea Research Program; T. Schaff*



## How can the **octopus' arm** be described?

The arm of most octopus species have about 240 suckers—usually in double rows. The suckers vary in size from fractions of an inch to 2.8 inches (a few millimeters to 7 centimeters) in diameter. To show how well the suction works, it would take about 6 ounces (170 grams) of pressure to break the hold of a sucker almost 1 inch (2.5 centimeters) in diameter. Imagine what it would take to break the hold of the average octopus! (Assuming 240 suckers at 1 inch, or 2.5 centimeters, each, it would take about 90 pounds, or 41 kilograms, of pressure to break free of this animal's grip.)

## Do any **octopuses** have the ability to **produce light**?

Yes, scientists have recently discovered a bizarre deep-sea octopus with suckers that glow in the dark—the *Stauroteuthis syrtensis.* Researchers believe that in the past, the octopus used its suckers for gripping, much like octopuses in shallow water use suckers to grip prey and rocks as they scamper across the ocean floor.

But this "new" bright orange octopus with webbed arms is different. It lives in open water at depths of about 325 feet (900 meters). And because

Horseshoe crabs, like the pair mating here, are the descendants of crustaceans that evolved millions of years ago in the oceans; crustaceans are among the most highly successful groups of animals that have ever lived. *NOAA/Mary Hollinger, NODC Biologist*

it has nothing to cling to, and feeds on prey too small to grab with its arms, it has developed some other tricks. The small, rounded pads on the inside of the octopus' arms look like suckers, but they lack the muscles needed for a good grasp. Instead, the suckers have photocytes, or light-producing cells, that emit a blue-green light. And because most octopuses have good eyesight and use visual signals to communicate, this light may be a good way to attract prey, threaten enemies, or contact others of their species in the darker environment. Scientists call this an evolutionary transition—an evolutionary change that is currently occurring in a species, usually to fill a niche (specific habitat). The only other known example of such a change from muscle to a light organ was in a fish.

## Where are **crustaceans classified** in the animal kingdom?

These familiar creatures, including the crab and lobster, are members of the class Crustacea in the phylum (group) Arthropoda. In fact, since there are more than 26,000 species of crustaceans they are the largest class of arthropods. The crustaceans, most of which are marine, have chitinous (fibrous) and/or calcareous (shell-like) exoskeletons; in other words, they wear their somewhat soft skeletal structures on the outside of their bodies. These animals have segmented bodies, usually with a pair of legs (or appendages) per segment. They also have two sets of antennae. Other marine crustaceans include the sea spiders, krill, and barnacles. (Their land relatives include wood lice.)

## What is **chitin**?



Chitin is the key component (a tough polysaccharide) of the exoskeletons of arthropods and the shells (also really exoskeletons) of crabs, lobsters, and insects. It is also present in the hard structures of some coelenterates, hard corals, and sea anemones. Chitin is actually a natural polymer that is structurally related to the sugar called glucose; it has been made into water-resistant paper and edible food wrapping. Scientists have even developed a technique to manufacture string-like strands of chitin—for use as dressings for cuts and burns, as the chitin apparently has antifungal properties and promotes healing.

**Using a gland on its head, the barnacle will "glue" itself to any surface, spending its life there in an upside-down position. Here, gooseneck barnacles have attached themselves to a dead sponge.** ***NOAA/OAR National Undersea Research Program; J, Moore***



## What are **sea spiders**?

Sea spiders are somewhat similar to spiders on land, but these 4- to 6-legged arthropods—also called whip scorpions—live in the oceans. The smaller specimens measure about 0.01 inches (2 to 3 millimeters) and live in shallow waters; the larger ones, some measuring more than 20 inches (50 centimeters) in length, live in deep-ocean waters. They are carnivorous, some feeding on other invertebrates, sucking them dry, or tearing apart prey with their legs.

## What **marine animal** spends its life **standing on its head**?

The crustaceans called barnacles spend their entire adult lives on their heads: The barnacle larva is free swimming, but when it finds a suitable spot to live, it glues itself to the surface using a gland on its head. The barnacle then builds shell plates around its upside-down body in a volcano-like shape—thus as an adult, the animal becomes immobile.

Many barnacle species live in intertidal zone, where they are alternately covered and uncovered by the tides. When the tide is out, the barnacle plates are shut tight, holding in water; when the tide comes in, the plates open at the top, and a feathery "hand" (remnants of the arthropod's leg) grabs the passing food particles, then passes the food into its stomach.

## Where do **barnacles attach** themselves?

Not all barnacles attach themselves to rock, as many boat owners can verify. Some species of barnacles are parasites of other crustaceans or of corals; others are independent species that live in communities attached to whales, turtles, and other marine organisms. Others have a negative impact on commercial marine enterprises, covering the bottoms of ships, piers, and offshore installations.

## What are the **"edible" crustaceans**?

The edible crustaceans are names familiar to most of us, such as krill, crab, lobster, shrimp, crawfish, and crayfish. But these animals really only represent about 3 percent of the total worldwide marine catch (fish total about 90 percent of the catch).

Ancestors of today's crustaceans evolved millions of years ago in the oceans, and are considered some of the most highly successful groups of animals that have ever lived. Of these species, only a relatively few are used as food sources for humans, food or bait for other marine animals, or fertilizer for crops.

## What are **krill**?

Shrimp-like krill are the largest plankton in the rich upper layer of the ocean, and some of the smallest crustaceans. Most krill are found in the South Atlantic Ocean off Antarctica; in the Norwegian Sea; and in other cold ocean waters around the world.

The krill differs from the shrimp in one important way: It has bristles at the end of its tail. Krill are often found in huge groups, numbering almost 100 million individuals in one region. Krill have an important role at the bottom of the marine food chain: They are food for certain fish, seals, penguins, seabirds, herring, and sardines; giant blue and baleen whales sometimes feed exclusively on krill. In fact, it is estimated that baleen whales consume about 33 million tons annually; penguins, about 39 million tons; and fur seals, about 4 million tons.

Krill is also harvested by a number of countries. Japan has been known to produce krill meat and protein concentrate. Russia has also been involved in krill catches, making such products as krill butter and cheese.



## Can **krill populations** indicate a **climate change**?

Yes, krill populations may indicate climate changes—and other environmental changes as well. The tiny animals are extremely sensitive to fluctuations in temperature, salinity, and ultraviolet radiation. Scientists have been watching the changes in the krill populations, especially as any reductions correlate to recorded decreases in the Southern Hemisphere's ozone layer—the layer that protects the Earth's organisms from the ultraviolet rays of the Sun.

## Where are the **major crab fisheries** around the world?

Major crab fisheries are found along the Asian and North American coasts bordering the central and northern Pacific Ocean (where king and Dungeness crabs dwell) and along the Atlantic coast of North America. The Bay of Biscay, off the northern coast of Spain and the western coast of France; the North Atlantic, off the southern coast of Ireland; and the North Sea are also prime spots for crab fishing.

The oldest crab industry in the United States began with the blue crab (*Callinectes sapidus*), with records mentioning the crab in the Chesapeake Bay as early as the 1630s. Today, the blue crab is often found off Florida, Maryland, North Carolina, and Virginia.

## What are **lobsters** and where are they **found**?

Lobsters are crustaceans, and include the true, spiny, Spanish (or slipper) and deep-sea lobsters. They are part of the order Decapoda, which, in addition to lobsters, includes the other crustaceans that are most often eaten by humans—crabs and shrimp. All lobsters have

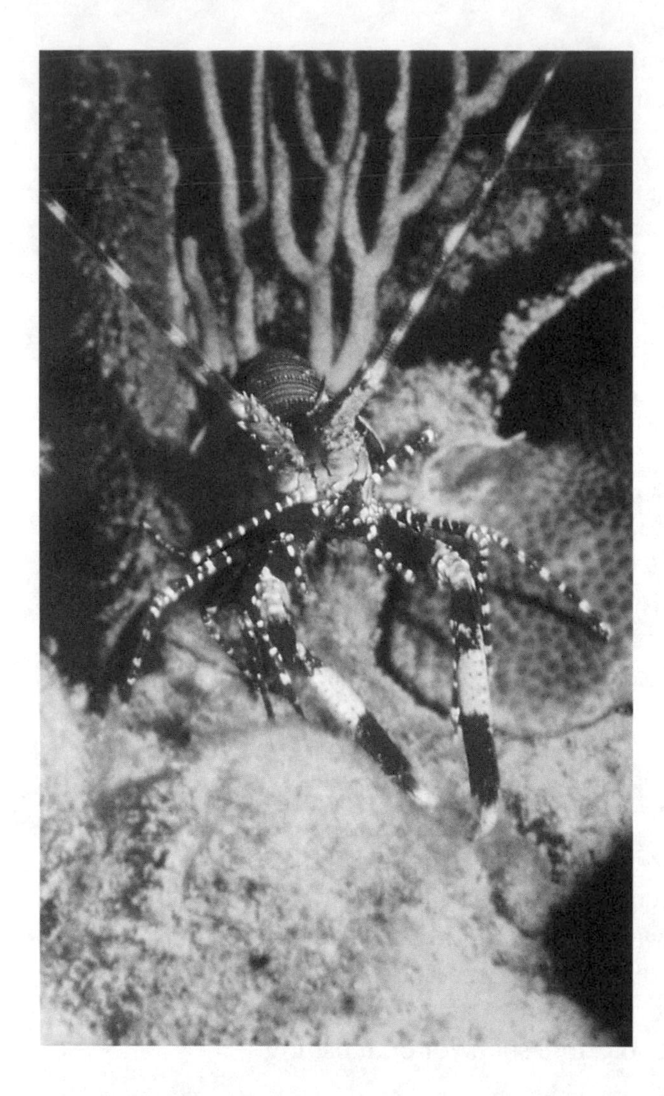

**Lobsters are most abundant in about 1,000 feet (300 meters) or more of water; this rock lobster was found on a Pacific reef.** ***NOAA/OAR National Undersea Research Program; E. Williams***

stalked eyes; chitinous (fibrous), segmented exoskeletons (on a three-part body); and five pairs of walking legs. One pair of walking legs is the chela, or claws—and in most cases, one claw is larger than the other.

Lobsters are very long-lived animals, maturing at about age 5 and living more than 50 years. They usually survive that long because they have few predators; if they can stay away from starfish or rays, their only predator is man. Lobsters are most abundant in about 1,000 feet (300 meters) or more of water; they live on carrion (dead animals), or even live fish when they can catch them.

## What is known about **lobster behavior**?

Lobsters seem to be their own animal: No one has been successful "growing" and harvesting lobsters commercially, which is why lobster cages are still found all along the coasts of the world. The behavior of this crustacean often seems bizarre, too. For example, North Atlantic lobsters have been seen traveling in columns along the ocean floor—but no one knows why. Some scientists have theorized that the marches are somehow associated with the animal's rise in abundance.

## Which **lobsters** are **harvested**?

There are three species of lobsters that are harvested the most: the American lobster (*Homarus americanus*), representing about 50 percent of the catch; the European lobster (*H. gammarus*), about 30 percent; and the Norwegian lobster (*H. norvegicus*) making up most of the rest.

## What is a **shrimp**?

Shrimp, of which there are more than 2,000 known species, are among the smaller crustaceans (hence the association of the word with anything that is small). They inhabit environmental niches (habitats) from freshwater rivers to the deep oceans, but most shrimp are ocean-dwellers. The body is somewhat flat, and has a tail fan attached to a rather long abdomen. Most shrimp spend their time walking around the bottom of the ocean on their 5 pairs of walking legs, or they use the limbs to dig up sediment to loosen food particles. They usually feed at night on other small crustaceans, worms, and mollusks.

Most shrimp live an average of 3 years. But they make up for a short lifetime with their prolific reproduction: In 3 years, a female shrimp can produce more than 20,000 offspring. Shrimp are also more resilient than are other crustaceans—they can tolerate drastic changes in the temperature and salinity of local waters.

## Why are **shrimp** called "**bandits of the coral reefs**"?

Not all shrimp are bandits; this nickname refers specifically to the 2-inch (6-centimeter) pistol shrimp, which dwells in and around coral reefs. It

## Do sea urchins throw their spines?

No, it's a myth that sea urchins throw their spines. They also do not leap onto helpless people passing by. And most sea urchins can be picked up and held, if it's done carefully. But there are exceptions. For example, the long-spine sea urchin of south Florida and the Caribbean has spines that can easily penetrate human skin, breaking off (like a splinter). And because the spines' barbs are slightly toxic, they can be quite painful.

Sea urchins are often found in tidal pools or just below the low-tide line—thus they are prey of seabirds, sea stars, lobsters, and terrestrial animals such as foxes. The globe-shaped body of a sea urchin is called a test; the mouth is on the bottom and the anus on top. The test is divided into 10 sections, 5 of which have holes through which tube feet protrude (in rows). A sea urchin eats using a structure in the mouth called Aristotle's lantern; at the center are 5 teeth that come together like a bird's beak. These teeth allow the animal to scrape algae off rocks; as the teeth wear down, they grow back (maintaining their size). The spines that cover the body are used for cleaning and defense, and some contain poison. Each spine is connected to the test with a ball-and-socket arrangement, allowing the spines to move. When a sea urchin dies, the spines drop off and the body inside the test decomposes, leaving only the shell.

sports a large right pincher claw that has a peg and matching hole. When a small fish wanders by, the tiny crustacean runs out of hiding, aims its "pistol," and snaps its large pincer shut. This creates a shock wave, stunning the fish—and allowing the shrimp to move in for the kill.

## Do some **fish** really **protect shrimp**?

Yes, a certain type of goby fish, the *Cryptocentrus coeruleopunctatus,* stands guard over a snapping shrimp at the entrance to their shared bur-

At low tide in central Oregon, a sampling of sea life gets stranded on the beach: kelp, a crab, a sea star, and a sand dollar. *CORBIS/Brandon D. Cole*

row on the ocean floor. As the tiny shrimp digs and cleans the burrow with its claws, the goby stands guard, one of its antennae touching the shrimp. If there is danger, the goby wiggles. The shrimp, feeling the movement in the antenna, runs for cover—with the goby not far behind.

## What is the difference between **crayfish** and **crawfish**?

Crayfish are crustaceans that usually live in freshwater; crawfish live almost exclusively in the oceans (or in brackish waters). Because crawfish resemble their crustacean cousin, the lobster, they are sometimes called "spiny lobsters" or "false lobsters"; but crawfish are not really lobsters, though many crawfish are just as large—and certain species are considered a delicacy.

## What is a **sand dollar**?

A sand dollar is actually a flat version of a sea urchin. It has very tiny moveable spines on the test (body) that give the sand dollar a smooth, felt-like appearance and touch. These spines allow the animal to dig into the sand. The flower-like shape on the top of the sand dollar corresponds to the sea urchin's 5 rows of tube feet. The tube feet that pop out of these petals are used for respiration (breathing).

## What is the **difference** between a **starfish** and a **sea star**?

There really is no difference: Starfish is an antiquated and inaccurate name that is now being replaced with sea star (since a starfish is not a fish at all). Sea stars, members of the phylum (group) Echinodermata, most often live in rock tidal pools, and measure 6 to 12 inches (15 to 30 centimeters) in diameter, although some have been found up to 26 inches (65 centimeters) in diameter. Most sea stars have 5 arms, although some have up to 10. The top of the sea star consists of horny (chitinous) plates; the undersides of the arms have tubed feet, or a series of small, flexible suction cups. It uses its arms to move and grab prey; it can also regenerate its arms. The mouth of this predator is found at the center of the star, on the bottom. Worms, crustaceans, and bivalves (especially oysters) are among the foods consumed by the sea star.

## TYPES OF MARINE COMMUNITIES

### What is a **marine community**?

A marine community is an area with certain marine organisms and/or physical features that distinguish it from other areas. Within the community, there is a unique collection of organisms that are mainly dependent on the food sources available in the area. Marine communities are located wherever there is a coast or open ocean.

### What are some **examples** of **marine communities**?

There are many examples of marine communities. They can be based on physical features, such as a rocky shore, sandy beach, or intertidal community. Or they can be based on and defined by organisms that dominate the community, such as a mangrove or a coral reef community. (Mangrove communities are discussed at length in this chapter; *see* pages 365–67. Coral reef communities are discussed in the chapter Beyond the Shore, beginning on page 168.)

### What are the **characteristics** of **intertidal communities**?

The intertidal (or littoral) zone, which is the part of the shoreline that is periodically covered by the highest tides and exposed by the lowest tides, is home to various plants and animals that can tolerate the alternating environment—in which they are exposed to air, then to water. During the high tides, organisms are exposed to stable water temperatures, although during storms they can be battered by waves and debris carried in the water. During low tides, organisms are exposed to the air, and the accompanying changes in temperature and light. In areas where there is snowmelt, intertidal organisms are also exposed to fresh water.

### Which **organisms** live in the **intertidal zone**?

It is difficult to categorize and generalize the varied marine life of the intertidal zones, mainly because of the different types of shorelines, such as rocky or sandy. But we do know there are complex combinations of flora and fauna: In the upper intertidal zone, the animals and plants resemble land organisms; in the lower intertidal zone, the flora and fauna resemble ocean organisms. The amount of land or ocean animals in each zone depends on the amount of time each area is exposed to the air or to the water.

### Are all **intertidal zones the same**?

No, not all intertidal zones are the same. Tidal ranges vary dramatically: For instance, in the Baltic Sea, there is virtually no tide; in the Bristol Channel, the spring tides vary more than 33 feet (10 meters). In addition, intertidal zones are located in various climatic regions—from the warm waters along the equator to the cold waters in the northern climates.

## SHORE COMMUNITIES (ROCKY AND SANDY)

### How do the **tides** influence the **organisms living along the shore**?

Many organisms—both animals and plants—that live along the shore are controlled by the rise and fall of the tides: For about 12 hours (part

Sea stars and other animals living in the intertidal zone are exposed at low tide at Burnaby Narrows, British Columbia, Canada. *CORBIS/Raymond Gehman*

of the usual tidal cycle) the surroundings are marine; the rest of the time the area is exposed, producing an often dramatic change in temperature, exposure to the rays of the Sun, and the drying effects of the wind and Sun.

## Do **currents** affect **organisms living along the shore**?

Yes, some organisms living along the shore can be affected by currents—but mostly by the surface currents, which are driven by winds. These currents move nutrients and organisms (such as plankton) around the shore's waters, providing food for coastal organisms. But these same currents can also wash nutrients out to sea, making the conditions in the shore community highly changeable.

## What is the biggest **difference** between **rocky and sandy shore organisms**?

There is one big difference between rocky and sandy shore organisms—whether plants or animals: Along the rocky coasts, organisms can attach themselves to the rock; along a sandy shoreline, there is no way for the organisms to attach themselves to the sand.

## What is a **rocky shore community**?

In general, a rocky shore community, like all shorelines, has distinct zones where organisms live—each one changing constantly. The major zones are:

**Upper beach/splash zone:** On the rocky shore, this is the area farthest landward that marine organisms can dwell; here water from crashing waves wets the rocks. The top zone is the rocky shore; below is the splash zone.

**Intertidal zone:** The intertidal zone (which occurs along every shore, rocky or sandy) is the most challenging for marine life, since it changes so dramatically: When the tide comes in, or when storms

King penguins making themselves at home along a rocky shore in St. Andrews Bay, South Georgia Island (in the South Atlantic Ocean). *CORBIS/Wolfgang Kaehler*

occur, the organisms in this range are hit by crushing waves; when the tide goes out, the organisms are exposed to the drying sun and air.

**Subtidal zone:** The subtidal (or below low-tide) zone is the area below the low-tide line, in which organisms are exposed to the air but only for very short periods.

## What is the **black zone** along a **rocky shore**?

Scientists subdivide the three main zones (splash, intertidal, and subtidal zones) found on a rocky shore into highly specific zones; a black zone is one of these—and the farthest landward of the marine zones. It is actually considered the transition zone between land and the marine environment. It can be thought of as the top part of the splash zone (the area where the rocks are splashed with water), or the highest points on shore rocks. Of the highly specific zones on a rocky shore, the black zone is the one that has the least exposure to water: it is only wet during the high spring tides.

## What **organisms** are found in the **black zone**?

The rocky shore's black zone contains a crust of cyanobacteria (formerly called blue-green algae). Other plant organisms, such as other types of algae and periwinkles, are in the black zone. These organisms have mechanisms that resist drying: For example, the algae are enclosed in gelatinous covers, and periwinkles are tightly sealed in conical shells, allowing them to retain the water they need to survive.

## What is the **white zone** along a **rocky shore**?

Another specific zone along a rocky shore is the white zone, the area just below the black zone. It is the area in which the rocks are exposed during low tide, and covered with water during high tide. It can be thought of as the bottom of the splash zone (the area where the rocks are splashed with water).

## What are some **organisms** found in the **white zone**?

Most of the organisms in the white zone exist by clinging to the rocks; these include barnacles, mollusks (such as limpets), and dogwhelks (a type of snail). During low tide, the creatures are tightly sealed within shells to resist drying out. For example, barnacles close four movable plates at low tide and pull open the plates during high tide, when they are exposed to water.

## What **other organisms** are found along a **rocky shore**?

There are more organisms (members of the intertidal community) below the white zone. For example, there are brown algae, with some species growing more than 8 feet (2 meters) long, and bladderwracks. Animals include mussels that attach themselves to rocks by using threadlike filaments secreted by a gland in the foot. In the lowest zone (which is uncovered only during the spring tides), there are starfish, sea cucumbers, limpets, mussels, and crabs.

Tidal pools, such as these along a rocky stretch of Mediterranean shoreline near Byblos, Lebanon, can be home to a variety of marine life. *CORBIS/Roger Wood*

## What is a **tidal pool**?

Tidal, or tide, pools are small collections of water below the tide line usually along rocky shorelines—and often called miniature seas. They are usually depressions in the rock or an area surrounded by high rocks that hold seawater. These small pools are colonized by many organisms (both plants and animals) that either live in the pool permanently or seek out such pools when the tide is out. Although the pools are found in nooks from the high- to low-tide lines, the ones that are the most occupied are found in depressions closer to the low-tide line.

## What **animals** live in **tidal pools**?

A multitude of animals can live in tidal pools—it mainly depends on where the tidal pool is located. They hold various species of limpets, mussels, crabs, sea slugs, sea urchins, starfish, hydroids, sea spiders, sea scorpions, anemones, and sponges. Fish, such as blennies, sticklebacks, and gobies, along with other sucker fish that can hold fast to rocks as the tide flows out, are also present—and many pools also contain thriving shrimp and prawn populations. These pools also attract other animals, including birds (such as gulls) and mammals (such as racoons) that live nearby; they come to the tidal pool in search of prey.

## What **plants** live in **tidal pools**?

Many different kinds of plants can live in tidal pools—and again, it depends on where the pool is located. The most prolific tidal pool plants are seaweeds, especially the red and green algae. These pools can also hold sea lettuce and sea oaks.

## What is a **sandy shore community**?

Sandy shore (or beach) communities do not have any large rocks or tidal pools; they also lack the algae found in a rocky shore community. Most of this area is covered with sand, and offers little protection in the form of tall plants. The action of waves and currents form and change the beach constantly—nothing stays in one place for very long. Animals and plants in the sandy shore communities either live above the high-tide line or below the high-tide or low-tide lines. The upper sandy beach is the transitional area from land to sea; true marine life appears in the intertidal zone, where the tides rise and fall.

## What **animals** live in the **sandy shore community**?

If you have walked a beach, you can probably identify a number of the many animals that dwell in sandy shore communities. Like the rocky shore community, this habitat is divided into various zones, each of which attracts its own unique animal life.

**Upper beach/splash zone:** Animals that live in the upper sandy beach (including the backshore and the backbeach) are usually more terrestrial than marine; these include ghost crabs and beach fleas. The lower part of this area is the splash (or spray) zone, which is affected by the spray from breaking waves.

**A sea anemone and other plants grow in a tidal pool on Asilomar Beach, Monterey, California.** ***CORBIS/Mike Zens***

**Intertidal zone:** Just below the upper sandy beach, in which there is more water available, the division of animals is based on how they obtain their food and oxygen. For example, fish need water; but other creatures, such as a sandhopper, can breathe air and live in the debris that washes up on the strand line (the detritus and driftwood that forms a dark line along the high-tide mark on a sandy beach). And the sand itself holds food: Among the sand grains live small crustaceans and worms that feed on microscopic algae, bacteria, and organic matter. Birds that live around the intertidal zones of sandy beaches include gulls, sandpipers, curlews, plovers, oystercatchers, and turnstones.

**Subtidal (below low-tide) zone:** Animals living in this area, which is below the low-tide line, include cockles, bivalves, lugworms, clams, coquina, snails, crab, sand dollars, starfish, and ostracods. At high tide, the lugworms, which burrow through the sand and feed on organic matter, are most active; one species of coquina clam travels up the beach at high tide, and retreats as the tide goes out.

## What **plants** live in a **sandy shore community**?

Sandy shore communities have few plants; this is because beaches have few "anchors" (natural, physical features) to hold plants to the sand. But there are some plants that live in this community. Above the high-tide line and around sheltered areas, organic matter often accumulates. Here, the sand mixes with mud, providing an anchor for beds of cord and eel grasses, beach peas, dusty millers, salt spray roses, and sea lettuce. Below the high-tide line, the variety of plants are few—mostly the usual photosynthetic plankton that live in the upper waters.

# COASTAL WETLAND COMMUNITIES

## What is the **nature** of a **coastal wetland**?

Coastal wetlands are the low-lying areas near the shore, including the estuaries, tidal (or coastal) marshes, and mud flats. The divisions of a

**Florida's Crystal River estuary is home to a variety of life—both sea- and land-based.** ***CORBIS/David Muench***

coastal wetland are complex—and many of the plant and animal organisms living in the various wetlands overlap with each other, making it difficult to say that specific animals and plants exist in any one region.

An estuary is the area in which the ocean tides meet a river current, with the river's freshwater diluting the saltwater; salt marshes are large, flat areas of land protected from the wave action of the tides, but still inundated with brackish to salty tidal waters; mud flats (or tidal mud flats) are relatively flat areas covered with very fine-grained silt (in a sheltered estuary), and alternately covered and uncovered by the tides; and the tidal marshes are found on the landward side of the salt marshes and mud flats.

## Are all **estuaries the same**?

No, not all estuaries are the same. Each one is dependent on the nature of the rivers and tides that feed it, and on the climate. For example, along the West Coast of North America, estuaries are narrow because the rivers travel down the steep slope of the continent; on the East

Coast, the estuaries tend to be wider and extend farther inland because of the more gradual slope of the land toward the ocean.

## Why are **estuaries** so **fertile**?

As estuary plants die and decay, they are carried seaward by the river's current and the ocean's retreating tides. Although some animals from the area feed on the nutrients, much of it is swept away, to enrich the coastal waters. However, the meeting of the river's freshwater and ocean's salt water creates a kind of nutrient trap—causing the waters of the estuary to be about 30 percent more fertile than the water of the open oceans.

## What **organisms** live in **estuaries**?

Estuaries are home to great numbers of plants and animals. But the diversity of life—even when the organisms living in nearby mud flats and salt marshes are taken into account—is not as great as that found in a forest or lake community. Estuarine plants include photosynthetic plankton, and various grasses and seaweeds; animals include plankton (dinoflagellates and diatoms), snails, birds (such as herons), jellyfish, and millions of baby fish.

## Why are **estuaries** called **natural filters**?

The vegetation growing in estuaries works as a natural filter: Grasses, seaweeds, and other plant life slow fast-flowing waters and remove certain pollutants from them as the tide rises and falls. But not all estuaries can filter water—especially if there is an excess of pollutants, or if silt covers the vegetation, smothering the plants' ability to filter the water.

## Why are **salt marshes** some of the most **dynamic and rigorous environments** on Earth?

Salt marshes are thought to be some of the most dynamic and rigorous environments because of the conditions set up by the tides. As the tides

**Egrets are among the animals that make their homes in salt marshes such as those in the Darling National Wildlife Refuge, Sanibel Island, Florida.** ***CORBIS/Raymond Gehman***

move in and out of the marsh, the animals and plants have to shift from being terrestrial (land) to oceanic organisms in a few short hours. During this time, the water levels, salinity, temperatures, and exposure to the air vary greatly—making it a challenge for the organisms that live here. To add to the pressure, there are also periodic tropical storms and spring and summer floods that must be endured.

Even though they are harsh environments, salt marshes are also one of the most productive ecosystems on Earth. The marshes are filled with rich organic nutrients: As the grasses rot, bacteria break them down; as the tides move in and out, they mix and spread the nutrients all over the marsh.

## What types of **animals** are common in **salt marshes**?

There are many types of animals commonly found in salt marshes. Larger animals, usually visiting from inland, include deer, raccoons, and muskrats. Birds, especially ducks, herons, egrets, and hawks, often frequent the salt marshes. In the shallow waters of the marsh are the scallops, sea urchins, mussels, clams, shrimp, crabs, worms, and smaller jellyfish. And one of the few salt marsh reptiles, a diamondback terrapin, eats plants and cracks open crustaceans and mollusks with its horny jaws.

## What types of **plants** are common in **salt marshes**?

The most common plants in salt marshes include salt marsh grass (and other grasses that grow in the brackish water), bulrush, and glassworts. In the shallow water are the usual photosynthetic algae.

## What **organisms** are found on **mud flats**?

Mud flats contain moving fresh and salt water, making it difficult for mobile organisms to live—but other organisms anchor themselves in the mud and thrive there. For example, cord, widgeon, and eel grasses root in the shallow waters. In calmer areas, sea lettuce grows, one of the favorites of ducks and geese. Small animals (many of which are burrowers) include clams, snails, mollusks, crustaceans, and the fiddler crab; these organisms are part of the diet of many larger creatures, such as wading birds, large gulls, and herons—all of which can be found in mud flats. Mud flats are also visited by mammals, including weasels, otters, and raccoons, searching for crabs.

## What **animals** are found in a **tidal marsh**?

Because the tidal marshes are blanketed with reeds and rushes (and farther inland, with trees and shrubs), these areas are filled with birds, including mallards, pintails, widgeons, herons, bitterns, snipes, and woodcocks. Mammals also frequent the area—looking for smaller animals hiding in the rushes (such as martens, otters, water voles, various mice, and in some places, wild cats) and in the mud (crabs, snails, and worms).

## How does a **tidal marsh** change during the **seasons**?

Many tidal marshes experience seasonal changes that are similar to those that can be observed on the prairie: In the spring, it is green; in the summer, it contains purple (sea lavender), yellow (goldenrod), and pink (rose mallow); in autumn, the colors range from white and purple (asters) to dark green (rushes); and in the winter, it all turns brown.



# MANGROVE COMMUNITIES

## What is a **mangrove community**?

A mangrove community (also called a mangrove swamp) is a collection of partially submerged salt-tolerant trees and shrubs; around the world, there are about 50 species of mangrove trees. In the United States, they are the basic vegetation found along Florida's southern and western shores, and also the Florida Keys. Mangroves are also found in the Philippines, Ecuador, India, Bangladesh, and Indonesia.

The roots of these trees are the key to the mangrove community: Not only do they funnel oxygen into the mud below, but the massive root system also holds onto the mud in order to build up surrounding land—creating an environment that teems with life.

## Which **kinds of mangroves** grow in **Florida**?

There are three main types of trees found in the Florida mangroves: red, black and white. Although they can be found together, many of the mangrove regions contain only one type of mangrove tree.

**Red mangroves:** The leaves of the red mangrove are leathery; this adaptation allows them to reduce the amount of freshwater the trees lose due to evaporation. The seeds of these plants are cigar-shaped seedlings called propagules; the seeds drop, sometimes rooting in the mud below the tree. Other seeds may float away to colonize another area downstream, and still others are carried elsewhere by animals or on floating debris. The roots of the red mangrove descend from their branches—in what is called aerial roots.

**Black mangroves:** Black mangroves have breathing roots called pneumatophores. These roots stick straight up in the mud like small

Mangrove islands off the coast of Belize form a forest in the Caribbean. *CORBIS/Kevin Schafer*

chimneys, giving the underground roots the aeration they need to survive. The leaves of the black mangroves excrete the excess amount of salt from the saltwater; this is why salt crystals are often seen on the leaves.

**White mangroves:** White mangroves usually live with the black mangroves or inland. Two small holes, or pores, in the leaves get rid of the excess salt as the plant takes in seawater to survive.

## Why is so much **life** found in the **mangroves**?

The mangroves contain a rich and varied food web. Their long roots provide a perfect hiding place for creatures that swim or crawl; the mud trapped by the many roots also make a sheltered nursery for young fish and prawns; the tops of the trees are ideal for nesting birds; and the decaying mangrove leaves provide nutrients to the water (with the temperate mangroves containing more nutrients than those in tropical and subtropical areas), attracting fish who feed here.

## What types of **animals** are commonly found in the **mangroves**?

A crab visits a mangrove swamp in Palawan Island, Philippines. Mangrove roots, which provide shelter for small fish and larvae, can be home to a rich ecosystem, attracting fish from the open sea, which come here to feed. *CORBIS/Arne Hodalic*

From the top of the trees down to the elaborate root system, and including the surrounding waters, there are many animals that live in the mangroves. Some representative species, and their homes, include the following:

**Tree tops:** Birds, such as the brown pelican, white ibis, and great white heron live in and next to the tree tops. Around the trees themselves are larger land animals, including certain types of deer; for example, Key deer are frequent visitors to the Florida Key mangroves.

**Upper water levels:** In the upper water levels (including at high tide) the nurse shark, yellow stingray, and mangrove snapper (a fish) may be found swimming.

**Lower water levels:** The lower water levels teem with animal life. One reason is that the roots provide shelter and a good place to hide from predators. Another is that decaying mangrove leaves provide nutrients, making them excellent for feeding. The animals in this area include toadfish, trunicates, barnacles, sea urchins, worms, anemones, shrimp, sponges, and crabs.

## What types of **plants** are commonly found in the **mangroves**?

Besides the dominant mangrove trees themselves, there are very few plants around these communities. The most profuse are the small, photosynthetic algae in the water; larger algae, such as certain seaweeds, may also be found.

## What are the **zones** of the **open ocean**?

There are five zones in the open ocean, but they do not equate with communities, as organisms overlap between the zones. (The last three zones—the bathyal, abyssal, and hadal—can be grouped together for a total of three major zones.)

**epipelagic (or euphotic):** The epipelagic zone of the upper ocean includes the upper sunlit layer, which receives enough sunlight for the process of photosynthesis to take place in many plant organisms; it begins at the surface and extends to 656 feet (200 meters).

**mesopelagic (or disphotic):** The mesopelagic zone is where there is minimal light and it fades into darkness. It ranges from about 656 to 3,281 feet (200 to 1,000 meters); some scientists call the area between 650 to 1,650 (200 to 500 meters) the deep-water zone.

**bathyal or bathypelagic (part of the aphotic):** The bathyal zone is also called the twilight zone of the open oceans—or the beginning of "the deep"—in which there is little light. It ranges from 3,281 to 13,124 feet (1,000 to 4,000 meters).

**abyssal or abyssalpelagic (part of the aphotic):** The abyssal zone is the area of total darkness; it ranges from 13,124 to 19,686 feet (4,000 to 6,000 meters).

**hadal or hadalpelagic (part of the aphotic):** This zone occurs only in the deepest parts of the oceans, beginning at 19,686 feet (6,000 meters) and extending to the lowest point of the ocean, the Mariana Trench (which is 36,198 feet, or 11,033 meters, deep). There is no light in this region.

## Do **currents** affect **organisms living in the open oceans**?

Yes, many organisms in the open ocean are affected by currents—even more so than are those living along the shores or in the intertidal zone. Each current has its own density, temperature, and salinity (saltiness). For example, the Gulf Stream runs along the East Coast of the United States, from Florida to the Mid-Atlantic states, and is one of the strongest known currents. But because it carries warm waters as far north as Cape Cod, many marine organisms—from plankton to whales—live within or around the current's waters. Currents also bring up sediment and nutrients to the surface, creating rich fisheries in many spots in the ocean.



## What are some **organisms** that live in the **epipelagic** zone of the open ocean?

The epipelagic zone (which begins at the surface and extends to 656 feet, or 200 meters) is an important part of the marine food web. The vast bulk of the organisms living here are the microscopic phytoplankton and zooplankton; small fish, jellyfish, and some larger animals (for example, baleen whales) feed on the plankton; in turn, these small fish are eaten by large fish (such as porpoises), as well as by seals, birds, and other animals. When organisms living in this open ocean zone die, their remains fall to the seafloor.

## Which **organisms** live in the **mesopelagic** and **bathypelagic**?

The mesopelagic and bathyal zones of the open ocean have no real plants, but are home to many marine animals. The areas (from ranges from about 656 to 3,281 feet [200 to 1,000 meters] and from 3,281 to 13,124 feet [1,000 to 4,000 meters], respectively) are home to shrimp, squid, octopuses, and a few specialized fish. Some animals also live in either zone during the day, then migrate vertically at night to feed on prey in the euphotic zone above.

## Which **organisms** live in the **abyssal zone**?

There are no photosynthetic plants known to live in the abyssal zone of the open oceans—between the depths of 13,124 and 19,686 feet (4,000 and 6,000 meters). And no one really knows the true extent of animals living in this area of total darkness. Besides the animals that live around volcanic vents (see farther on, page 370), most of the animals known to live deep in the oceans, such as the hatchet fish, lanternfish, and snipe eel, were found by fishermen who dragged them up from the depths.

### Which **organisms** live in the **hadalpelagic zone**?

Like the abyssal zone, the hadalpelagic zone (deeper than 19,686 feet, or 6,000 meters) has few animals—at least that we know about. The actual number and types of animals living here is unknown since the depths of the hadalpelagic—the deepest points on the surface of the Earth—remain to be explored by humans.

## COMMUNITIES AT THE OCEAN VENTS

### Why is a **hydrothermal vent** likened to a **desert oasis**?

The features on the deep-ocean floor known as hydrothermal vents are likened to oases since life thrives around them—in the otherwise lifeless "deserts" of the deep oceans. These vents are actually volcanic openings where super-hot water escapes from the depths of the Earth. (These plumes of water are referred to as black smokers.) The ecosystem around the hydrothermal vents is the only one on Earth that does not rely on the rays of the Sun as its foundation.

### What **animals** live near **volcanic ocean vents**?

Colonies of life forms, including bacteria, crabs, clams, fish, and 10-foot-(3-meter-) long-worms have been found in the deep oceans around volcanic ocean vents. Of the approximately 300 species found around the vents, about 97 percent are new species—that exist no where else in the ocean.

One such volcanic hot spot was discovered in 1977 by explorers aboard the scientific submersible *Alvin,* about 500 miles (805 kilometers) west of Ecuador—and about 8,000 feet (2,438 meters) down.

### What is the nature of the life found around hydrothermal vents?

There is a wide variety of strange life forms that live around hydrothermal vents. For example, there are colonies of worms that live in intertwined white tubes; these worms can be up to 10 feet (3 meters) long and 4 inches (10 centimeters) in diameter. There are also 10-inch- (25-centimeter-) long clams, bizarre limpets, crabs, sea anemones, sea spiders, squat lobsters, mites, shrimp, and even some types of fish.

## How have **organisms adapted** to **hydrothermal vent** environments?

The biggest adaptation that organisms have made to volcanic vent environments is how they produce energy. Plants near the surface of the ocean use photosynthesis to convert sunlight into energy (food); these plants are the basis of the food chain in the upper ocean. But at the deep-ocean vents, the organisms at the bottom of the food chain don't have any sunlight. Instead they live off the superheated seawater (erupting from the vents), which is loaded with chemical and mineral particles—including hydrogen sulfide and metallic sulfides. Microorganisms thrive off these nutrients—and provide food for other hydrothermal vent–dwellers.

But how do they live under such conditions? Scientists discovered part of the answer to this question by studying large tube worms at the vent. These creatures, with no mouth or gut, had special bacteria living inside their tissues. The special bacteria harvest energy from the sulfides released by the hot vents using a process called chemolithoautotrophy (a self-feeding process based on mineral chemicals); in this way the tube worm converts nutrients in the water into food Some mats and giant clams around the hot vents also have these bacteria; other animals around the vents eat the bacteria directly. Still others (larger organisms) eat animals that live off the bacteria.

## How do **hydrothermal vent–dwelling animals travel** between vents?

Organisms that live off hydrothermal vents have a bizarre way of traveling from vent to vent: They are carried. Several hundred feet (a few hundred meters) above the ocean floor, giant whirlpool-like plumes, fed by the fluids erupting from the vents below, can break free—and "fly" through the underwater landscape. These masses of released water carry heat, chemicals, and even animals between the vents.

The "traveling plumes" work in this way: When hot seawater from the vent hits colder, denser water above, the hot water increases in buoyancy (since heat rises); as the plume of hot water rises, it eventually reaches water of the same density, and it then spreads out sideways. Researchers assumed that the plumes would just keep spreading out, but this assumption was not valid since it did not take into account the fact that ours is a rotating planet. Once the known data about the plumes were input into a computer model, scientists discovered that the plumes rotate, too, becoming coherent bodies distinct from the water around them (in other words, bodies of water that have their own properties).

In conducting further studies, researchers discovered that some of these plumes can form distinct rotating vortices (eddies or whirlpools)—and some can break off from the plume. From this information, scientists concluded that if the eddies occur at hot vents in the oceans, they can also "roam" the ocean floor—transporting dissolved chemicals, mineral particles, and living creatures from vent to vent.

## How is **life at the hydrothermal vents important** to the discovery of **life in other parts of our solar system**?

Scientists are looking for life on other planets—by looking for life on Earth! In a giant kelp forest at the Monterey Bay Aquarium, researchers are testing a new scientific probe (built by the National Aeronautics and Space Administration) that may one day look for life on Jupiter's icy moon Europa. Here the instruments of the probe can be tried out in a controlled, watery environment rich with organic material.

NASA's Jet Propulsion Laboratory's Loihi Underwater Volcanic Vent Mission Probe will eventually be sent to an underwater volcanic vent near Hawaii, where it will be placed inside the hydrothermal vent by a robotic arm controlled from a nearby submersible.

It's hoped that this exercise will lead to future missions to explore harsh environments—including Antarctica's Lake Vostok, which is under 2.5 miles (4 kilometers) of ice, and eventually Jupiter's icy moon. The main questions scientists want to answer are: Do volcanic vents exist on Europa? If so, can and do simple biological species exist within the hot water vents? And if they do, are there any temperature limits or chemical conditions they need to grow elsewhere in our solar system?

## COMMUNITIES OF THE DEEP OCEAN

### Why don't **scientists know** much about the **deep ocean**?

The main reason scientists know little about the deep ocean is its great depth. Because water pressure increases, light decreases, and temperatures drop with depth, these waters cannot be explored without specially built equipment. It is estimated that less than 5 percent of the oceans have been explored—and (by volume) the dark, cold deep oceans represent 80 percent of the available living space on Earth!

### How do conditions in the **deep ocean differ** from those of the upper levels?

There are many different conditions in the deep waters compared with the upper levels of the ocean: There is little life in the deep ocean; there is little food; the pressures are as much as a thousand atmospheres; there is no light; the oxygen concentrations can be less than a tenth than the oxygen concentration at the surface; and it is extremely cold, around 36° F (2° C).

### How many organisms are living on the deep-ocean floor?

Since people have only recently begun to explore this realm, no one really knows the number of organisms living in the deepest oceans. But it's estimated that there are one million or more species yet to be found.

## Why is it **difficult** for scientists to **collect creatures** from the **deep ocean**?

The main reason is that the conditions at the sea surface are very different than those the animals are used to (they're acclimated to the high-pressure, low-temperature, low-oxygen environment of the deep). It is therefore quite common for the creatures that are collected from the depths to decay quickly once they have been removed from their natural environment.

## Why are there **no plants** in the **deep waters** of the ocean?

There are no plants living in the deep waters of the ocean because plants live by photosynthesis, the process of converting sunlight to energy, and sunlight can only reach several hundred feet (a few hundred meters) into the oceans. In short, the deep-ocean waters are simply too dark for plant life. However, plant particles may be found in the deep ocean: When plants living at upper levels die, their decayed remains fall through the water column. (An exception to the rule about plants not living in the deep ocean is, of course, the plant life found at the hydrothermal vents, where plants rely not on photosynthesis to survive but rather on the nutrients that gush out of these deep-sea openings in the ocean floor.)

## What are some of the **animals** found in the **deep waters** of the ocean?

One of the stranger animals found in the deep waters of the ocean is the *Lipogenys gilli*: Using its toothless mouth, this fish sucks in great quantities of deep-ocean ooze and, using its very long intestines, it extracts what little organic nourishment there is in the depths. The deep ocean is also visited by residents of the mesopelagic zone (above): Many of these creatures are vertical migrants, such as the lanternfish, shrimps, krill, squids, and arrow worms, which move from one level of the ocean to another in search of food.

## What is the **greatest challenge** to **animals** in the deep ocean?

While it might seem obvious to assume that the high pressure exerted by the deep-ocean waters would pose the greatest challenge to animals living in the depths, this is not the case: These animals have adapted to the high-pressure environment—and not by making any special structural changes. Because most fish have balanced internal and external pressures, a gradual readjustment was all they needed: The internal system of most fish is equal to the external pressure of the water, and therefore, when these fish migrated to deeper waters, they only needed to make an incremental change to their internal pressure in order to adapt to their new environment.

The real challenge at the depths is the scarcity of food. Here, successful predators take advantage of whatever prey wanders by. Some rely on food that drifts slowly down from the more hospitable ocean waters above, including carcasses, marine snow (small organic particles such as feces, molts, etc.), and plant detritus (debris or decayed plant life). In this competitive world, small fish even take on larger fish in order to survive; for example, the head of the otherwise small gulper eel is massive, allowing the animal to open its gaping jaws to grab larger fish. The black dragonfish has long spiked teeth that it uses to trap prey in its mouth. And a creature called the great swallower can accommodate large prey, since its stomach and body walls are elastic enough to expand.

## How have **deep-ocean animals** adapted their **coloration** to suit their environment?

Deep-ocean fish tend to be dark (black or red)—allowing them to blend in with their murky environment, but they can also be white; invertebrates are often red or colorless. In addition, many deep-ocean dwellers have developed bioluminescence (luminescence produced through a physiological process).

## How have **deep-ocean animals** adapted in **size** to suit their environment?

Deep-ocean animals are not the creatures portrayed in movies; unlike the huge and threatening beasts of the big screen, the real-life residents of the deep ocean are mostly small. There are always exceptions—including the giant squid, isopods, shrimp, and urchins (and around the hydrothermal vents, there are giant clams and tube worms). The reason mostly smaller animals live at this level is the scarcity of food; there simply is not the abundance that is needed to support larger animals. Also, deep-ocean dwellers have a slower metabolism, and thus they develop much more slowly.

## How have **deep-ocean animals** adapted their **vision** to suit their environment?

Animals in the deep oceans usually have very small eyes or are completely blind. Some are only blind when compared with how we see. For example, certain species of deep-ocean shrimp are blind, but they can "see" hydrothermal vents.

## How have **deep-ocean animals** adapted their **bodies** to suit their environment?

Most deep-ocean fish have weak, flabby muscles and some have no air bladder (or have only a smaller air bladder) to help them swim; instead buoyancy is provided by the water's density. Some animals have evolved so their jaws open wider to take on larger prey; some also have expandable stomachs to accommodate larger prey. They also have poorly ossified skeletons (in other words their bones are not strong), no scales, and carry high levels of water in their tissues. And many have developed bioluminescence to lure mates and prey.

## How have **deep-ocean animals** adapted their **reproduction** to suit their environment?

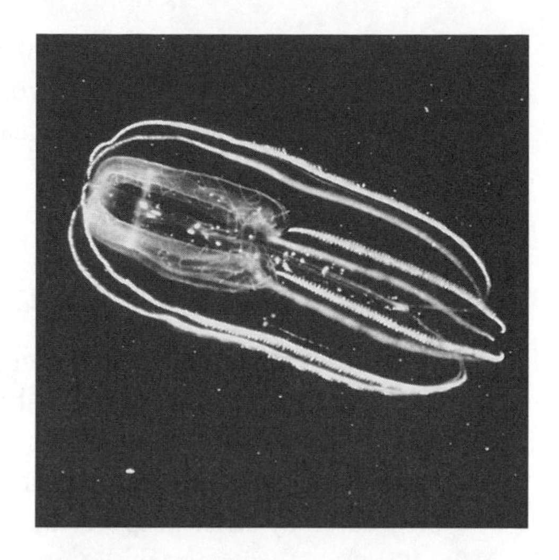

This bioluminescent fish, a lobate ctenophore, emits a blueish glow. *NOAA/OAR National Undersea Research Program*

There are several ways in which deep-ocean animals have evolved their own type of reproduction. The basics are the same, but some of the practices differ. For example, many take on one mate for life, because of the chance of not finding another one; some use their bioluminescence (physiological ability to glow in the dark) to locate a mate. Others use a kind of parasitism; for example, some males become attached to a female, then control the female's sexual functions through her hormones.

## Why do some **deep-ocean animals glow**?

Another way deep-ocean animals have adapted is by bioluminescence, or the production of natural light. In most cases, the light is produced by bacteria in the animal's tissue. Scientists believe organisms glow in the dark for a number of reasons, including:

to break their silhouette so they are not as easily seen by predators
to avoid predators
to attract prey (which are attracted to the light)
to communicate
to allow a mate, or potential mate, to see them
for species identification

## Why are **whales** important to the **deep ocean**?

Some scientists believe that whale carcasses could be one of the keys to how animals survive in the deep oceans. They theorize that the decaying whales could provide important nutrients—even "stepping stones" that allow organisms to migrate across the deep-ocean floor. If this is true, it may also give scientists a clue as to how organisms evolved in the deep ocean.

## Do **bacteria** live in the **deep oceans**?

Yes, some bacteria live in the deep oceans. Scientists know that there are bacteria that live around the hydrothermal vents (volcanic openings where super-hot water escapes from the depths of the Earth); they also know some marine animal carcasses that fall to the ocean floor are not only eaten by other deep-ocean animals, but are decomposed by bacteria.

## What **discovery** of a **deep-ocean fish surprised scientists** in the past century?

During the twentieth century, scientists were very surprised to learn that a fish that was thought to be extinct is actually alive and swimming in our oceans: The "living fossil," called the coelacanth (SEE-la-kanth), is a fish with a three-lobed tail and fins that have arm-like bases. The first was found in 1938 off the coast of Madagascar; many have been discovered since, including a possible new species in 1999.

# HUMANS AND OCEANS

# MAKING USE OF THE OCEANS

## OUT TO SEA IN SHIPS

### Why did **people first sail** the oceans?

Humans have sailed the seas since ancient times. Sea tales and legends are among the oldest elements of our folklore; to wit: the ageless saga told by the Greek poet Homer in the *Odyssey*, which dates to the 8th century B.C.E. And many of the reasons people have been drawn to the oceans remain true today. For example, the ocean, which teems with fish and other edible marine animals and even plants, has long been a source of food. Historically, the ocean also provided humans with other raw materials, such as oil (particularly from whales), shells, tusks, fur, and even skins. Though these are no longer particularly sought-after materials, other ocean products are of interest, such as kelp, which is a source of nutrients (particularly in the Japanese diet); reef corals and sponges, which may have chemicals that could be used in pharmaceuticals; and minerals (found in deposits on the ocean floor).

Since Roman civilization, merchants have sailed the known waterways to ship products from one place to another—literally to take them to market. For the merchant marine, sailing was (and is) their livelihood—and for some, a source of great wealth. Ambitious traders were always eager to find new routes over which they could transport their goods. This desire led to exploration—and the discovery of new lands. But there were also religious reasons for ocean exploration: As both Christianity and Islam spread, religious leaders became eager to claim new lands and convert native peoples. Even Christopher Columbus's first voyage to the New World in 1492 was religiously motivated; his adventure was sponsored by Spain's King Ferdinand and Queen Isabella—known as the Catholic Monarchs.

Since ancient times people have also used the world's waterways for their own transportation. Much of this has been driven by our natural curiosity to see new lands and learn more about the world we live in. Until the advent of the airplane (early in the twentieth century), traveling the seas was the only way for humans to move between landmasses that are separated by unbridgeable bodies of water. Even after the airplane came into use, ships remained a primary mode of oceanic transport for many years: Between 1881 and 1920, tens of millions of immigrants made transatlantic and, to a lesser extent, transpacific crossings to find new opportunities and settle in North America.

As people moved around the globe and modern nations were established, sailing the seas also became a necessary mode of defense: Countries established navies, a seafaring military, to protect their own borders as well as their interests in other lands.

Though the map of the world may change (when country lines are redrawn or even geological changes occur), there are no new lands to discover on our planet. And yet this has not stopped ocean exploration. On the contrary; beginning in earnest in the 19th century, ocean exploration took a new twist—the probing of the vast and mysterious underwater world. Today's scientists have sophisticated equipment and technologies available to them; these have proven critical to the human quest to understand the world ocean, much of which still remains to be discovered and studied.

## How did early humans **travel** the **seas**?

Most early humans used rafts made of wood and reeds or canoes made of dugout wood, bark, or taut animal skins to travel the seas. A more advanced boat had a wooden framework of ribs covered with a layer of thin wooden boards.

Whaling was once an important industry—but harpooning one was dangerous business, as depicted in this old painting by J. S. Ryder. The animals were valued for their oil and blubber. *NOAA Marine Fisheries Service*

## Why does a **boat float**?

A properly designed boat or ship floats because of buoyancy. This is a force exerted upward by the water on the vessel; the amount of this force is equal to the weight of the water displaced by the vessel.

## What are **port** and **starboard**?

Port and starboard are nautical terms used to indicate direction. If you are on a ship and are looking forward, toward the bow, port is the left side and starboard is the right-hand side. The left side was named port because ships in a harbor were traditionally tied up (at port) on that side so that the steering oar, on the right, would not be crushed by the dock (this was in the days before rudders were used to steer ships). The starboard was named because the ship's right side was where the steering oar was set: 'star' is a form of the Middle English word , *stēor,* which means steering oar; and 'board' is from the Middle English word for the side of the ship, *bord.* An easy way to remember these terms is by their length: port and left are both short words; right and starboard are both longer words.

## What is the **difference** between a **boat** and a **ship**?

There is really no precise line of demarcation between these two terms, but many people do differentiate between them based on size: Ships are considered large vessels that can carry goods and people over water; while the term boat is usually applied to smaller vessels that carry fewer goods and people. But the terms can be used interchangeably.

## When did **sailboats** become a reliable means of **travel**?

Although many early galleys had a single square-rigged mast, sails became a reliable means of travel about 1000 C.E. And although the rudder was developed around 100 B.C.E. by the Chinese, it didn't catch on until about 1200 C.E. The rudder, along with the sails, gave sailing ships the maneuverability needed to make them a reliable form of sea travel.

## What **types of sailing ships** have been used on the oceans?

There have been many different types of sailing ships used on the oceans over the centuries—and many types are still in use today. Here are a few examples:

*Arabian dhow:* The Arabian dhow usually has one mast (although there may be two). The main sail of the dhow is triangular and very large in order to catch whatever wind is available. This allows it to tack (beat against the wind), particularly in areas that have gentle winds.

*Chinese junk:* A junk is a flat-bottomed, slow-moving sailing ship used by the Chinese and Southeast Asians. It has a square bow, rudders, and a built-up, squared-off stern (the rear part of the ship). There are normally two to five masts that support large and small sails, each strutted with long horizontal rods of bamboo.

*European galleon:* The galleon was the main sailing ship used by the Europeans from 1500 to 1700. They were large trading ships built by many nations, including Spain, to bring back riches from the New World. Most galleons had four masts and were square-rigged, but also carried some triangular sails. They also had a large, castle-like structure toward the rear; galleons are no longer in regular use today.

*Hybrid:* A hybrid sailing ship, called the *Shin Aitoku Maru,* was launched in 1980—the first sail-assisted commercial ship in 50 years. This ship had two sets of rectangular canvas sails stretched over steel frames; the sails were controlled by a computer that monitors the wind speed and direction, making adjustments as needed. The ship also had a diesel engine for additional propulsion.

## What **kind of ships** did **early humans** build?

The first significant ships were the galleys built by the Phoenicians, which became the standard for travel on the Mediterranean Sea. Galleys were in existence before 3000 B.C.E., and remained in use until the 18th century. The most well-known examples were used by the Greeks, and measured about 130 feet (40 meters) long and 19 feet (6 meters) wide. Oars were the principle means of propulsion, although there was often a single square-rigged mast. Early galleys had a single line of oars (and the men to row the oars) on each side, while later versions employed two and three banks (or decks) of oars on each side. A ship with two decks of oars was called a bireme; one with three decks was called a trireme.

## Where were the **earliest ships built**?

Some of the earliest ships were built in ancient Egypt—and probably in other areas of the Mediterranean. The ships were made of tapered bundles of papyrus stalks bound together, and measuring about 35 feet (11 meters) long.

## Why were the **Phoenicians** known as the **Sea People**?

The Phoenicians were known as the Sea People because of their close ties to, and use of, the sea. They were seafaring traders and explorers who did not exist as one cohesive nation, but rather as a series of independently ruled port cities all along the coast of the Mediterranean. By about 1200 B.C.E. the Phoenicians had become the leading traders of the ancient world, establishing colonies along the north coast of Africa, along the south coast of Spain, and on the Mediterranean islands of Cyprus, Sicily, Minorca, Sardinia, Majorca, and Ibiza. The Sea People established Carthage, a major Mediterranean port for centuries, around 800 B.C.E.

There is evidence that the Phoenicians, who were a Semitic people, were the first to circumnavigate Africa, around 600 B.C.E.— a full 2,000 years before Portuguese navigator Vasco da Gama did so (in 1497–98). They are also thought to have sailed as far as England, the Azores (in the Atlantic), and Brazil.

## What is one of the **oldest surviving vessels**?

One of the oldest surviving vessels belonged to the Egyptian pharaoh Cheops (or Khufu), who ruled in the 26th century B.C.E. The approximately 4,600-year-old ship was split apart and buried near the Great Pyramid of Cheops at Giza. The pieces were eventually pulled out of the sand and put together. The vessel is on display in a museum in Egypt.

## What were some **historic ocean voyages** of **discovery**?

Throughout our history, many cultures and countries have embarked on ocean voyages of exploration. Some of the more well-known expeditions were:

c. 30,000 B.C.E.: Aborigines use a seagoing craft to reach Australia.

c. 1500 B.C.E.: The Polynesians begin to explore and colonize the Pacific Ocean.

600 B.C.E.: Kaleus sails through the Straits of Gibraltar (then called the Pillars of Hercules), between Africa and Europe (Spain).

c. 310–306 B.C.E.: Greek navigator Pytheas is reputed to have sailed around Britain, and is thought to have found Iceland (Thule) to the north.

300 B.C.E.: Hanno of Carthage sails down the west coast of Africa.

C.E. 100: Chinese explorers sail to India.

c. 450: The Polynesians, in the command of chief Hawaii-Loa, sail some 2,400 miles (3,900 kilometers) of open ocean to reach the Hawaiian Islands.

c. 700: Arab sailors reach the Spice Islands (the Moluccas).

800: Vikings explore the North Atlantic.

874: Norwegians colonize Iceland.

1001–02: Norseman Leif Eriksson, son of Eric the Red, sails west from Greenland, reaching a place he calls Vinland, probably Nova Scotia or Newfoundland, Canada.

1405: The first of six Chinese expeditions sets sail to East Africa, Aden (present-day Yemen, on the Arabian Peninsula), and the Philippines.

1427–31: Portuguese navigator Diogo de Sevilha discovers the Azores islands.

1487: Portuguese navigator Bartolomeu Dias sails around the Cape of Good Hope, at the southern tip of Africa.

1492–1502: Italian-born explorer Christopher Columbus, sponsored by Spanish monarchs Ferdinand and Isabella, sails west in hopes of finding a trade route to Asia (the Indies); he voyages across the Atlantic four times between 1492 and 1504, discovering the New World and claiming many lands for Spain. He made landfall throughout the West Indies, including Cuba, Hispaniola (Haiti and the Dominican Republic), Jamaica, Trinidad, Martinique, and Honduras; he reached the South American mainland at Venezuela, where he discovered the mouth of the Orinoco River.

1497: Portuguese navigator Vasco da Gama sails around Africa to India.

1497: Genoese-born John Cabot, with the backing of the English, makes two voyages in search of a northwest passage to China; he lands at Labrador, Newfoundland, or Cape Breton Island. On second voyage (1498), he is lost at sea.

1498: Vasco da Gama, sailing on behalf of Portugal, reaches the western coast of India after rounding the Cape of Good Hope and sailing through the Indian Ocean; it is the first western European voyage around Africa to the East.

1499–1501: Italian Amerigo Vespucci sails with the Spaniards and explores the coast of South America and unwittingly lends his name to the Americas; the homage is the result of an early geographer who believed Vespucci was the first European explorer to actually realize he was in a new land (and not in Asia).

1519–22: A Spanish expedition, led by Ferdinand Magellan (who does not return), is the first to sail around the world. Though Magellan is killed in the South Pacific, navigator Juan Sebastian de Elcano captains the ship home; he and 18 sailors claim the distinction of being the first to circumnavigate the globe.

1535: Frenchman Jacques Cartier sails up the St. Lawrence River to present-day Montreal.

1584–85: Englishman Sir Walter Raleigh leads an expedition to colonize North America; it lands in a region he names Virginia (actually the coastal region north of Florida).

1592: Juan de Fuca discovers British Columbia, Canada.

1603: Frenchman Samuel de Champlain makes the first of 11 voyages to explore Canada; he will discover Lake Champlain (in Vermont), the Ottawa River, and the Great Lakes.

1609: Henry Hudson explores the Delaware Bay, Hudson River, and Hudson Bay (Canada's great inland sea).

1766–69: French navigator Louis de Bougainville sails to the Pacific and discovers Tahiti; he commands the first French expedition to circumnavigate the globe. A woman, Jeanne (Jean) Bare, also makes the trip—disguised as a man. She becomes the first female to travel around the world.

1768–79: Three around-the-world expeditions are led by English mariner and explorer James Cook (better known as Captain Cook); he maps the coasts of New Zealand, eastern Australia, and Papua New Guinea, discovers many islands in the North and South Pacific, and crosses the Antarctic and Arctic circles.

1791: George Vancouver explores the west coast of North America.

1893–95: Norwegian Fridtjof Nansen explores the Arctic in the *Fram.*

1895–98: Joshua Slocum becomes the first person to single-handedly sail around the world.

## How were **sailing ships rated** in the 18th century?

During the 18th century, when sailing ships were in heaviest use—and were used in battle, the vessels were classified by a rating system that was based on the number of guns it had. For example, a first-rate ship had more than 100 guns while a fourth-rate ship had only 50 mounted guns.

## What was the **clipper ship**?

The pinnacle of the sailing ship was the clipper ship— an extremely fast sailing vessel that could handle small, expensive cargoes. To keep up with increasing overseas trade, in the mid-1800s North American shipbuilders developed these swift ships, which had sleek, sculptured hulls and numerous sails (as many as 35). A typical clipper was large, with three or more masts rigged with square sails and triangular sails set on the bowsprit (the leading sail support). Designed to reach speeds of up to 20 knots, the vessels were said to "clip off the miles."

The high-water mark of the clipper ships came between 1845 to 1860, when these vessels carried their cargoes around the world, setting record travel times. The first true clipper ship debuted in 1845 with the *Rainbow,* designed by American naval architect John W. Griffiths.

Clippers also carried the first tea leaves of the season from China, which led to the establishment of the "tea race" as an annual event. Although clipper ships were still built and used well into the twentieth century, by that time the construction of canals around the globe had shortened most sea trade routes, eliminating the need for the swift clippers. They were replaced by steamships, which by the mid-1800s had begun carrying raw materials and finished goods across the Atlantic Ocean.

## What **time records** did the **clipper ships set**?

The clipper ships set many time records. The fastest clipper sailed from Hong Kong to New York in 74 days, and from San Francisco to New York, by way of Cape Horn (the southernmost tip of South America), in 89 days. In 1851 the *Flying Cloud,* launched by Canadian-American shipbuilder Donald McKay, sailed from New York's East River, around the tip of South America, to San Francisco in just under 90 days—a

## When did iron become an important material in ship-building?

For centuries, the main material used to construct oceangoing ships was wood, but during an economic depression in 1857 the British began building so-called "composite ships," large vessels planked with wood over iron frames. At the same time, iron ships were also being built and used to carry freight around the world.

The new material was soon put to use in the construction of naval vessels; wooden ships could no longer withstand the assault posed by explosive shells (developed in the 1820s). "Ironclads" were warships built of wood or iron and covered with thick plates of iron. The first battle between two ironclads was staged during the American Civil War (1861–65): On March 9, 1862, the Union's *Monitor,* originally built as an ironclad and equipped with a revolving gun turret, faced the Confederacy's *Virginia,* which was made by raising the sunken federal boat the *Merrimack* and covering the wooden vessel with iron plates. The ships met at Hampton Roads, Virginia, a channel that empties into Chesapeake Bay. Though the outcome was indecisive, the *Monitor*'s performance in the battle was enough to warrant the U.S. Navy's production of a fleet of ironclad ships, signaling the beginning of modern naval warfare.

record. The next year, McKay's *Sovereign of the Seas* sailed from New York to Liverpool, England, in 13 days, 14 hours. And his *Champion of the Seas* (1854) covered 465 miles (748 kilometers) in 24 hours. It took 25 years for steamships to break that record.

## What are **steamships**?

The steamship, or steamer, is a ship propelled by steam engines, which drive paddle wheels (either along the boat's side or on the stern) to move the vessel through water. Steamboats were first developed in the late

1700s and became commercially viable in the early 1800s. There were two types of steam-driven vessels—those designed for the deep coastal waters along the eastern seaboard of the United States and those designed to navigate the shallower inland rivers of the nation's interior. The first workable steamboat was demonstrated by Connecticut-born inventor John Fitch (1743–98) on August 22, 1787, on the Delaware River.

The world's first commercially successful passenger steamboat service began in 1807—the *Clermont,* built by Robert Fulton (1765–1815). It traveled up the Hudson River from New York City to Albany in 30 hours, and then returned. The first transatlantic steamship was developed in 1838—the British *Great Western,* which sailed from Bristol, England, to New York City in just 15 days.

### What was the **progression of freight ships** after the steamship?

Freight ships, or those that carry large amounts of goods across the oceans, progressed rapidly through the twentieth century. The steam engine was followed by the steam turbine; the steam turbine was followed by the diesel engine in the early twentieth century; and by the 1950s, nuclear marine engines were introduced. Modern freight ships include the largest generation of oil carriers ever—the supertankers. These are some of the largest ships ever put to sea.

## NAVIGATING THE SEAS

### What is **navigation**?

Navigation at sea is the process of determining where a vessel is located on the ocean and plotting a course to get from one place to another. It combines observation, science, technology—and even a little art.

### How did **early sailors navigate** the oceans?

Early sailors had it tough compared with their modern counterparts—they navigated by simply moving from one point on the visible shore to another. Their technology was probably limited to such "instruments" as wind direction indicators.

Sailors eventually attempted to travel the open oceans, a task made difficult because there were no landmarks to steer by, just a flat expanse of water. The method of "dead reckoning" was used initially: A sailor would use an established point to determine his position, then aim toward his destination along a straight-line course. There were problems with this method: First, the surface of our planet is curved, so a straight-line course may actually end up being a great circle. Second, without any fixed objects to steer by, holding a straight-line course to a destination was almost impossible—and getting lost on the wide expanse of the ocean was easy and deadly.

But there were still some "sea-smart" early cultures that could navigate open ocean. For example, during their early voyages in the Pacific, the Polynesians used clues such as the presence and patterns of birds, clouds, color of the sea, or even the smell to guide them to land.

## What **method** did **early sailors** eventually use to **navigate** the open ocean?

Early navigators eventually turned to celestial navigation to travel the oceans. With this method, the sailor would determine the position of a ship on the ocean relative to the position of the Sun, Moon, planets, or a star on the celestial sphere—the imaginary, hollow sphere that surrounds the Earth and has a coordinate system similar to our longitude (hour angle) and latitude (declination); it also makes one rotation on its axis daily. The stars are fixed on this celestial sphere, while the planets, Moon, and Sun are not—but all can be used by the ocean traveler to determine their position. The great seafaring cultures, including the Phoenicians, Norsemen, and Polynesians, all knew and used the sky to navigate.

Celestial navigation was the mainstay of ocean navigation for centuries and continues to be used to some extent today. But there are drawbacks—the biggest being visibility: If a celestial object is covered by clouds, fog, mist, or haze, it's almost impossible to use it as a navigation tool. And, of course, the system is of little help to the sailor during daylight.

## What were some of the **early** navigation **instruments** used by sailors?

One of the earliest instruments used in navigation was the astrolabe, developed by the ancient Greeks and used until about the 18th century. The astrolabe typically consisted of a wood or metal disk suspended from an attached ring; the circumference of the instrument was marked off in degrees. A movable pointer, called the alidade, pivoted at the center of the disk. Sightings toward a celestial object using the pointer were translated into angular distances by taking readings off the circumference of the disk.

The compass, developed in Europe during the 12th century, finally gave sailors a firm directional reference. This instrument consists of a freely suspended magnetic needle, which always aligns itself with the magnetic north and south poles of our planet. It is still in use today, albeit in more refined forms.

More sophisticated instruments were developed between the 15th and 17th centuries, such as the backstaff, the quadrant, and the octant. But one invention made a great impact: the sextant. Around 1730, the sextant was invented independently by American inventor Thomas Godfrey (1704–49) and English mathematician John Hadley (1682–1744). This instrument combined an adjustable index mirror, a half-silvered horizon mirror, telescope eyepiece, and a degree scale. With it, a sailor could accurately measure the altitude of a celestial body—such as the Sun, Moon, or a particular star—from the reference point of the horizon. Commonly used stars were the pole star in the Northern Hemisphere, and the dominant star in the constellation of the Southern Cross in the Southern Hemisphere. At a particular time on a specific day, these objects are at a certain degree above the horizon. Tables were formulated that translated the data read from the sextant into position, in degrees north or south of the equator. Although instruments like the sextant were a great improvement over their predecessors, they were only able to fix the latitude of a ship.

## What is **latitude**?

Latitude is the angular distance north or south of the Earth's equator and is measured in degrees, minutes, and seconds of arc. The equator is defined as 0 degrees; the north and south poles are 90 degrees north and

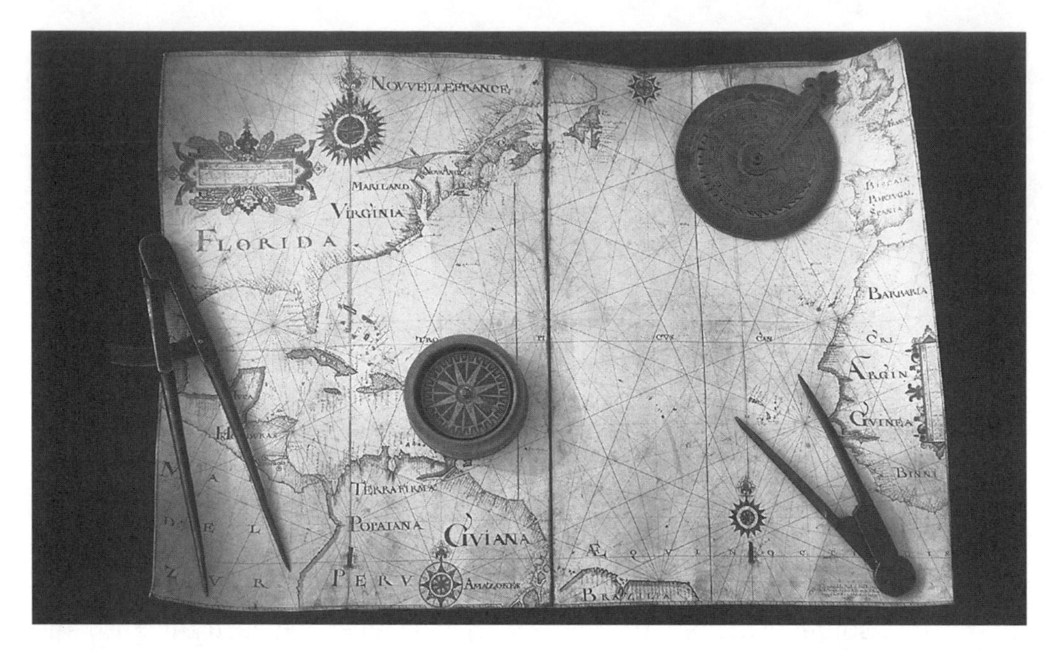


**The tools of early navigation: Compasses and a map of the new world dating to the seventeenth or eighteenth century are on display at the Jamestown Museum, Virginia.** ***CORBIS/Richard T. Nowitz***

south, respectively. Lines of identical latitude form horizontal circles around the globe, and are called parallels of latitude, with one degree of latitude equal to 60 nautical miles (111 kilometers).

But the location of a ship cannot be completely known just using latitude. This merely gives the ship's position in degrees, north or south of the equator. To firmly fix the location, the east or west position, called the longitude, must also be known. The combination of latitude and longitude can determine the exact location of a ship on the ocean.

## What is **longitude**?

Longitude is the angular distance east or west from a reference circle, and is measured in degrees, minutes, and seconds. Lines of identical longitude, called meridians, form vertical circles passing through the poles. The reference meridian, called the Prime or Zero Meridian, passes through Greenwich, England; other longitude positions are located east and west of this reference. The International Date Line is on the 180 degrees (east or west) meridian and is located in the Pacific Ocean. Most world maps clearly show lines of latitude and longitude—the parallels and meridians.

## What **instrument** allowed sailors to **determine longitude**?

Although the concept had been understood for many years—as well as the need to determine it—longitude could not be accurately measured on a moving ship until the invention of the chronometer. This was (and still is) an instrument designed for keeping highly accurate time. Like a watch, a mechanical marine chronometer is spring-driven, but its parts are more strongly built and include devices to compensate for temperature-driven changes in spring tension.

The invention of the chronometer is a story in itself: Around 1712, the British Board of Longitude offered a prize of 20,000 pounds to the first person who could invent a method of determining longitude at sea—with accuracy within 30 nautical miles and for a 6-week journey. English clockmaker John Harrison (1693–1776) took up the challenge, and spent most of his life perfecting a marine chronometer. He finished his first in 1735; the 4th was finished around 1760—and was smaller and more accurate. Harrison won the prize in 1763, but received only small amounts of the money until 1773, when King George III intervened on his behalf. Harrison's story was told in full in the best-selling book *Longitude,* written by Dava Sobel and published in 1995.

Harrison's chronometer method was based on the premise that by using an accurate clock set to Greenwich Mean Time in England (today's site of the prime meridian), a sailor could read it at noon local time and determine the ship's longitude. Since the Earth spins 15 degrees per hour, the sailor would just have to multiply the difference in time by 15. So, for example, if the local time is 5 A.M. and the chronometer (set to Greenwich Mean Time) reads 1 A.M. , the ship is located at 60 degrees longitude, also called the 60th meridian, which runs through the Arctic and Indian oceans. (The latitude, determined using a sextant, would tell the sailor the horizontal location.)

Improvements continued to be made on chronometers. For example, a few years after Harrison's 4th chronometer was built, French clockmaker Pierre Le Roy (1717–85) invented a chronometer resistant to changes in temperatures. The establishment of the prime meridian at Greenwich, England, in 1884 also added to the chronometer's accuracy.

Modern versions of the chronometer use quartz movements to insure accuracy; they are also mounted on gimbals, devices that keep the chronometer steady even if the ship is being tossed by high waves or storms. Although they are still being used (mostly as backup), most chronometers were replaced in the early twentieth century by more sophisticated instruments—especially with the establishment of the radio; and later, the advent of satellite technology.

## What **instruments** are used in **modern ocean navigation**?

The gyrocompass, radar, loran, and GPS (global positioning system) are all important instruments in ocean navigation today. With the exception of the gyrocompass, which uses the Earth's magnetic poles, these instruments use either radio waves or satellite imagery and data to pinpoint locations.

**gyrocompass:** The gyrocompass is an improved version of the venerable navigation compass. It consists of a rapidly spinning, electrically driven rotor; the rotor is suspended so its axis automatically points along the geographical meridian. This makes the gyrocompass more accurate and less prone to external influences than the compass.

**radar:** Radio waves transmitted from a fixed station can be detected and used as a directional aid—a modern version of dead reckoning. Radar (an acronym, it stands for radio detection and ranging) is a ship-based instrument that sends out ultrahigh-frequency radio waves. The time needed for these waves to be reflected back to the ship is used to detect the position, motion, and nature of any object within detection range.

**loran:** Loran (an acronym, it stands for long range navigation) is a radio-based navigation system. Geographically separate radio stations send out pulsed transmissions that are received by instruments on a ship. By measuring the time interval between the arrival of a pair of these signals, a sailor can plot a position line on which the ship is located; a 2nd measurement, using another pair of transmitting stations, produces a 2nd position line. The intersection of the lines is the precise location of the ship.

GPS: The NAVSTAR Global Positioning System, or GPS, is a series of Earth-orbiting satellites—with each of the 24 satellites continuously broadcasting time and position messages. These signals are detected by receivers and translated into precise, real-time location information. The GPS system allows position to be determined anywhere on the planet to within about 300 feet (90 meters) for civilian purposes, and to within a few feet for military users—although many civilian GPS receivers will one day have the same precision.



## What are **nautical charts**?

Nautical charts are maps that are used to plot the course a ship will follow. The charts are based on Mercator projection—the Earth is projected on a cylinder tangent to the surface at the equator, then "flattened" to give the appearance of neatly spaced longitude and latitude lines (the longitudes, or meridians, are drawn parallel to each other and the lines of latitude, or parallels, appear as straight lines, whose distance between each other increases with their distance from the equator). These nautical charts also show surface features (of islands or along a coast) and underwater topography.

## How are **today's nautical charts** changing?

The United States' National Oceanic and Atmospheric Administration (NOAA) has made the transition from hard copies of nautical charts to more accurate, digital navigation charts. The agency took its huge nautical chart databank—an accumulation of more than 150 years of ocean surveys—and converted the charts into a digital format for the computer. In this way, charts of various ocean areas are readily available to people involved in ocean work (researchers, fishermen, and ship navigators); the maps are more accurate and can be updated much quicker than the old, paper nautical charts. In addition to digitized nautical charts, most ships now use a Global Positioning System (GPS) to determine their location in the oceans.

## Are there **less-technical aids** to help with **ocean navigation**?

Yes, besides computer-generated nautical charts and Global Positioning Systems, navigators also use less technical aids. These include navigation buoys, visual landmarks, and radio beacons. Each one of these aids is listed on the nautical charts, assisting mariners in their navigational tasks.

# PERIL AT SEA

## How do **shipwrecks occur**?

Most shipwrecks occur along rocky or shallow coastlines, where rocks can be hidden just under the surface of the water. For example, the rocky coast of the Scilly Isles in southwestern England have claimed many ships through the years. But wrecks also occur during storms (the *Monitor,* a Civil War ironclad ship, sunk off Cape Hatteras, North Carolina, during a storm on December 31, 1862); as casualties of war (the British liner the *Lusitania* sunk after it was hit by a German submarine on May 7, 1915); and from colliding with other vessels or even large objects in the water (the oceanliner *Titanic* struck an iceberg and sank on April 14–15, 1912).

## Where was a **Phoenician shipwreck** recently found?

An ancient shipwreck, thought to be a Phoenician vessel, was recently discovered lying on the bottom of the Mediterranean Sea, about 3,000 feet below the surface. This shipwreck has been named *Melkarth,* after the Phoenician god of sailors.

A remotely operated vehicle (ROV) showed that the wreck contained many large amphora, ceramic jars used as shipping containers for olive oil, wine, fish, honey, and other products used by the Phoenicians. The distinctive style of amphora at this site indicates that the ship was Punic (western Phoenician) in origin and sunk sometime in the 5th century B.C.E.. The depth at which the wreck is located—combined with the low oxygen levels and frigidity of the water—have kept the site well preserved.

**Disaster at sea: Seven destroyers shipwrecked in 1924 in the Pacific, off Point Conception, southern California.**
***NOAA/C&GS Season's Report, Lukens 1925***

## What are some of the **oldest shipwrecks** ever found?

One of the oldest shipwrecks that has been found was discovered in 1954 in the Mediterranean Sea: It was a Bronze Age cargo vessel that sank around 1200 B.C.E. The shipwreck was found by accident: In talking to American journalist Peter Throckmorton, who was also a diver, a Turkish sponge diver named Kemal Aris happened to mention something about seeing a wreck. The journalist pinpointed the wreck 90 feet (27 meters) down, and the University of Pennsylvania became involved in its excavation, which brought up copper ingots, stone balance weights, and artifacts from the modest merchant vessel.

Another ancient shipwreck, which is considered by some to be the oldest yet found (but there are disagreements about the vessel's cultural origins, and therefore, its dates), was discovered in 1982 by sponge-diver Mehmet Cakir. The vessel was found 140–170 feet (43–52 meters) below the surface of the Mediterranean Sea, near Uluburun, in southern Turkey, and is believed to date to the fourteenth century (late Bronze Age). The recovery was completed in 1994 and the artifacts are housed at Turkey's Bodrum Museum of Underwater Archeology.

## How did **Florida's Treasure Coast** get its name?

Florida's Treasure Coast, an area of reefs and beaches off Sebastian and Ft. Pierce, was so named for the shipwrecks offshore there, in the Atlantic Ocean. They were caused by what is called the 1715 Plate Fleet disaster, one of the most colorful sagas in Florida's coastal history.

In 1715, the Terra Firma Fleet (loaded with Peruvian silver, and emeralds and gold from Columbia) and New Spain Flota (loaded with silver and gold from Mexico, and silks, porcelain, and spices from China) met in Havana, Cuba, to sail back to Cadiz, Spain, together. They were hoping the combined fleets—11 Spanish and 1 French ship—would dissuade pirates and raiders from attacking, which often happened to Spanish ships in the Florida Straits.

But the pirates were not the problem: The ships reached the Bahamas by July 24, 1715, but by then, it was well into the hurricane season. The ships were all thrown back by the winds, pushed into the reefs and beaches along what is now called the Florida Treasure Coast. Some vessels crashed along the coast; 5 of the ships were never seen again and more than 700 sailors lost their lives. Those who survived had to trek about 200 miles (322 kilometers) to the north to the nearest Spanish settlement in St. Augustine—and were subjected to both natural exposure and harassment by the locals.

Because of the wealth of cargo lost, Spaniards and others who heard of the shipwrecks scavenged the area—and not long after the wrecks, there were close to 10,000 people living on the beach in search of the lost cargo. Today, coins and other fragments of the cargo sometimes wash up on the beaches or are found in the sands of this region—especially after major storms. Many of the relics are now in museums—testimonies to Florida's ocean history.

## What are some **famous shipwrecks** of **recent history**?

Among the most well known and well-documented shipwrecks are those that have occurred since the mid-1800s, including:

1833: On May 11, *Lady of the Lake* was bound from England to Quebec when it struck an iceberg and 215 people perished.

1853: On September 29, the *Annie Jane,* an emigrant vessel, sunk off the coast of Scotland and 348 people perished.

1898: On November 26, the *City of Portland* sank just off Cape Cod, Massachusetts, killing 157 people.

1912: On April 15, the ocean-liner *Titanic* sank after colliding with an iceberg; 1,513 people died.

1915: With World War I (1914–18) under way, on May 7, the British liner the *Lusitania* sunk off the coast of Ireland after it was hit by a German U-boat; 1,198 civilians were killed.

1917: The French munitions carrier the *Mont Blanc* was struck by another ship and exploded in the harbor at Halifax, Nova Scotia, claiming 1,635 lives and severely injuring more than 1,000. The severe destruction was due to the power of the explosion (the ship was laden with thousands of tons of TNT and acid): it was so terrific that it laid waste to much of Halifax and generated a huge wave that swept through the city.

1928: On November 12, the *Vestris,* a British steamer, sank in a gale off the Virginia coast, claiming 110 lives.

1939: On May 23, the *Squalus,* a submarine carrying 59 men, sank off Hampton Beach, New Hampshire; only 33 lives could be saved.

1942: On October 2, the *Queen Mary* rammed and sank a British cruiser; 338 people aboard the cruiser died.

1948: On December 3, the *Kiangya,* a Chinese passenger ship carrying refugees fleeing Communist troops during civil war in China, struck an old mine, exploded, and sank off Shanghai. It is thought that more than 3,000 people were killed.

1952: On April 26, the *Hobson,* a minesweeper, collided with the aircraft carrier *Wasp* and sank during night maneuvers in the mid-Atlantic Ocean; 176 people perished.

1953: On January 31, the *Princess Victoria,* a British ferry, sank in the Irish Sea, taking 133 people to watery graves.

1956: On July 25, the Italian liner *Andrea Doria* collided with the Swedish liner *Stockholm* off Nantucket Island, Massachusetts. It sank the next day; more than 1,600 were rescued but 52 people, mostly passengers on the Italian ship, were dead or unaccounted for.

1962: On April 8, a time bomb on the *Dara,* a British liner, exploded in the Persian Gulf; the ship sank. 236 lives were lost.

1963: On April 10, the *Thresher,* an atomic-powered submarine belonging to the United States, sank in the North Atlantic, killing 129.

1968: In late May, the *Scorpion,* a U.S. nuclear submarine, sank in the Atlantic Ocean 400 miles (644 kilometers) southwest of the Azores, killing 99 people. The actual wreck was finally located on October 31 of that year.

1987: On March 9, a British ferry capsized in the North Sea after leaving the Belgian port of Zeebrugge with 500 people aboard. The event was probably caused by water flowing into its open bow, and 134 people drowned. And on December 20, more than 4,000 passengers were killed when the ferry *Doña Paz* collided with the oil-tanker *Victor* off Mindoro Island, 110 miles south of Manila, Philippines.

1990: On April 7, what was believed to be an arson fire aboard the Danish-owned North Sea ferry *Scandinavian Star* killed at least 110 passengers in Skagerrak Strait off Norway.

1991: On December 14, a ferry carrying 569 passengers struck a coral reef and sank in the Red Sea off the coast of Safaga, Egypt—causing more than 460 people to drown.

1993: On February 17, the *Neptune,* a triple-deck ferry, capsized off the southern peninsula of Haiti during a squall; more than 1,000 passengers lost their lives.

1994: On September 28, the *Estonia,* a passenger ferry, capsized off the coast of southwest Finland and sank in the stormy Baltic Sea. Of the approximately 1,040 passengers aboard, only about 140 survived.

## How did the **C.S.S. *Alabama* sink**?

The C.S.S. *Alabama* was one of the most famous Confederate States Naval vessels during the American Civil War (1861–65)—and its wreck

## Have any ships mysteriously disappeared in the oceans?

Yes, there have been many mysterious disappearances of ships in the oceans, especially before the advent of advanced communications. For example, the Danish sail-training *Köbenhavn,* left the River Plate for Melbourne, Australia on December 14, 1928. The ship had a crew of 75 that included 45 boy cadets learning to sail. The last contact with the ship was made on December 22—and all was well. No one ever heard from the ship or its crew again.

Another well-known mystery involved the disappearance of a ship's crew and passengers: On November 5, 1872, the *Mary Celeste,* a sailing ship, left the New York harbor for Genoa, Italy—carrying its captain, his family, and a 14-member crew. On December 5, another sailing ship, the *DeGratia,* discovered the *Mary Celeste* floating as a derelict—with no one onboard. There was no sign of violence or trouble and there were plenty of supplies. To this day, no one knows what happened—although there have been a number of theories.

was recently found by underwater explorers in the English Channel, off the coast of Cherbourg, France.

The *Alabama* was the most sought-after vessel of the war, wreaking havoc on the Union states' commerce. It was built in Liverpool, England, for the Confederacy, and was launched under the name *Enrica* in 1862, then renamed the *Alabama.* Its crew, under the command of Rear Admiral Raphael Semmes, was entirely British. The month after it was launched, it captured and burned some 20 vessels (including a dozen whaling ships) in the North Atlantic Ocean. As she came closer to the United States, the *Alabama* attacked many more ships, including the Union vessel the *Hatteras.* The Confederate ship became a legend—a phantom that could never be found. It never even went into ports to resupply; rather, the crew used the supplies and fuel (coal) from captured ships.

By 1864, the Union ship U.S.S. *Kearsarge*—matched almost evenly to the *Alabama*—was sent to search for the Confederate vessel off the coast of Europe. Learning that the *Alabama* was at Cherbourg, the captain of the *Kearsarge,* John A. Wilson, took up position just off the nearby coastal breakwater. (The news of the impending battle also brought sightseers from Paris and other areas to watch the fight from the beach.) On June 19, 1864, the *Alabama* steamed out of Cherbourg to face the *Kearsarge.* Wilson managed to draw the *Alabama* out of French waters, about 7 miles (11 kilometers) offshore; then the *Alabama* engaged the *Kearsarge.* During the fierce fighting, the *Alabama* attempted to head for shore, but by that time, she was taking in too much water. It was evident the boat was sinking, and Semmes hoisted the white flag of surrender.

The *Alabama* sunk fast into the channel. Captain Semmes and 41 others from the *Alabama* survived—and in all the confusion, managed to escape back to England. In all, during her 21 months at sea, the *Alabama* reportedly destroyed 60 vessels, capturing supplies and material thought to have totaled about $6 million.

## How was the **S.S. *Oregon* sunk**?

The S.S. *Oregon* was a Scottish luxury-liner—the biggest and fastest ship afloat at the time—and was called "the queen of the Atlantic." On her maiden voyage in 1881, she crossed the Atlantic Ocean in 6 days, 10 hours, and 40 minutes. One night in 1886, while coming into New York after a 7-day journey from England, the *Oregon* was struck by the 3-masted schooner, the *Charles R. Morse.* The hit penetrated the *Oregon's* hull and made 3 large holes in the liner's left side. The schooner and the *Oregon* were temporarily locked, but then drifted apart, and the *Morse* quickly sunk, along with its 9-member crew. The *Oregon* floated for more than 8 hours after the collision, allowing rescue ships to reach the sinking ship. Amazingly, no crew members or passengers on the *Oregon* were killed—a textbook rescue.

Not only was the ship the fastest in its day, it also gained notoriety for being the largest ship that ever sank off Long Island. We know, too, that the ship sits just about 22 miles (35 kilometers) from the Fire Island Inlet—the hull still in one piece. Down with the ship went all the passengers' baggage, more than 300 mail bags, and cargo worth about a million British pounds.

## How did the **R.M.S. *Titanic*** sink?

The R.M.S. *Titanic* is one of the most famous ocean-liners of all time—and one of the worst maritime disasters in history. The British ship was on its maiden voyage from Southampton, England, to New York City when it struck an iceberg just before midnight on April 14, 1912. Of the 2,220 people onboard, 1,513 perished in the freezing waters of the North Atlantic; the others were saved by the ocean-liner *Carpathia.* But the iceberg isn't entirely to blame for the dear loss of life: To save space, there were lifeboats for only about half the passengers and crew, the ship may have been going too fast for such dangerous waters (it was a known iceberg lane), and the radio operator was off duty (and asleep) when the warnings about icebergs were sent to the ship.

## How was the **R.M.S. *Lusitania*** sunk?

The *Lusitania* was a Royal Mail Ship (R.M.S.) that was sunk by a German U-boat in 1915. The liner was a civilian passenger ship, and the idea that someone would sink an unprotected ocean-liner was unthinkable—even though, with World War I (1914–18) raging, the Germans had already threatened this was a possibility.

The Germans should have been taken for their word. On May 7, a U-20 submarine under the command of Kapitan-Leutnant Walther Schwieger hit the *Lusitania* with one torpedo at a range of 700 yards (640 meters). Twenty minutes later, the liner sank, killing 1,198 men, women, and children—including 128 Americans.

The sinking of the *Lusitania* is considered one of the most notorious milestones in World War I—and for some people, it was the catastrophe that shifted American public opinion in favor of U.S. entry into the conflict. But the real consequence was the German reaction: The Kaiser (the head of Germany), worried that the U.S. would in fact join the conflict, halted all submarine warfare in British waters. The hiatus allowed Britain to recoup its losses and build up its naval and merchants marine—and eventually convince the United States to join the war. Many view the sinking of the *Lusitania* as the deciding factor in the war: The Kaiser's mistake cost Germany victory.

## Why is the **Truk Lagoon famous**?

The Truk Lagoon, in the Federation States of Micronesia (in the western Pacific Ocean, east of the Philippines), is renowned for its shipwrecks from World War II (1939–45). In 1944, U.S. bombers surprised a fleet of 60 Japanese merchant vessels that had taken refuge in the lagoon, which is about 250 feet (76 meters) deep and 40 miles (64 kilometers) wide. The bombers sunk every ship, and 1,000 men died. Today, the shipwrecks form an enormous artificial reef, filled with corals, algae, sponges, and other reef life; it is also a popular diving spot.

## How do explorers and archaeologists know **where to find shipwrecks**?

In the past, most shipwrecks were found by word of mouth (reports passed from person to person about a ship that had gone down), accidental discovery, or observing first-hand a ship going down. In the past few decades, the introduction of advanced sonar and electronic sensing devices has helped underwater explorers and archaeologists detect previously unknown sunken ships.

## What **recent expeditions** have found **shipwrecks** in the deep oceans?

Thanks to advanced technology, which allows the sites of shipwreck to be pinpointed, researchers, scientists, and adventurers have conducted many recent expeditions in the deep oceans to hunt for shipwrecks. These expeditions have recovered—or at least recorded—details about some of the most famous shipwrecks of our time.

Many of the most well-known expeditions have been led by underwater explorer Robert Ballard and his team, who discovered the wrecks of the R.M.S. *Titanic* and the German battleship *Bismarck.* Ballard takes great advantage of new technologies, such as remote operating vehicles (ROVs), to find and take close-up photos of the wrecks. One of his most famous expeditions was to the *Titanic,* the ocean-liner that sunk tragically in 1912 after hitting an iceberg.

In 1998 Ballard discovered several wrecks in the Mediterranean, lying along an ancient trade route off the coast of Tunisia. Using an ROV named *Jason,* Ballard found the remnants of eight ships. The vessels span 2,000 years of history, with five of them from ancient Rome; the oldest vessel is thought to have sunk about 100 B.C.E. The ships were found about 2,500 feet (762 meters) under the ocean surface, with their debris—in the form of artifacts and cargo—scattered over 20 square miles (52 square kilometers). Some of the cargo included pre-cut stone blocks and columns, probably for a temple.

An interesting side effect of such deep-ocean exploration has been a boon to scientists: Most shipwrecks found in shallow water (less than 200 feet or 61 meters) are usually pounded by waves, stripped by looters, and covered with sediment, debris, or even corals. Wrecks discovered in the deep oceans are usually more pristine and not broken up. There is also less debris covering the vessel and artifacts—and the wrecks usually have not been looted. With the sophisticated technology that is available to find and explore deep-sea shipwrecks, some underwater scientists and explorers believe there should be international laws or conventions to protect and preserve these historical vessels.

### How do **ship captains know** where **shipwrecks** are located in shallow water?

The Automated Wreck and Obstruction Information System (AWOIS) is a collection of data on submerged wrecks and obstructions. Not all shipwreck positions are known, but those that have been found and charted by the National Oceanic and Atmospheric Administration's (NOAA) surveys in the coastal waters of the United States are noted. The latitude, longitude, and historical description are listed on specific NOAA nautical charts, allowing ships and boats (commercial and sports fishermen) to avoid any possible collisions with the submerged shipwrecks. The AWOIS is also of interest to historians, scuba divers, researchers, and salvagers.

## VENTURING UNDER THE WAVES

### When did **people** first venture **into the ocean waters**?

The first individual to go into the oceans with a supply of air—allowing a prolonged stay—may have been Macedonian king Alexander the Great

The diving bell, a dome supplied with a few minutes of compressed air by a hose, was the first apparatus that humans used to aid in underwater exploration. *NOAA/OAR National Undersea Research Program*

(356–323 B.C.E.). It is thought that he was lowered into the water in a bubble of glass, a kind of primitive diving bell. The great Greek philosopher Aristotle (384–322 B.C.E.) also mentioned a certain device that allowed a person to breathe underwater. But the first practical attempts to stay underwater using special apparatus began in the 1700s.

## Why is **diving without apparatus bad** for **humans**?

Diving, or descending beneath the surface of the ocean (or any water), means entering a realm to which humans are not adapted. Because we cannot breathe underwater, a human is limited to the length of time they can hold their breath in the water—which is anywhere from 1 to 3 minutes. Also, humans are limited to the depths they can dive without artificial means. This is because the pressure of the water forces the air out of our lungs at approximately 300 feet (100 meters).

There are also physiological problems associated with diving under the ocean's surface, especially in deeper waters. Most of these problems stem from the effects of pressure and gases on our tissues. The following problems can affect human divers:

**Lung tissue damage** is caused by the increased pressure on the lungs exerted by the inhaled gases from the breathing apparatus.

**Nitrogen narcosis** is an intoxication effect caused by an increased amount of nitrogen dissolved in the blood, which impairs reflexes and judgment.

**Decompression sickness**, better known as "the bends," is caused by nitrogen bubbles that form after emerging from compression too quickly (in other words, from rising too quickly to the surface), which causes localized pain in joints and muscles, labored rapid breathing, paralysis, or even death.

**Increased oxygen and carbon dioxide in the tissues**, caused by the increased pressure at increased depths, can lead to convulsions, labored breathing (due to too much oxygen), and intoxication effects (from too much carbon dioxide).

## What were some of the **earliest attempts to dive** underwater?

The diving bell, a dome supplied with compressed air by a hose, was the first diving apparatus humans used. British astronomer Edmond Halley (1656–1742), best known for his discovery of the periodic comet named for him (Halley's comet), devised the first practical diving bell in 1717. The wood chamber had glass windows on top and was open at the bottom. Leather tubes were hooked to air casks; as water displaced the air in the casks, the air was pushed into the bell.

The hunt to find a good diving helmet and suit began even earlier—in the 1600s; but it took until about 1819 before German-British inventor

**State of the art in 1915: The diving suit had come a long way since its invention in 1819. Here, the diver (outfitted in J. Peress' 1-atm dive suit) prepares to be lowered into the water to explore the wreckage of the *Lusitania,* the British luxury liner downed by the Germans during World War I. *NOAA/OAR National Undersea Research Program***

Augustus Siebe (1788–1872) devised the first practical diving suit, which consisted of a leather jacket and metal helmet. Air was pumped into the helmet from the surface through a hose, and although the helmet was not watertight, the air's pressure kept water below the diver's chin. In 1830 Siebe invented another suit, complete with airtight and controllable air pressure; it was the first modern (closed) diving suit.

## How **deep** do **divers dive**?

With or without suits, humans have dived to various depths. Without any external devices to aid them, humans can dive to about 60 feet (18 meters). Some pearl and sponge divers have been reported to reach 100 feet (30 meters), but normally, they are only underwater to depths of about 40 feet (12 meters). For dives with the standard mixtures of compressed air and oxygen, the limit is about 250 feet (76 meters); with special mixes (such as oxygen with helium or hydrogen to replace nitrogen), divers can reach depths of about 500 feet (152 meters).

## What **equipment** has **enhanced diving** capabilities?

In order to stay underwater for longer times, humans developed breathing equipment such as scuba (Self Contained Underwater Breathing Apparatus) gear. This gives divers more time to explore and work in the watery realm. Scuba gear is especially used for skin diving, and is designed for swimming as opposed to walking on ocean floors. However, this technique is still limited to certain depths.

## Who **invented scuba** equipment?

The most popular form of scuba, or self-contained underwater breathing apparatus, was invented during World War II (1939–45) by underwater explorer (who was then a French naval officer) Jacques-Yves Cousteau (1910–97). His aqualung includes 1 to 3 cylinders, which the diver carries on his or her back; the compressed air from the cylinders is fed through a hose to the diver's mouthpiece. This air passes through several valves that allow a constant flow at a pressure that automatically adjusts to the surrounding water pressure.

## How can **divers dive deeper** into the oceans?

In order to dive deeper, humans now use submersibles. These small vessels are built to withstand the tremendous pressures of deep dives, carry their own atmosphere, and remain underwater for extended periods of time.

## What was the **first deep-diving submersible**?

The first deep-diving submersible was the bathysphere—a heavy steel, spherical, non-maneuverable vessel that was lowered by a cable from a ship. In 1934, American naturalist and explorer Charles William Beebe (1877–1962) and engineer Otis Barton (1899–?) descended to a then-record depth of 3,028 feet (923 meters) in the Puerto Rico Trench off Bermuda.

Because the bathysphere was not maneuverable—and if the cable broke, there was no way back to the surface—Swiss physicist Auguste Piccard (1884–1962) designed the first drive-able bathysphere. This vessel could be maneuvered in the deep sea, and was kept afloat by a large container filled with gasoline. In 1954, Piccard reached a depth of 13,125 feet (4,000 meters); in 1960, his son Jacques (b. 1922) reached a depth of about 35,810 feet (10,915 meters) in the submersible *Trieste,* in the Mariana Trench off Guam.

## What are **modern submersibles** like?

Modern submersibles are small, maneuverable research vessels that can function at great depths. Typically, they have a compartment for the crew contained within a pressure-proof hull, with view-ports to the outside, life-support and power systems, and sensors such as sonar

The DVS *Turtle* is a diving submersible; a state-of-the-art vessel that has allowed humans to investigate previously unseen areas of the ocean. Here it is being taken aboard the U.S. Navy support ship *Laney Chouest* in California's Monterey Bay. *NOAA/OAR National Undersea Research Program; J. Butler*

hydrophones. A variety of instruments are attached to the outside of the submersible, including external lights, cameras, and mechanical arms that are used to collect samples and perform a variety of tasks. Modern submersibles are used for exploration, repair work, collecting samples, taking photographs, recovery work, and as a base for deep-sea divers.

## How many **kinds of modern submersibles** are there?

There are many types of modern submersibles, which are capable of diving to great depths. There are rigid suits, which function as mini-sub-

**The *Johnson Sea Link* is an underwater sphere that allows pilot and observer to explore the depths. *NOAA/OAR National Undersea Research Program***

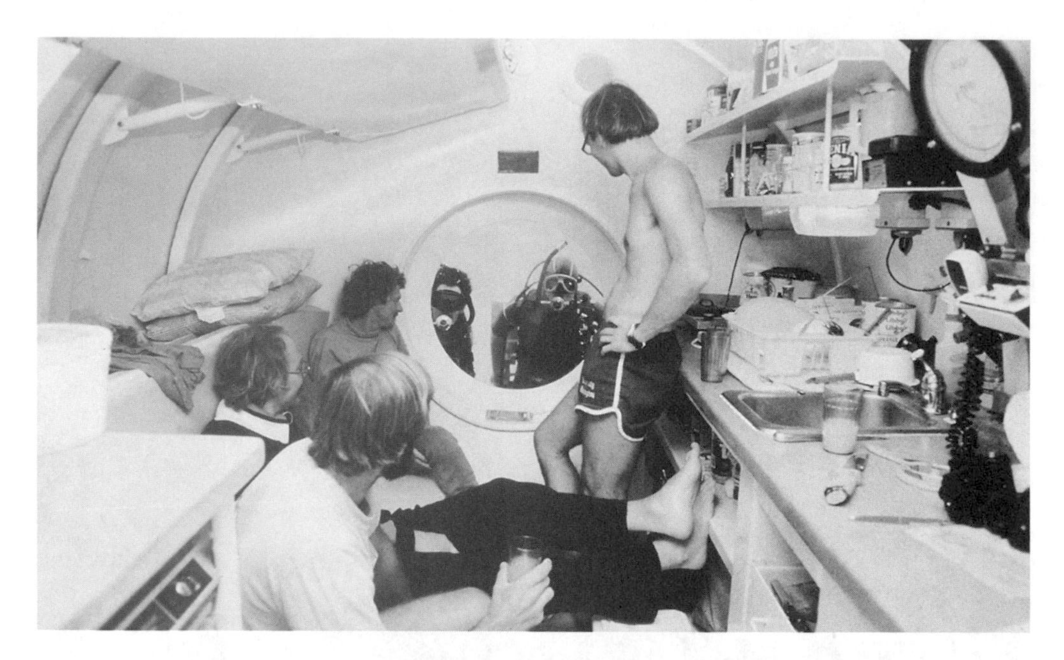

**The insider's view: Aquanauts relax in the tight quarters of the underwater research facility *Hydrolab,* while other divers (visible through porthole) are at work just outside. The vessel held four divers at a time, but only slept three. *NOAA/OAR National Undersea Research Program***

mersibles, carrying one person; these include the JIM and Spider. For deeper dives there are diving saucers or spheres, which can hold more than one person. These include Jacques Cousteau's *Soucoupe Plongeante*; bathyscaphes such as the *FRNS* (Fonds pour la Recherche Nationale Scientifique) series and the *Trieste,* designed by Auguste Piccard; the *DSRV* (Deep Submergence Rescue Vehicle), used to rescue sunken submarine crews; the *Alvin,* a 3-man submersible capable of diving to 13,000 feet (3,960 meters), which was used to explore the wreck of the *Titanic*; and the *Aluminaut,* a 6-person submersible made of high-strength aluminum alloys and capable of operating at 15,000 feet (4,570 meters).

## What do **deep-sea submersibles, whale corpses**, and **laundry** have to do with each other?

Although the connection between deep-sea submersibles, whale corpses, and clean laundry seems remote, they really might be related! Scientists from the University of Hawaii and a biotechnology company are collabo-

Divers work off the undersea research facility *Hydrolab* (near the Bahamas, 1966). Submersibles such as these have allowed divers to explore the ocean's depths. *NOAA/OAR National Undersea Research Program; D. Clarke*

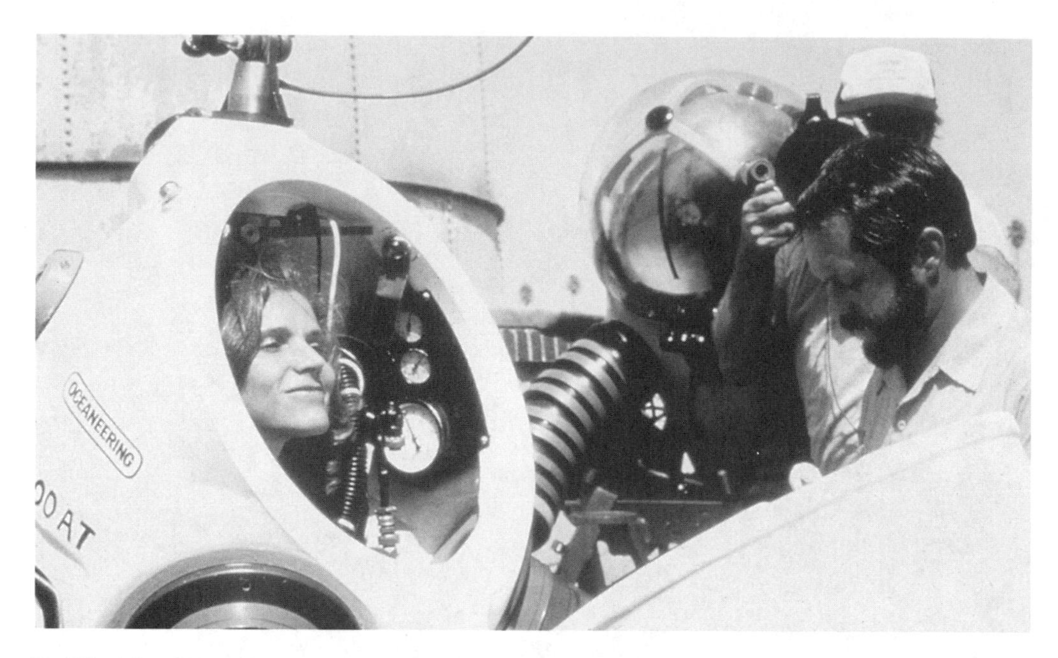


**The JIM suit is a rigid suit that functions as a mini-submersible, carrying one person underwater for exploration. Here Dr. Sylvia Earle prepares for a dive.** ***NOAA/OAR National Undersea Research Program***

rating to gather and investigate bacteria that live on whale corpses found rotting deep in the Pacific Ocean off southern California. The purpose of their work is to improve laundry detergent.

Scientists recently used the *Alvin,* a submersible owned by the United States Navy and operated by the Woods Hole Oceanographic Institution, to study and retrieve the creatures living on rotting whale corpses. The bodies normally consist of unusual bacteria, crabs, and worms that eat away the flesh. The bones then become coated with slimy mats of bacteria, which subsist on the oils that leak out of the carcass. To accomplish this trick, these bacteria have developed unusual enzymes that readily digest oils and fats at cold temperatures—around 41° F (5° C).

The enzymes found in conventional biological washing powders are only efficient at temperatures higher than 104° F (40° C). Scientists have found that the enzymes from the bacteria found on whale corpses can be made more active if they are warmed slightly to between 59° and 68° (15° and 20° C)—which is ideal for a cold-water detergent.

The DVS *Alvin,* a diving submersible, in 1978—one year after it was used for the first exploration of hydrothermal vents in the deep-ocean floor. *NOAA/OAR National Undersea Research Program*

## What is a **submarine**?

A submarine is a naval vessel that can operate underwater for extended periods of time. But submarines cannot dive as deep as the deepest-diving research submersibles. However, they are much larger and can carry more people.

The earliest version of a submarine was built around 1620 by Dutch inventor Cornelis Van Drebbel (1572–1633). It was a leather-covered rowboat with tubes that brought air from the surface (the tubes were held on the surface by floats)—and may have been able to stay under water for up to 15 hours.

The first submarine used in combat was invented in 1775 by American engineer David Bushnell (1742?–1824); the egg-shaped vessel was called the "Bushnell Turtle." During the American Revolution (1776–83), the submarine unsuccessfully attacked a British ship in New York Harbor. Robert Fulton (1765–1815), the American inventor who also built the first steamboat, put together the *Nautilus,* a submarine with many innovations—such as rudders for horizontal and vertical control, and compressed air to supply oxygen. And the Confederates used several versions of submarines during the American Civil War (1861–65).

The modern version of the submarine was initially developed around the beginning of the twentieth century—and these underwater vessels were used extensively during both World Wars. The addition of nuclear power to submarines was a major step, as vessels with this capability can remain submerged for great lengths of time. Although modern submarines are designed for combat, they are now increasingly used to study the oceans.

# FAMOUS OCEANOGRAPHIC EXPEDITIONS AND FAMOUS OCEAN SCIENTISTS

## What was the **H.M.S. *Beagle*?**

The H.M.S. *Beagle* was a survey ship sent out by the British Navy in 1831 to chart the coast of South America and circumnavigate the globe. On its 5-year voyage, the crew collected geological, biological, and meteorological data of the South American coast, the southern islands, Australia, and Asia. The voyage was also significant because of a young naturalist on board: Charles Darwin (1809–82) used his observations from the voyage to develop his theory of evolution, which he published as *On the Origin of Species by Means of Natural Selection* (1859). Wherever the *Beagle* stopped, the 20-something-year-old Darwin gathered data on flora, fauna, and geology. He also made a study of fossils and species he found in the Galapagos Islands. In addition to his work on natural selection, Darwin published many other important works based on what he learned from the *Beagle* expedition; these included *The Structure and Distribution of Coral Reefs* (1842) and *Geological Observations on Volcanic Islands* (1844).



## Who was **Matthew Fontaine Maury**?

Matthew Fontaine Maury (1806–73) was an American oceanographer whose wind charts, published in 1848, established the first "sea lanes"—ocean areas with the most favorable winds. His work resulted in shorter sailing times. In addition, Maury attempted soundings of the Atlantic Ocean, which he did in preparation for the laying of the first transatlantic cable (to enable telegraph messages to be sent between North America and Europe). To avoid at-sea collisions, Maury also recommended different travel lanes for eastbound and westbound steamers. In 1855, he published *Physical Geography of the Sea,* which became a standard in the field and has had many subsequent editions.

## Why is the **voyage of the H.M.S. *Challenger*** so famous?

The 4-year voyage (1872–76) of the British ship H.M.S. *Challenger* is often thought of as the beginning of oceanography; it was the first expedition to circumnavigate the globe for the purpose of scientific study. The ship was commanded for most of the voyage by British naval officer George Nares (1831–1915), and the lead scientists were British zoologist John Murray (1841–1914) and Scottish naturalist C. Wyville Thomson (1830–82).

During the voyage, deep-sea investigations were made at more than 300 spots—depth soundings were taken; temperature and salinity measured; and water and sediment samples were collected, as were plants and animals.

## What **discoveries** were made by the **H.M.S. *Challenger***?

There was a wealth of new discoveries made during the H.M.S. *Challenger* voyage (1872–76). For example, many new species of plants and animals were collected and classified for the first time. The data accumulated during the 4-year journey also confirmed earlier work done on currents and winds; underwater mountains were mapped and the circulation of cold Antarctic water was charted; and the ship discovered the Mid-Atlantic Ridge that cuts north-to-south through the Atlantic Ocean.

## Who was **George Strong Nares**?

George Strong Nares (1831–1915), a British admiral and surveyor, was the ship's captain for most of the voyage of the H.M.S. *Challenger.* Nares had previously served on numerous scientific voyages, surveying the coast of Australia and the Mediterranean Sea, and charting the currents in the Strait of Gibraltar, the narrow passage of water between Spain and Africa, which connects the Atlantic Ocean with the Mediterranean Sea.

## Who was **Charles Wyville Thomson**?

Charles Wyville Thomson (1830–82) was a Scottish naturalist and a professor of natural history and zoology at Belfast, Ireland (1854–68), and Edinburgh, Scotland (1870–82). Thomson made a special study of deep-sea life. He also described sounding and dredging expeditions, writing a book on the subject, *The Depths of the Sea* in 1873. He is most well known for being a lead scientist aboard the H.M.S. *Challenger,* from 1872 to 1876; his account of the famed expedition was published as *The Voyage of the Challenger* (1877).

## Who was **John Murray**?

Canadian-born John Murray (1841–1914) was a British oceanographer who became an assistant to Scottish naturalist Charles Wyville Thomson, the head of the H.M.S. *Challenger* expedition (1872–76). Murray was on the voyage for 3 and a half years, and did much of his work on the smallest organisms in the top layers of the oceans—the single-celled protozoans. He also studied sea bottom sediments. Murray sorted hundreds of the specimens that were gathered during the *Challenger* expedition, compiling a multi-volume report on the voyage. Later in his life, Murray explored the North Atlantic Ocean (1910).

## Who was **Fridtjof Nansen**?

Fridtjof Nansen (1861–1930) was a Norwegian explorer, scientist, and statesman who was the first to explore the Arctic. His expedition, from 1893 to 1896, took the ship *Fram* to the most-northerly point humans had yet ventured—86° 14' North. Between 1910 and 1914, he made oceanographic expeditions in the North Atlantic. Once World War I (1914–18) got under way, the scientist turned his attention to matters of state, a dedication that would eventually earn him the Nobel peace prize (1922). Among his published works were *Norwegian North Polar Expedition* (1893–96) and *Northern Waters* (1906).

## Who was **William Maurice Ewing**?

William Maurice Ewing (1906–74) was an American geophysicist who was a proponent of German geophysicist Alfred Wegener's (1880–1930) theory of continental drift. He also took the first seismic recording in the open ocean in 1935; suggested a relationship between earthquakes and undersea rifts (mid-ocean ridges); and in 1939, was the first to photograph the deep sea. He was also the first director of the Lamont-Doherty Geological Observatory at Columbia University, New York.

## Who was **Edwin Link**?

American Edwin A. Link Jr. (1904–81) was an explorer and an inventor of both flight and oceanographic equipment. He was born in Huntington, Indiana, but also spent part of his life in Binghamton, New York, where his father owned a piano and organ factory. In 1929 he combined his keen mechanical skills, which he learned at his father's factory, with his love of flying to build a pilot trainer. This flight simulator launched a completely new industry—and helped train many World War II (1939–45) pilots, among whom the "blue box" was famous. More elaborate simulators based on Link's ideas are still used today to train aviators as well as astronauts.

But Link's accomplishments did not end in the air, he also focused some of his attention on underwater archaeology and exploration. After 1954, he designed and developed unique diving systems and submersibles—

French underwater adventurer Jacques Cousteau was also a filmmaker, author, environmentalist, and inventor of the aqualung. *Associated Press/The Cousteau Society*

and did so until his death in 1981. His work improved diving equipment, allowing divers to go deeper, stay underwater longer, explore more safely and efficiently, and return to the surface with less risk.

## What were **Jacques Cousteau's** accomplishments?

The world-renowned marine explorer Jacques Cousteau (1910–97) is credited with no less than revealing the mysteries of the deep to the public at large. In 1943, along with French engineer Émile Gagnan, Cousteau (who was then in the French navy) developed a practical aqualung,

Deep-sea explorer Robert Ballard, shown in Mystic, Connecticut, in May 1999; the research submarine *Turtle* is behind him. *AP Photo/Steve Miller*

enabling a diver to stay underwater for several hours. It was the first regulated compressed-air breathing device—or, the first self-contained underwater breathing apparatus (scuba). After World War II (1939–45), he helped create an underwater research program and in 1950 he outfitted the now-famous ship *Calypso,* a former minesweeping vessel, for oceanographic research. With *Calypso* as his base of operations, Cousteau and his crews explored the depths, often filming their underwater adventures. It was with this footage that Cousteau created numerous made-for-TV movies and produced three full-length feature films—*The Silent World* and *World Without Sun* (both of which won Oscars for best documentary), as well as *Voyage to the Edge.* He also wrote and contributed to more than 50 books. The highly recognizable captain died at the age of 87, having spent the last two decades of his life heightening public awareness of the importance of ocean conservation. He was also active in various United Nations initiatives for environmentally sustainable development. These activities garnered him numerous awards, including the highest civilian honor that the United States government bestows, the U.S. Presidential Medal of Freedom (in 1985). His vision continues to be carried out by the Cousteau Society (www.cousteausociety.org).

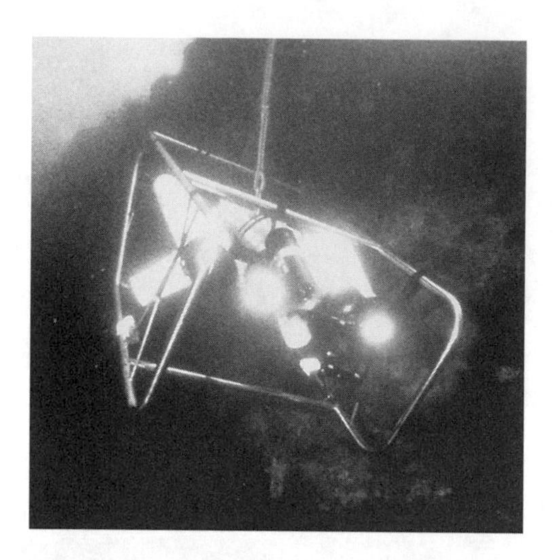

Tethered robots such as the *Phantom S2* (shown diving the face of a reef wall) are among the new technologies that allow humans to explore the ocean depths, including otherwise inaccessible wreckage on the seafloor. *NOAA/OAR National Undersea Research Program*

## Who is **Robert Ballard**?

Robert D. Ballard (*see* photo on previous page) is a deep-sea explorer and scientist. He discovered the sunken *Titanic,* the *Bismark* and the *Yorktown,* and has led or participated in more than 100 deep-sea expeditions using diving submersibles, including the *Alvin, Archimede, Trieste II, Turtle, Cyana,* and the *Ben Franklin.* These expeditions included the first manned explorations of the Mid-Atlantic Ridge, the discovery of warm water springs and their unusual animal communities in the Galapagos Rift, and the discovery of high-temperature black smokers, the super-heated water plumes escaping from hydrothermal vents in the seafloor. Ballard also discovered 11 warships from the lost fleet of Guadalcanal and participated in the explorations of the luxury liner *Lusitania* and the hospital ship *Britannic.* After searching for 12 years, Ballard led the expedition that discovered the *Titanic* in 1985.

Ballard is also the founder and the head of the Institute for Exploration (IFE) at Mystic Marinelife Aquarium in Mystic, Connecticut. Specializing in deep-ocean archaeology, IFE's goal is to establish a new field of underwater research that uses state-of-the-art technology such as advanced mapping and imaging systems, underwater robotics, and manned submersibles.

## What is the ***Jason* Project**?

Deep-sea explorer and scientist Robert Ballard founded the *Jason* Project in 1989 after receiving thousands of letters from school children wanting to know how he discovered the wreck of the R.M.S. *Titanic.* The mission of the Jason Foundation for Education, founded in 1990 to administer the program, is to get students interested in and excited about science and technology and to motivate and provide professional development for schoolteachers.

## Have any **recent expeditions** brought **black smokers** from volcanic vents to the surface?

Yes, for the first time, in the summer of 1998 oceanographers brought up large chunks of volcanic chimneys or vents (called black smokers) from a mid-ocean ridge; some of the chimneys were still steaming when they reached the deck of the ship. Scientists from the University of Washington (Seattle) and the American Museum of Natural History (New York) hauled up the smoking chunks using the research vessel *Tully,* just off the northwest coast of the United States. These thick chimneys form where magma rises to the ocean floor—in areas where the crust is pulling apart because of the movement of the continental plates. A specially designed cage dragged up sections of the smokers from almost 1 mile (1.6 kilometers) down—some of them weighing up to 2 tons.

# HARVESTING THE OCEANS

## What do people **harvest** from the **oceans**?

Almost everything living and non-living in the ocean is harvested by humans—from fish and other animals to plants, minerals, and materials.

## What is **fishing**?

Fishing is catching fish and other marine and freshwater animals for food or sport. It can be done with lines and hooks, rods and reels, nets (trawls), spears, or cages.

## What **fish** make up **most of the world's catch**?

Approximately half of the world's catch of fish is make up of five kinds of fish: herring, cod, jack, tuna, and mackerel.

## What kind of **fish** is **consumed most**?

Herring is the world's most popular fish—at least when it comes to eating.

## How did **early humans** fish?

The methods used by early humans to fish were quite simple, and are still in use today in many parts of the world. Early fishermen used small boats and didn't wander too far from the shore, so their choice of fish was limited. The harpoon was used to spear fish, while hooks and lines were used to snag them and bring them ashore or aboard; fish were sometimes caught by hand. The invention of nets allowed larger numbers of fish to be caught at one time; these nets could be set by hand in shallow water, or dragged from a small boat (this method is called trawling).

Traps were also set to catch fish, while animals such as crabs and clams were collected by digging on shorelines at low tide, or by hand in shallow water. In Asia, cormorants were (and still are) used for fishing. These birds dive for fish while attached to a small boat by means of a line. They can't swallow the fish they catch because of rings or cords fastened around their necks.

## What is **commercial fishing**?

Commercial fishing is the practice of harvesting large amounts of marine organisms—mostly for human consumption. Commercial fishing fleets have huge ships that go to the sea—sometimes for months at a time. Some of the larger vessels are called factory ships, which process and refrigerate their catch. Technology has also helped commercial fishing: sonar and satellite data can be used to pinpoint schools of fish.

## How **much** fish is caught **each year**?

The amount of fish that people catch in a year is staggering: It is estimated that commercial fishing enterprises bring in more than 90 million tons annually, and individual fishers catch an additional 24 million tons annually—for a total of more than 114 million tons of fish.

## Where are the world's **most fertile fishing grounds**?

The richest fishing grounds are at the continental shelves and areas of upwelling (where cold water wells up from deep in the oceans).

## What are the world's **top fishing nations**?

The top fishing nations are Japan, Russia, and China. According to national law, all coastal countries have the right to fish within 200 miles (322 kilometers) of their shores.

## What **types of nets** do commercial fishermen use?

There are many types of nets used by commercial fishermen, including the trap net, used to herd and trap fish in shallow waters; drift or gill nets, nets stretched for hundreds of miles to catch such fish as tuna in the open oceans; and purse seines, an upside-down mushroom-shaped net used to catch shoal species such as anchovies.

## How might **fishing** be done **in the future**?

One day, like the cowboys of yesterday, who rustled the cattle across the open plains of the American West, there may be "fishboys" who herd schools of fish in the oceans. Using natural submarine seawalls, large schools of fish could be contained in giant "ranches." Sonar and miniature devices attached to the fish could be used for herding. The reason for the interest in fish ranching and herding is the population growth: By the year 2150, the United Nations estimates that the world will hold close to 11.5 billion people. This huge population will have to rely on the bounty of the oceans as part of their food source.

## What is a **"no-take" zone**?

In some oceans, there are "no-take" zones—reserves in which commercial and recreational fishermen are prevented from taking any fish or other marine animals from the waters. This action preserves organism biodiversity and prevents over-fishing in certain areas.

## How is the problem of **shark over-fishing** being addressed?

Scientists believe that sharks are being over-fished in some areas—with many of the larger coastal sharks being killed twice as fast as they can reproduce (most species typically reproduce every 2 to 3 years). Now the National Marine Fisheries Service has adopted new restrictions to help stop the over-fishing of large coastal sharks in the Atlantic Ocean. The rules call for major cuts in both the commercial and recreational shark-fishing quotas. There is also a moratorium on 19 shark species, including the longfin mako, Caribbean reef, and dusky sharks; and a commercial fishing quota for porbeagle and blue sharks. For those fish not mentioned in the rules, recreational fishermen can only take one shark larger than 4.5 feet (1.4 meters) per vessel per day. Even with these strict rules, it is estimated that it will still take about 40 years for sharks to return to even one half their original numbers in the coastal Atlantic.

## What is **aquaculture**?

Aquaculture is the farming of aquatic organisms in freshwater, brackish water, or saltwater—and, in the majority of cases, this means fish. The fish are stocked, fed, and provided protection from predators in a controlled environment. This method cuts down on disease among the fish. Aquaculture is becoming more important around the world, as the oceans' wild fish stocks are declining. Overall, fish farming (the more well-known term for aquaculture) produces more than 10 million tons of fish, crustaceans, and mollusks each year.

## What **marine plants** are **food sources for people**?

The most popular marine plants used as a food source for humans are the algae commonly called seaweeds. For example, a green algae called sea lettuce has the consistency of wet wax paper. It can be added to salads or dried, powdered, and used as a seasoning. Another edible seaweed is the red algae called purple laver or nori. It is transparent and paper-thin—and looks like sea lettuce. Many people eat nori in sandwiches; it can also be seasoned and eaten with rice.

## What are some of the **materials harvested** from the oceans?

There are many materials harvested from the oceans (or from deep beneath the ocean floor) for human use: oil and gas, precious corals, and deep-sea nodules. Some scientists are also hoping to develop more drugs using materials from the sea; substances from abundant marine invertebrates and plants may have the potential to help humans.

## What are **deep-sea nodules**?

Deep-sea nodules are large lumps of rock—composed mainly of manganese and iron oxides—that are found in abundance over expanses of the deep-ocean floor. The nodules vary in their composition: Some are nickel-copper combinations, while others are mostly cobalt or mostly manganese. These mineral deposits form when elements in seawater accumulate around a central object, such as a fish bone or a shark's tooth. It is estimated that about 1,500 trillion tons of deep-sea nodules are lying on the ocean floor and can be harvested for their minerals. They can be brought up to the surface in a variety of ways; one method is to use special vacuums that pull the nodules up to a ship.

## What **ocean minerals** were **recently discovered near Japan**?

A huge volcano 250 miles (400 kilometers) south of Tokyo hides a huge mineral deposit in its caldera (the collapsed center). The deposit lies at a depth of 4,000 to 4,500 feet (1,220 to 1,372 meters), and contains pyrite, galena, arsenic, and other minerals—and even gold and silver. The researchers who examined the volcano believe it may be the first of many such volcanoes that could be harvested for their useful minerals.

# GLOBAL OCEAN CONDITIONS

## POLLUTION

### What is **pollution**?

Pollution is the contamination of the environment by waste or other materials. The vast majority of pollution on our planet is a direct result of human activity.

### Where does **marine pollution** along our **coasts** originate?

The pollution along our coasts comes from many sources, including sewer overflow, storm water runoff, polluted water runoff, and sewage treatment plant malfunctions. Approximately 160,000 factories dump 68,000 tons of toxic metals and 57,000 tons of toxic organic chemicals into the coastal waters of the United States each year. And about 5.9 trillion gallons of wastewater from sewage treatment plants are also dumped into these waters annually.

It is estimated that the world's oceans receive about 3.25 million tons of oil each year—with the majority of the oil coming from street runoff along the coasts rather than tanker spills. And more than 70 percent of all the oceans' pollution comes from land-based sources, with the coasts being the most polluted.

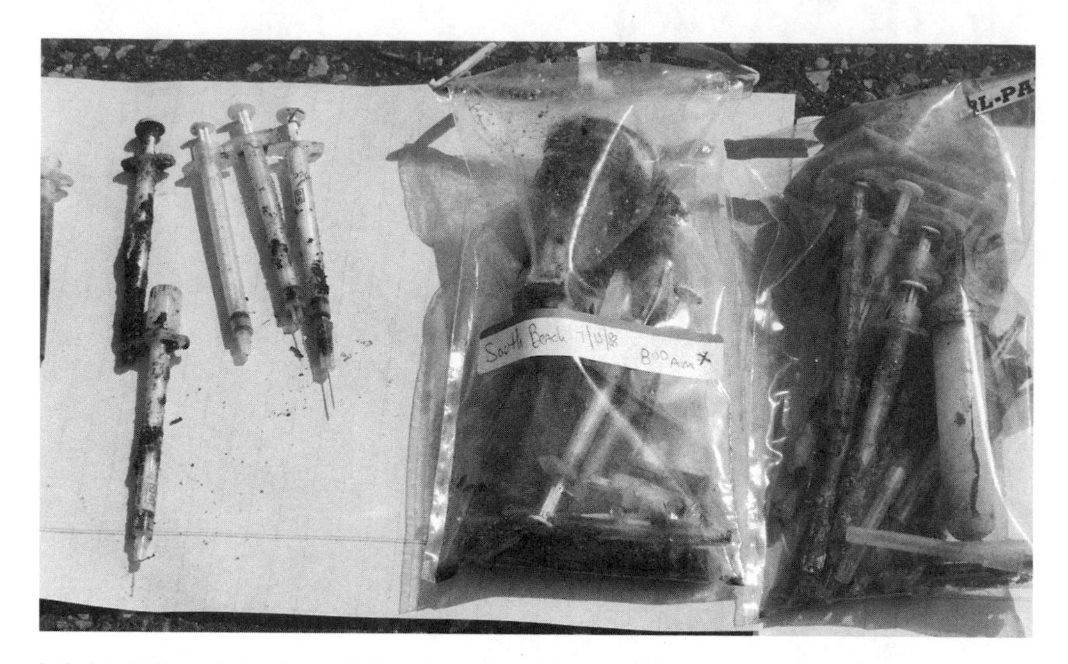


In the late-1980s medical waste repeatedly washed up along Atlantic beaches; these syringes came ashore on South Beach, Staten Island, New York. Area hospitals were found to be dumping their waste into the ocean—a practice that was soon banished. *CORBIS/Bettmann*

## What are some of the **ways humans pollute** the oceans?

Pollution entering the oceans takes on many forms, and comes from many sources. Industries, large cities, even farms can generate a wide variety of waste materials that eventually find their way into the seas.

For example, in their manufacturing processes, some industries use water to remove chemicals. If the facility is on the coast, this contaminated water may be dumped directly into the ocean or into rivers, eventually reaching the sea. Major cities, with their large populations, produce huge amounts of garbage and sewage, some of which may find its way into neighboring rivers and oceans. And the fertilizers and pesticides used on farms can mix with rain runoff, eventually making their way to the sea. Other pollution can come directly from ships at sea, such as oil tanker spills and dumping of waste. Finally, airborne pollution by land-based sources can eventually settle into the oceans; this pollution can include small particles of metals, wood pulp, and manufacturing waste, or gases such as sulfur and nitrogen oxides.

**Ocean-dumping of refuse accounts for about 10 percent of the pollutants in the world's oceans; pollutants originating on land account for most ocean pollution.** ***Reproduced by permission of Greenpeace***

## How are the **sources** of **marine pollution** distributed?

The following are the major sources of ocean pollution, listed according to the percentage the source contributes (by weight):

| Source | Weight Percentage |
|---|---|
| discharge and run-off from land | 44% |
| airborne land emissions | 33% |
| accidental spills and shipping | 12% |
| ocean dumping | 10% |
| offshore oil/gas drilling and mining | 1% |

## Where is there a **massive amount** of **atmospheric pollution**?

A massive amount of atmospheric pollution—covering an area approximately the size of the continental United States—has recently been found overlying the tropical Indian Ocean. It was discovered by scientists participating in the Indian Ocean Experiment (INDOEX), whose purpose is to investigate how tiny pollutant particles (aerosols measuring less than a few microns) are transported through the atmosphere, and how they affect climate. Data are gathered through the use of balloons, satellites, surface stations, research aircraft, and oceanographic ships.

The large area of atmospheric pollution found over the tropical Indian Ocean is present as a dense, brownish haze reaching from the water's surface to a height of 0.5 to 2 miles (1 to 3 kilometers). It extends almost 1,000 miles (1,600 kilometers) off the southern coast of Asia, covering most of the northern Indian Ocean, including the Arabian Sea, and a good part of the Bay of Bengal. The haze layer reaches as far south as the equatorial Indian Ocean, extending to about 5 degrees south of the equator. The haze itself consists of large concentrations of aerosols, including soot, sulfates, fly ash, mineral dust, organic particles, and nitrates; gases are also present in high concentrations, including sulfur dioxide, carbon monoxide, and various organic compounds.

What concerns scientists is that this haze appears to scatter incoming solar radiation and reduce the amount of energy absorbed by the ocean surface by as much as 10 percent. This reduction will disturb the hydrological cycle, resulting in a lower amount of moisture evaporating from the water's surface; this could lead to changes in rainfall amounts on a regional, or even global, level. In addition, a lessening of the amount of sunlight reaching the surface might also have a negative impact on marine plant life, such as plankton, which rely on photosynthesis to produce energy. And this would, in turn, directly affect the marine food chain.

## What is the **International Coastal Cleanup**?

The International Coastal Cleanup is a yearly event during which volunteers around the world spend a day cleaning up shorelines. The first event took place in 1986, and was sponsored by the Center for Marine Conservation (CMC); 2,800 people removed trash from along approximately 122 miles (almost 200 kilometers) of Texas shoreline during this inaugural event. The cleanup has grown every year since, with more volunteers and support from corporate donors. The program went national in 1988, and international in 1989 when Canada and Mexico joined in. Now, during each cleanup event, approximately 350,000 people around the world—from all the states and territories of the U.S. and 80 countries—pick up millions of pounds of debris from more than 4,500 beaches, river banks, underwater sites, and lakeshores. (To volunteer to help with a cleanup, call 1-800-CMC-BEACH.)

The International Coastal Cleanup has also taken on other missions, such as collecting data on the amounts and types of debris found during the cleanup event, educating people about the problem of coastal pollution, and supporting policies and legislation to protect marine habitat and wildlife. The most recent International Coastal Cleanup in the U.S. involved approximately 160,000 volunteers, who removed more than 3 million pounds of debris from 6,887 miles (11,081 kilometers) of coastline. Included in this large amount of debris were 1,317,092 cigarette butts, 162,046 straws, 337,342 pieces of plastic, and 299,154 plastic food bags and wrappers.

## What **items have been found** during the **International Coastal Cleanup**?

Some of the most common items found during the U.S. cleanup include cigarette butts, straws, metal bottle caps, and plastic bottles. But tires, glass bottles, lumber, car parts, fishing line and syringes have also been found and removed. Many types of marine life, including seals, turtles, fish, otters, and seabirds, which died after becoming entangled in the debris, also had to be removed. Some of the more unusual items found during the latest International Coastal Cleanup were a baby carriage filled with adult books, a chemical gas mask, a female mannequin, and a BMW sedan.

Nearly 60 percent of the debris collected from the shores during the latest International Coastal Cleanup originated on land. This is in direct contrast to the popular perception that most beach trash originates with the maritime and cruise industries and eventually washes ashore. The normal litter seen on land, such as candy wrappers and cigarette butts, can eventually find its way to the shore, littering a beach.

## Do any **U.S. beaches** have pollution **problems**?

Yes, many beaches in the United States have pollution problems. Although only about 10 percent of the coastline is regularly tested, there are more than 2,500 beach closures each year because of polluted water.

**Sign of the times? Ocean pollution can close beaches such as this one at Point Loma in southern California where millions of gallons of sewage contaminated the waters.**
***CORBIS/Rick Doyle***

## How is **pollution of U.S. coastal waters** monitored?

An annual National Health Protection Survey of Beaches has recently been instituted to monitor the quality of United States coastal waters, with government agencies collecting information on local beach health. The results are summarized and published by the Environmental Protection Agency (EPA). These can be viewed on their website (www.epa.gov).

In the second annual survey, the EPA sent 322 questionnaires about local beaches to beach health protection agencies in 33 states; 193 usable responses were received containing information on 1,403 beaches. The vast majority of respondents were local government agencies from coastal counties, cities, or towns. The areas they covered included coastal areas along the Atlantic and Pacific oceans, the Gulf of Mexico, and the Great Lakes.

The results of this survey show that there is a wide variety of standards used in monitoring the quality of beach waters. Some 935 beaches have water quality monitoring programs of which 67 percent are checked at least once a week. However, 33 percent are monitored less than once a week, and 2 percent of the local agencies only monitor water quality after special events or precipitation.

## Why is the **Mediterranean Sea** so **polluted**?

The Mediterranean Sea is very polluted for a number of reasons. It is bordered by 18 countries with approximately 120 major cities that send their industrial wastes and sewage into this body of water. Compounding the problem, the Mediterranean Sea is almost landlocked, so the pollutants cannot be washed away. Instead, they stay in place, building up over time in this relatively small area.

A beach at Manorbier Bay, on the Welsh coast, covered with crude oil that spilled from the *Sea Empress* oil tanker in 1995. *CORBIS/Chinch Gryniewicz; Ecoscene*

## What are **oil spills**?

Oil spills occur when tankers release crude oil, also known as petroleum, into the ocean environment. Although accidental spills from oil tankers are highly publicized, they only account for approximately 20 percent of the oil that is released. The other 80 percent of oil from tankers is the result of their routine operations, such as emptying ballast tanks.

## What can happen when **oil** is **released into the ocean**?

There are a multitude of things that can happen when oil is released into the ocean. For example, in a few cases of accidental spills from tankers, some of the oil has caught fire. This burning releases gases and large quantities of toxic ash, which may be blown over large areas, producing allergic reactions and breathing problems in people and other animals.

Even without combustion, the lighter components of the spilled oil will eventually evaporate into the atmosphere, where they may react with sunlight and oxygen to form greenhouse and acid gases, contributing to global warming and acid rain, respectively.

The heavier components of the crude oil form a "mousse," a sticky mixture of oil and water. This mixture may sink to the ocean floor, where it combines with sediments to form a tar-like mass. This mixture destroys the habitat of bottom-dwellers, as well as ruining spawning areas for shellfish and fish. It may also float, drifting with the currents and tides, eventually washing up on shore. Spills that occur close to a coastline may contaminate local wells: The oil washes up on the shore, then works its way into fresh groundwater reservoirs that extend under the beaches.

## What **effect** can an **oil spill** have on **marine organisms**?

Oil spills can be very harmful to marine organisms. In addition to destroying habitats, oil can injure or kill marine life through direct contact or poisoning.

Marine birds and mammals are the most visible victims of direct contact with spilled oil. The feathers of birds become clogged with the oil, making it impossible to fly. This added weight can also cause birds to sink and drown rather than float. As if that wasn't enough, the oil also destroys the insulating properties of the birds' feathers; if the weather is cold, the birds may die of significantly reduced body temperatures, a condition called hypothermia. Marine mammals in cold waters are also susceptible to hypothermia, as their fur loses its insulating ability when covered with oil.

Marine organisms can also be poisoned after ingesting spilled oil. Although the negative effects of ingesting oil are not well known for microorganisms such as plankton, larger organisms such as fish and marine mammals are visibly affected. Fish take in large amounts of oil through their gills, which is why large fish-kills are often associated with spills. Marine mammals that are covered with oil attempt to clean

**After the Exxon *Valdez* oil tanker spilled her load in Prince William Sound, Alaska, in 1989, the clean-up effort was extensive. *Reproduced by permission of Greenpeace***

themselves, leading to the ingestion of large amounts of oil. This oil can destroy the internal organs of the animals and interfere with their reproductive processes. For example, upward of 15,000 otters died as a result of poisoning associated with oil spilled from the Exxon *Valdez* tanker in 1989 in Alaska's Prince William Sound. In addition, carnivorous animals and birds that scavenge the dead can also end up ingesting fatal amounts of oil.

## Where did the **first major commercial oil spill** in the **oceans** occur?

The first major commercial oil spill in the oceans occurred on March 18, 1967, when the tanker *Torrey Canyon* was grounded on the Seven Stones Shoal off the coast of Cornwall, England. The spill sent 124,000 tons of Kuwaiti oil into the ocean.

Although it was the first commercial tanker accident, World War II (1939–45) still holds the record for the most oil ever spilled in the oceans by tankers over a short period of time: Between January and June

1942, German U-boat attacks on tankers off the East Coast of the United States sent 590,000 tons of oil into the ocean. Another unfortunate record was set in January 1991, when Iraqi troops began deliberately dumping oil into the Persian Gulf from Sea Island, Kuwait, sending 1,450,000 tons into the water.

## Which **oil-tanker spills** have been the **biggest**?

The following table lists some of the major oil spills that originated from tankers.

| Size of Spill (tons) | Vessel | Year | Site |
|---|---|---|---|
| 257,000 | Atlantic Empress | 1979 | West Indies |
| 239,000 | Castillo de Beliver | 1983 | South Africa |
| 221,000 | Amoco Cadiz | 1978 | Atlantic, France |
| 132,000 | Odyssey | 1988 | Mid-Atlantic |
| 124,000 | Torrey Canyon | 1967 | English Channel |
| 123,000 | Sea Star | 1972 | Gulf of Oman |
| 101,000 | Hawaiian Patriot | 1977 | Hawaiian Islands |
| 95,000 | Independenta | 1979 | Turkey, Bosphorus |
| 91,000 | Urquiola | 1976 | North Coast of Spain |
| 82,000 | Irene's Serenade | 1988 | Greece |
| 76,000 | Khark 5 | 1989 | Mediterannean, Morocco |
| 68,000 | Nova | 1985 | Gulf of Iran |
| 62,000 | Wafra | 1971 | South Africa |
| 58,000 | Epic Colocotronis | 1975 | West Indies |
| 41,000 | Burmah Agate | 1979 | Gulf of Mexico, USA |
| 36,000 | Exxon Valdez | 1989 | Alaska, USA |
| 28,000 | Argo Merchant | 1976 | East Coast, USA |
| 21,000 | Ocean Eagle | 1968 | Puerto Rico, USA |

## How are **oil spills cleaned** up?

There are four methods often used to clean up an oil spill:

**Containment:** The initial method used after an oil spill is called containment and recovery: Long, floating barriers called booms are placed around an oil slick to contain it and prevent its spreading or migrating.

Skimmers are then used to physically remove the oil. Vacuum skimmers suck up the oil into storage tanks, while floating disk and rope skimmers are passed through the oil, which adheres to them. Materials such as straw, sawdust and talc, which absorb oil, can be also be added to the slick, and then these materials can be removed.

**Dispersal:** Chemical dispersants can be added to the oil slick to break it up into millions of small globs, which are easily dispersed and carried out to sea. However, this method doesn't remove the oil from the environment, it just spreads it over larger areas of the ocean.

**Bio-remediation:** The bio-remediation method uses microbes and fertilizers to enhance the biological degradation of the spilled oil. This works best when the oil has washed up on shore.

**Burning:** Burning is rarely used, since it spreads the pollution (into the atmosphere), rather than removing it from the environment.

## What new method is being tested to **clean birds' feathers** after an oil spill?

A new method to clean birds' feathers coated with oil is currently being tested in Australia, and entails dusting the feathers with a fine iron powder. The oil sticks more readily to the powder than to the feathers—and the iron and oil are removed when a magnet is used to comb the feathers. This new "dry cleaning" method is less stressful to the birds than are other methods of removal, which rely on detergents that can destroy the waterproofing properties of the animals' feathers.

## How are **sensors** being used to **prevent oil and chemical spills** from ships?

In Galveston Bay, Texas—home to the busy Houston Ship Channel—a series of electronic sensors has been installed to help prevent oil or chemical spills resulting from ship accidents. This series of sensors, installed in 1995 by the National Ocean Service, is called the Physical Oceanographic Real-Time System (PORTS). The real-time information collected by these sensors consists of wind speed, water level, weather conditions, water speed and direction, salinity, and water temperature—which the public may access using the Internet or telephone.

The Galveston Bay area is similar to many coastal bays around the world. The delicate ecosystem and many industries, such as tourism and fishing, are dependent on its continuing good health. This area is also a center for many oil refineries—with 50 percent of the United States' chemical production taking place here. Added to this, Galveston Bay is shallow outside the Houston Ship Channel, with frequent changes in water level due to the effect of prevailing winds on the tides. The normal currents and tides can also be influenced by the many rivers and floodways that feed into the bay.

The information provided by PORTS is critical to pilots and captains of large oil and chemical-carrying ships in the area. It helps them to avoid collisions and groundings—thus preventing their cargoes from spilling out and polluting the fragile bay.

## What **ocean feature** was once proposed as a place to **dump nuclear wastes**?

No one knows where to safely dump wastes from nuclear power plants, nuclear submarines, or other sources. At one time, scientists believed the best place to dump the wastes was the Mariana Trench (in the Pacific Ocean), the deepest part of the ocean floor. Because the trench is an area where one continental plate is subducting (moving underneath) another plate, they believed that the wastes would be "recycled" into the Earth. But there were problems with such a solution, including the methods of delivering the wastes and the ability to keep the wastes within the trench. The idea has since been abandoned.

## Is the **ocean's health** linked to **human health**?

Yes, there is an indication that the ocean's health is linked to human health in many ways. Several scientists recently looked at marine disease outbreaks in the Atlantic and Caribbean oceans— from Labrador, Canada, in the north to Venezuela, in the south. Going back to 1972, they found an increase in marine diseases—some known to infect people directly as they swim or consume contaminated fish and shellfish. The scientists also connected episodes of the El Niño phenomena—the periodic warming of ocean waters off the coast of western South America—with cholera outbreaks in both Peru and the Bay of Bengal (in the Indian Ocean).



## Are there any **noise pollution problems** in the **oceans**?

Yes, scientists recently discovered that noise from supertankers, oil exploration, and new military sonar equipment create "noise" in the oceans. Although it is yet to be proven, some researchers believe such noise may scramble the communication systems of sea life—possibly forcing changes in marine animal migration routes and breeding grounds.

# NATURAL MARINE PROBLEMS

## What is the **red tide**?

The red tide has occurred for centuries in many oceans (and some freshwater areas) around the world, and is either caused by natural or manmade events. The red tide is actually an increase in the production of single-celled, red algae called dinoflagellates, mostly of the two species *Gonyaulax* and *Gymnodinium.* These organisms multiply when there is a profusion of nitrogen and phosphorus, warm temperatures, and little competition. As the organisms grow and eat up the nitrogen and phosphorous compounds, they produce an algae "bloom"—a carpet that absorbs oxygen and cuts off the sunlight to the sea surface. Even after they die, the process uses up more oxygen and marine life is literally suffocated, causing what may be a familiar sight to some beach-goers: a multitude of dead fish, washed ashore.

The "red tide" was named after the intense discoloration of the water from the pigments within the algae. But when these algae proliferate, the water doesn't always become discolored; and sometimes the water can turn red-colored but the conditions are harmless. Therefore, scientists now use the term "harmful algal blooms" (HABs) to describe any damaging blooms—no matter what the color.

In April 1998 a red tide (or HAB—harmful algae bloom) occurred off Hong Kong Island, devastating the region's fishing industry. Here a vessel collects fish carcasses floating at the surface. Scientists believe land-based pollution contributes to HABs—which are toxic to marine life as well as to humans. *AP Photo/Vincent Yu*

## Why do some **shorelines** have more **harmful algal blooms** than others?

Harmful algal blooms (HABs) appear along some coasts mainly because of human activities. The dinoflagellates live off the nitrogen and phosphorus from sewage and rainwater runoff carrying chemicals from agricultural fields, residential lawns, and some industrial effluent. The results of the algal blooms (HABs) are devastating, mostly in the form of massive marine life kills. Not only do they cut off the other organisms' much-needed oxygen, but they can also cause the paralytic poisoning of shellfish—toxins swamp the bodies' defenses and paralyze the nervous systems. This kills off the shellfish and makes them inedible for humans.

## Why has there been a **recent increase** in **harmful algal blooms**?

Harmful algal blooms (HABs) have definitely increased over the last few decades, impacting more fisheries and wider areas—resulting in greater economic losses. The culprit generally blamed for this increase is pollution, since the species that make up HABs thrive on the nutrients present in sewage and other pollutants.

However, only some HABs have been definitively linked to pollution problems, so there are other factors at play here. Sometimes major storms or hurricanes carry algae species into new areas, where they grow and bloom year after year. For example, the area impacted by the New England red tide—caused by blooms of the dinoflagellate species called *Alexandrium*—greatly increased in 1972 as a result of Hurricane Carrie.

People can also cause an increase in the geographical expansion of HABs, sometimes carrying algae species to new areas in the ballast water of ships. The increase in aquaculture (fish-farming) may also contribute to the increase in blooms: Young stock imported to these marine farms, whether fish, seaweed, or shellfish, may carry species that lead to these blooms. The aquaculture facilities themselves may change the amount and type of nutrients found in their local areas, making conditions ripe for a bloom.

Another reason for the increase in HABs may actually be heightened awareness. Improved analytical instrumentation and communication have allowed both for the detection and widespread reporting of blooms, which otherwise might have gone unnoticed.

## What is ***Pfiesteria piscicida***?

*Pfiesteria piscicida* is the latest species of algae to be placed on the list of those causing HABs. This organism was discovered less than a decade ago in an estuary in North Carolina. Since that time, it has been found along with its close relations in estuaries in Maryland, Virginia, and Florida.

*Pfiesteria* is a dinoflagellate harmful to a variety of marine animals, particularly fish; it is also harmful to humans. This organism has been linked to polluted waters, found in areas downstream from hog farms in North Carolina and large chicken farms in Maryland. Scientists believe *Pfiesteria* may take advantage of pollutants both directly and indirectly: It may thrive on the increase in organic nutrients found in polluted waters, and also consume algae that use the inorganic nutrients found there.

## What **natural contaminant** was recently found in **seabird eggs**?

A previously unknown contaminant was recently found in the eggs of seabirds from both the Atlantic and Pacific oceans—and is thought to be of marine origin. To date, most contaminants that accumulate in wildlife originated from human activities, such as industries. However, this new natural contaminant seems to be from a marine source, and has been localized to the surface layer of the ocean. It contains high levels of bromine, chlorine, and nitrogen, and is similar to PCBs and dioxins. It may also be the first known contaminant to be naturally produced by a marine organism, and its effect on the birds' eggs is still unknown.

## What are invasive marine species?

Invasive marine species are organisms that are transported—usually by humans—in some way to an area different than their home. There, the "new" species often successfully establishes itself, invading the pre-existing and otherwise stable native ecosystem. Species that are suddenly transported to a new environment may not survive, but those that do thrive are called invasive, usually usurping other competing organisms within the habitat. The consequences of such invasions include a change in the habitat and the disruption of natural ecosystem links—changes that are often catastrophic for the native species. This process, which occurs in every part of the world, is now thought to be one of the major causes of extinction and ecosystem change over the past few centuries (one of the other leading causes is habitat destruction). Scientists are most concerned because such invasions cause an irretrievable loss of the native biodiversity.

### What **marine animal** recently **invaded** the coastal waters of **Australia**?

The northern Pacific sea star, *Asterias amurensis*—first found in 1986 in Tasmania (an Australian island, south of Melbourne)—has greatly expanded its range and numbers, threatening coastal waters from the state of New South Wales (in southeastern Australia) to the state of Western Australia on the Indian Ocean. This marine invader was probably brought to Australia on the hulls of ships or boats, or even in ships' ballast water. Recent estimates place the number of these sea stars at around 30 million in Tasmania's Derwent Estuary alone, with 12 million in Port Phillip Bay, Victoria, Australia.

Efforts to get rid of the northern Pacific sea star have failed so far. Some of the options being considered for their eradication include chemical control, physical removal, and biological control. However, it appears that any method will take time. For example, genetic methods might get rid of the marine invader without affecting native sea stars, but would take at least five years to develop. Biological control seems the most promising method, and Australian scientists are searching for natural parasites or diseases that will attack the invading sea star. Those that are promising will have to be thoroughly tested for their effect on the environment and native sea stars before they are widely used.

## What **marine pest** is **invading Australian coastal waters**?

The black-striped mussel has recently been found in three marinas in Australia—all now quarantined and one has been treated with chlorine. This mussel is a cousin to the zebra mussel, an invasive species that caused tremendous damage to the Great Lakes of the United States.

The black-striped mussel is a native of the tropical and subtropical waters around Central America. It is believed the species was transported to Australia on ships passing through the Panama Canal. This mussel can thrive in the warmer parts of the South Australian gulfs (along the Indian Ocean), as well as the tropical and subtropical areas of northern Australia. The effects of large populations of these mussels could include fouling of marinas, wharves, all types of vessels, equipment, and marine farms; the economic cost would be staggering. For example, it costs approximately $600 million dollars a year to clean and eradicate zebra mussels in the Great Lakes. And the costs to the overall ecosystem are beyond estimation.

## What **bottom-dwelling sea creatures** are found floating in **Georges Bank**?

Hydroids, tiny tentacled sea creatures normally rooted to the sea bottom, have been found floating free and in large concentrations—as high as 100 per gallon of water—in the area of New England's Georges Bank (a submerged sand bank in the Atlantic Ocean, east of Massachusetts). The approximately 0.10-inch- (0.25 centimeter-) long hydroids are voracious predators of copepods (tiny crustaceans), and are therefore in direct competition with the local fish larvae that also eat the copepods. In addition, the hydroids appear to attack and kill young fish—threatening the populations of cod and haddock, two commercial fish that are already decimated by years of over-fishing in this area (which was closed in 1994 so the waters could be replenished).

Scientists still don't know why the Georges Bank hydroids left the bottom and are now drifting in large numbers. These cousins to the jellyfish have two major life stages, sedentary (bottom-dwelling) and drifting. The hydroids found at Georges Bank are still in their sedentary stage—but something has interfered with their normal life cycle and caused them to drift. Some researchers speculate that seasonal storms contributed to the uprooting of the hydroids. Others believe this was caused by commercial fishing trawlers: The ships' heavy chains hold down trawling nets, stirring up the fish. These chains are dragged along the bottom, and may have uprooted the large number of normally bottom-dwelling hydroids.

## Why are **gray whales dying** along **North America's West Coast**?

Dead gray whales have been washing up along the West Coast in larger numbers than is normal, but this may not be caused by pollution. And it is not an indication that they are in danger of extinction. Rather, scientists believe the population of gray whales has increased dramatically in the last few years—and has outgrown its food supply.

By about fifty years ago, the gray whale had been extensively hunted in this area of the Pacific, leaving a population of less than 2,000 individuals. With commercial whaling banned in 1951 and endangered species status granted in 1973, the gray whale started a comeback. In fact, the latest figures show a population of 26,635—larger than has ever been recorded. As a result, the gray whale was removed from the endangered species list.

This large increase in population has some problems associated with it, such as finding enough food. Gray whales eat creatures living in sediment, such as ghost shrimp and other small crustaceans. Most of the whales washed ashore have been juveniles, with many showing signs of starvation. Whether there are too many gray whales now for the food supply, or there is just a temporary food shortage due to the effects of an event like La Niña (the periodic cooling of Pacific waters off western South America), is not yet known.

## What are **bottom-dwelling organisms** in the **North Pacific Ocean** experiencing?

According to a study conducted from 1989 to 1996, some bottom-dwelling organisms in the eastern North Pacific Ocean are experiencing long-term food shortages. There are numerous organisms that live on the bottom of the ocean, ranging from protozoa and bacteria to crustaceans and worms. These organisms rely entirely on material that drifts down from the ocean's surface for their food. The most likely reason for the decline in food is the increase in ocean temperatures that occurred during the same period.

## Are **California sea otter** populations **declining**?

Yes, the numbers of California sea otters are declining and the cause is unknown. Surveys of these marine animals are conducted each spring by scientists of the U.S. Geological Survey, California Department of Fish and Game, and the Monterey Bay Aquarium. About 375 miles (600 kilometers) of the central California coast are covered—from Santa Barbara to Half Moon Bay.

From a high of 2,377 sea otters in 1995, the population has shown a steady decrease. In spring 1999, the survey counted 2,090 sea otters, an overall reduction of 1.14 percent from 1998—and a sizeable 12 percent since the high (in 1995). The group of sea otters categorized as independents, such as adult and subadults, declined to 1,858, a decrease of almost 5 percent from the previous year (1998).

The cause of this decrease in population is still unknown. About 40 percent of the recorded sea otter deaths was due to disease, but there is no evidence that the rate of disease-related deaths has increased. Other factors that could contribute to this steady decline include the effect of contaminants, lack of food, and entanglement in fishing gear.

# THE OCEANS AND NATURAL HAZARDS

## What are some of the **natural hazards** associated with the **oceans**?

There are many natural hazards associated with the oceans—most of them having to do with the interaction of the ocean with the air. Among the more common natural hazards are:

Hurricanes: A hurricane is a tropical cyclone that can form in the Atlantic Ocean, Gulf of Mexico, Caribbean Ocean, or eastern Pacific Ocean. Hurricanes have sustained winds of more than 74 miles (119 kilometers) per hour in any part of their weather system; a clear, calm area at the center known as the eye; and bands that wrap around the eye extending up to 300 miles (483 kilometers) or more from the center. At the center, the air pressure is very low—the lower it drops, the stronger the winds around it and the stronger the storm.

Volcanoes: Volcanoes are blemishes on the Earth's surface, where hot molten rock called magma reaches the surface. Although most of us think of volcanoes as terrestrial, there are also those that form beneath the ocean's surface—rising up from the seafloor. These oceanic volcanoes increase the surrounding water temperatures; if the volcano builds to above the ocean surface, it can create a new landmass.

Tsunami: A tsunami is an often huge ocean wave that is formed by an earthquake, usually on the ocean floor (but continental earthquakes that occur near coastal areas may also generate tsunamis). These monster waves are mistakenly called "tidal waves"; in fact, they have nothing to do with the tides.

## How do certain **natural hazards** damage **ocean areas**?

Ocean damage by a natural hazard depends on the magnitude of the event. Strong hurricanes in the ocean do most of their damage along coastlines, where winds, waves, and excessive rains often change—or even destroy—natural habitats, not to mention humanmade structures. Most of this damage is in the form of erosion, especially of the coastal sands. For example, the Outer Banks, barrier islands off the coast of North Carolina are particularly susceptible to major hurricanes, as the

### What island area has the greatest concentration of active volcanoes in the world?

There are about 1,133 seamounts and volcanic cones around Easter Island in the South Pacific Ocean, about 2,000 miles (3,200 kilometers) west of the Chilean coast; this area is thought to have the greatest concentration of active volcanoes on Earth. Most of the volcanoes rise more than a mile (or 1.6 kilometers) above the ocean floor, and scientists estimate that two to three of the volcanoes are erupting at any given time. Some scientists also suggest that El Niño events (the periodic warming of the waters off the western coast of South America) may be precipitated by this volcanically active area nearby—but this theory is highly debated.

winds and water erode and deposit the sands—slowly "moving" the islands over time.

Major volcanic eruptions can cause localized weather changes, as the dust and ash spread high into the sky; they can also generate tsunami waves. Huge tsunamis usually do not affect the open ocean; but if they reach the shore, a tsunami can cause major damage to organisms' habitats and humanmade structures.

## Is there any way to **warn people** about an approaching **tsunami**?

It is not easy to determine whether or not an earthquake will generate a tsunami—or if the tsunami will reach a shoreline before it has dissipated. But in the Pacific Ocean, where the majority of tsunamis occur, there is the Pacific Tsunami Warning Center (PTWC) located near Honolulu, Hawaii. The PTWC, operated by the United States National Weather Service, is the headquarters of the Operational Tsunami Warning System, a group that works with other agencies to monitor seismic and tidal stations and instruments around the Pacific Ocean. This information is used to evaluate potential tsunami-producing earthquakes.

There is also an effort to warn the public about tsunamis: The Intergovernmental Oceanographic Commission (IOC), an international cooperative effort that involves many member states of the Pacific rim, which in 1968 set up the International Coordination Group for the Tsunami Warning System in the Pacific. This group meets every two years to review the progress of and update the warning system—and includes such members as Australia, Canada, Chile, China, Colombia, Cook Islands, Democratic People's Republic of Korea, Ecuador, Fiji, France, Guatemala, Hong Kong, Indonesia, Japan, Mexico, New Zealand, Peru, the Philippines, Republic of Korea, Singapore, Thailand, Russia, the United States, and Western Samoa. The IOC also maintains the International Tsunami Information Center, which works closely with the PTWC.

There are several tsunami warning systems in the Pacific Ocean. The Alaska Tsunami Warning Center in Palmer, Alaska, is responsible for releasing public watches and warnings, protecting Alaska and the Aleutian Island chain. The Pacific Tsunami Warning Center in Ewa Beach, Oahu, Hawaii, releases public watches and warnings for all U.S. territories and states bordering the Pacific Ocean.

Seismic stations around the Pacific collect data, detecting earthquakes and their magnitudes; tide stations in the system detect the tsunami and send back data on the nature of the tsunami-driven wave. The results are sent to the weather forecasting offices, state civil defense agencies, the military, and the Federal Emergency Management Agency (FEMA)—all in order to alert the public about an incoming wave. The Pacific Tsunami Warning Center also sends the information to countries and territories throughout the Pacific basin.

## What are some **safety rules** the **tsunami warning centers** give to the public?

The following are some safety rules put out by the tsunami warning centers to help people in coastal areas prepare for an incoming wave.

1. All earthquakes do not cause tsunamis, but many do. When you hear that an earthquake has occurred, stand by for a tsunami emergency.

2. An earthquake in your area is a natural tsunami warning. Do not stay in low-lying coastal areas after a strong earthquake has been felt.

**A huge tsunami washes over Hilo, Hawaii, in April 1946. It was one in a series of waves that originated off the coast of Alaska and hit Hawaii as 55-foot (17-meter) waves occurring at 15-minute intervals, causing great destruction. The event was the impetus for establishing a tsunami warning system. (Note man on pier at left.) *NOAA***

3. A tsunami is not a single wave, but a series of waves. Stay out of danger areas until an "all-clear" is issued by competent authority.

4. Approaching tsunamis are sometimes preceded by a noticeable rise or fall of coastal waters. This is nature's tsunami warning and should be heeded.

5. A small tsunami at one point on the shore can be extremely large a few miles away. Do not let the modest size of the tsunami make you lose respect for all other tsunamis.

6. The Pacific Tsunami Warning Center (PTWC) does not issue false alarms. When a warning is issued, a tsunami exists. The tsunami of May 1960 killed 61 people in Hilo, Hawaii, because they thought it was "just another false alarm."

7. Never go down to the shore to watch for a tsunami. When you can see the wave you are too close to escape it.

8. Sooner or later, tsunamis visit every coastline in the Pacific. Warnings apply to you if you live in any Pacific coastal area.

9. During a tsunami emergency, your local civil defense, police, and other emergency organizations will try to save your life. Give them your fullest cooperation.

## What are a **tsunami watch** and a **tsunami warning**?

According to the Pacific Tsunami Warning Center (PTWC), when an earthquake of sufficient magnitude to generate a tsunami occurs in the Pacific Ocean area, the PTWC personnel determine the location of the earthquake epicenter (the point on the Earth's surface right above where the deep earthquake occurred). If the epicenter occurred in or near the ocean, a tsunami is possible.

Based on this, the center issues a tsunami watch to participants: This indicates where and when the earthquake has occurred, and that the possibility of a tsunami exists. Because the tsunami follows specific physical laws, the tsunami center can produce accurate estimated times the tsunami will arrive at each spot around the Pacific Ocean.

A tsunami warning occurs when there is a positive indication that a tsunami has truly formed—based on data from tide stations nearest the earthquake. When this confirmation is received, the PTWC issues a tsunami warning, alerting warning system participants of the approach of a potentially dangerous tsunami—and repeats the arrival time for all locations.

## What happens if **people ignore tsunami** warnings?

In the past, many people have ignored tsunami warnings, traveling to a beach to "watch the wave." This is very dangerous. As a tsunami approaches, the water appears to retreat quickly, exposing marine organisms—and one of the most common reasons people die from a tsunami is because they gather shells and other marine organisms along the cleared shoreline, ignoring the warnings. But not long after—even as short as a few minutes—a huge wave will replace the dry shore, the initial wall of ocean water moving rapidly. Tsunamis kill people because they do not heed the official warnings.

## Have any **tsunamis destroyed** coastlines?

Yes, there have been many tsunamis that have caused a great deal of coastline destruction. In 1883, the August 26–28 eruption of the volcano Krakatau, a small, uninhabited island in the Sunda Strait west of Indonesia, destroyed most of the island. The eruption created a tsunami in the region that drowned more than 30,000 people on other islands. This volcano is still active today. In 1960, a powerful earthquake (unofficially measuring 9.6 on the Richter scale) originating in Chile, South America, created a huge tsunami that spread halfway across the world—killing 200 people when it reached Japan.

Some tsunamis strike close to shore, giving no chance of warning—even with today's advanced warning system. For example, in July 1998, a tsunami swept across Papua New Guinea, leaving only a scattering of boards and broken brush where a village once stood. In this case, there was no warning, as a magnitude 7.0 earthquake occurred just off the northwest coast. This quake was too close to the shore for monitors to issue warnings in time; scientists speculate that a landslide probably occurred offshore, causing the tsunami. Within 10 minutes after the quake, the tsunami struck—and more than 3,000 people died, making this tsunami one of the most deadly of the twentieth century. About ten such near-shore earthquakes occur every year, but they usually do not produce tsunamis.

## What is the connection between **tsunamis** and **asteroids and comets**?

There may be a connection between tsunamis and space objects called asteroids and comets. Researchers know that these objects have struck the Earth's surface in the past—on land and in the oceans. And they also now know that a strike in the oceans could be just as devastating as one on land: The body hitting the ocean creates waves that retain the enormous energies of the initial strike—in the form of an impact-generated tsunami.

Scientists recently calculated that a 3-mile- (5-kilometer-) wide asteroid striking the middle of the Atlantic Ocean would inundate the East Coast of the United States. On the other side of the ocean, such an asteroid-generated tsunami would swamp the coasts of France, Spain, and Portugal. But there is little need for concern: Such a huge asteroid is estimated to strike the Earth only once in about 10 million years. But that does not mean we should ignore the smaller asteroids: Those about 656 feet (200 meters) in diameter are much more common. One just a little larger, such as an asteroid 1,300 feet (400 meters) in diameter, could cause a football field-high wall of water to inundate a coastline.

# CHANGES IN THE OCEAN AND CLIMATE

## What is **global warming**?

Certain gases in the Earth's atmosphere influence how much of the Sun's energy is absorbed by the planet and how much is radiated back into space. These gases act like the glass in a greenhouse, trapping much of the energy emitted from the ground and making our Earth inhabitable. But if there is an excess of greenhouse gases, the atmosphere is thought to warm—a process scientists call greenhouse warming. And because greenhouse gases include many humanmade air pollutants, such as carbon dioxide, methane, and nitrogen oxides, many scientists believe we will continue to contribute to the overall increase in global temperatures.

## What is happening to the **ice shelves around Antarctica**?

Apparently, climatic warming over the past 50 years has changed some of the ice shelves (or floating glaciers) around the coastline of Antarctica. Two of these shelves are very large—each nearly the size of Texas. Scientists now believe global warming is causing the ice shelves to disappear, either by melting the shelves or keeping them from freezing. Still other scientists believe that melting water could form on top of the ice and drain into crevasses, causing the ice shelves to weaken and eventually disappear. Only time will tell if the disappearance of the ice shelves will continue—or if the melting will have a global effect.

## Is the **sea ice** rimming the **Arctic Ocean melting**?

Yes, data from satellites indicate that the Arctic Ocean sea ice is melting—but there is hesitation among scientists who believe it is difficult to draw

**Some scientists believe global warming is causing the ice shelves of Antarctica to disappear, either by melting the shelves or keeping them from freezing.** ***NOAA***

conclusions from such a short period of measurements. First noticed in 1979 in the far north, some data seem to indicate that the ice is slowly dwindling—some scientists believe this shows that temperatures are climbing across the region. Many climate scientists will be watching future satellite measurements of the sea ice closely, as they believe it could be a warning signal that global warming is becoming a reality.

## Will **global warming** affect the formation and power of **future hurricanes**?

Some scientists believe that rising ocean temperatures will increase the power and speeds of hurricanes (tropical cyclones)—especially in the northwestern Pacific Ocean where the most intense hurricanes occur (they are called typhoons in this area). Because hurricanes draw their power from the heat within the tropical ocean waters, scientists believe that greenhouse warming could increase the strength—in the form of higher winds and heavier rains—of these huge storms.

### What would happen if some or all of the world's ice sheets melted?

If the Greenland ice cap were to melt, the average global sea level would rise more than 24 feet (7 meters); the melting of the Antarctic ice cap would cause a rise in global sea level of about 210 feet (64 meters). And it is estimated that if all the ice sheets in the world melted, the average global sea level would rise by 250 feet (76 meters), engulfing or swamping landmasses.

Using a computer model, one study showed a 1 degree rise in the global ocean water temperature would raise the wind speeds 5 to 12 percent higher in the strongest storms. Some estimates show that an increase in winds (in the strongest storms) by 10 miles (16 kilometers) per hour could double the damage to coastal areas hit by the storms.

In the Atlantic Ocean, it may be a different story. Hurricanes generated in the warm tropical ocean waters off the western coast of Africa could occur less frequently as a result of global warming. Scientists believe that global warming increases the frequency of El Niño events—the periodic warming of the ocean waters off the western coast of South America in the Pacific—and may inhibit hurricane formation in the Atlantic Ocean.

But not all scientists agree. Some suggest that global warming will not have an effect on the strength of hurricanes in any oceans. Plus, hurricane behavior varies from year to year and decade to decade, making it difficult to determine true changes in hurricane strength. We will have to wait and see what effects global warming has on the Earth's climate and oceans.

## Could **iron filings** modify our **climate**?

Researchers from California's Moss Landing Marine Research Laboratories recently carried out an experiment near the Galapagos Islands (in the eastern Pacific Ocean, about 600 miles, or 965 kilometers, off the shore of Ecuador). The researchers dumped iron filings into the ocean near the islands to see if this would encourage the growth of photosynthetic plankton. The idea behind this experiment was that the plankton would grow with the addition of iron and pull more carbon dioxide from the air. They would also keep the carbon dioxide in their systems as they died and fell to the ocean floor, lowering the amount of this gas in the atmosphere. Scientists blame an excess amount of carbon dioxide (a greenhouse gas) for an increase in average global temperatures.

The researchers found that iron did increase the growth of the plankton—with an estimate of 100 tons of carbon removed from the atmosphere in the 7-day trial. But this idea, called ocean fertilization, has many problems associated with it—and many of the questions have yet to be answered, such as What are the side effects? Will other organisms suffer if the number of plankton increases? How do we control the amount of growth? What are the political consequences of such climate modification? Will we end up changing the climate in the wrong way? and, finally, Which countries have the "right" to make such decisions?

## Are scientists trying to pump **carbon dioxide** into the **oceans**?

Yes, scientists are attempting to pump carbon dioxide into the oceans to decrease the amount of this greenhouse gas in the atmosphere. At the Natural Energy Laboratory of Hawaii, a pipe that extends to depths of 3,300 feet (1,000 meters) will be used to send liquid carbon dioxide into the Pacific Ocean. This experiment will determine if it is safe to use the ocean as a "storage area" for large amounts of this gas.

The driving force behind this experiment is the increasing level of carbon dioxide in our atmosphere. This increase comes mainly from human activities, especially the accelerating rate of fossil fuel combustion. Some scientists believe the increase in carbon dioxide, along with other greenhouse gases, is contributing to global warming and climate changes. And some have further suggested that the rate at which forests are cut down and the slow advance in alternative (non–fossil fuel) energy methods make the goal of reducing global carbon dioxide emissions almost impossible. So, instead of reducing emissions, they are attempting to capture the carbon dioxide from power producers before it is sent into the atmosphere. This captured gas would be liquefied under high pressure and injected deep into the oceans using pipes from land or modified tankers. Because the gas would be in liquid form, it would stay in the oceans for several centuries before it finally bubbled to the surface. The oceans are already the largest reservoir of this gas, absorbing about a third of the amount produced by humans.

Advocates for this plan point out that it would have two positive effects: The amount of carbon dioxide reaching the atmosphere would be greatly reduced, which would result in a slowdown in global warming and climate change. This would give people time to find alternative sources of energy and reduce overall emissions. Detractors of this plan think the increased levels of carbon dioxide in the oceans would harm marine life, and they view it only as a temporary fix that does not address the real solution—finding and using alternative sources of energy.

# WATCHING THE OCEANS

## Why is it **important** to **watch the oceans**?

Many scientists and non-scientists alike argue that the oceans are the most important natural resource on our planet. Without the oceans, life as we know it would not exist. These immense bodies of water provide homes to incalculable numbers of animals and plants, which are a vital source of food for humans. The oceans are the engine for our weather and climate; no matter where we live on the planet, we are dependent on the moisture that originates with the interaction between the oceans and the atmosphere. We are, like all plants and animals on Earth, intimately intertwined with the oceans. We must carefully monitor them—for their condition of well-being, or disease, is also our own condition.

## How did **scientists used to monitor the oceans**?

Long ago, scientists watched the oceans the best way they knew how: By sailing out on ships and making careful surface observations. Most of the accumulated data were limited by their being constrained to the ocean's surface: They dredged samples off the bottom and took temperature measurements of the water; they collected plants and animals, and attempted to classify them; and depth measurements were used to determine the topography of the mysterious lands beneath the sea. Some dives were made, but the apparatus available just a century ago was cumbersome and dangerous. Nevertheless, scientists before the twentieth century accumulated a good deal of information about the oceans.

## How do **scientists monitor the oceans today**?

During the decades of the twentieth century, the technology used to gather information about the oceans has grown increasingly sophisticated. Scientists are no longer limited to long ocean voyages and taking samples from the surface. Those who study the oceans now have a wide array of methods to monitor the oceans, including nuclear submarines, safe and flexible diving apparatus, specially equipped aircraft and balloons, Earth-orbiting satellites, deep-sea drilling platforms, remote sensors on the ocean bottom and surface, deep-diving submersibles (both human and remotely piloted), and sophisticated computers and models. What has emerged from the use of all this technology is a picture of a global, holistic, dynamic ocean system. Scientists can now take the pulse of the ocean—and get an idea of its health over time.

# MONITORING OCEAN WEATHER

## What is **Doppler weather radar**?

Doppler weather radar detects the presence and location of an object by bouncing an electromagnetic signal off the object. By measuring the time it takes for the signal to return to the transmitter, the distance and direction of the object from the radar source can be determined. In the case of weather over the oceans, the "objects" are particles of water, ice, or dust in the atmosphere.

Also with weather, the radar signals are constantly moving; the radar signals reflected from the object undergo a change in frequency related to the speed of the object as it moves toward or away from the radar antenna. From this information, meteorologists can calculate both the speed and direction of motion in severe storms. For example, early detection of a developing tornado can often be made before the swirling column of air touches down on the Earth's surface. Thus with the help of conventional radar maps that pinpoint the location and intensity of storm systems, Doppler radar is often used to spot more details about storms over land and oceans.

## What is the **Automatic Surface Observing Systems**?

The Automatic Surface Observing Systems (ASOS) are currently being installed at more than 993 locations throughout the United States and will be the nation's primary surface weather observing network. The ASOS program is a joint effort of the National Weather Service (NWS), the Federal Aviation Administration (FAA), and the Department of Defense (DOD). With a large and modern complement of weather sensors, this network will more than double the number of full-time surface weather observing locations. ASOS will work non-stop, updating observations every minute, 24 hours a day, every day of the year. This system increases the National Weather Service's ability to send out timely weather forecasts and warnings—especially for the aviation community's runway touchdown zones and even to ships traveling along the coasts.

## What is the **Advance Weather Interactive Processing System**?

The Advance Weather Interactive Processing System (AWIPS) is actually a high-speed computer station and communication network that keeps people updated on warnings and forecasts—on land and on the oceans. With AWIPS, forecasters can quickly send out public warnings and forecasts on fast-breaking storms. The AWIPS computer workstations are the nerve centers of the Weather Forecast Offices and Regional River Forecast Centers. The computers receive and process the huge volume of weather data, including information from Doppler Weather Radar, environmental satellites, the hundreds of Automatic Surface Observing Systems (ASOS) sensors, and other data sources such as river gauges.

## Why are scientists **studying** the Indian Ocean **summer monsoon**?

Scientists are studying the Indian Ocean summer monsoon because it frequently disrupts weather patterns and climates around the world—not just in the area in which it occurs. The effects of the monsoon can lead to large amounts of rain and flooding along the coastal areas of the southeastern United States and South America, as well as droughts in Indonesia and Australia.

An international team will soon be participating in the Joint Air-Sea Monsoon Interaction Experiment (JASMINE). Participants in this study include the University of Hawaii, the National Oceanic and Atmospheric Administration (NOAA), the National Science Foundation, National Aeronautics and Space Administration (NASA), the University of Washington, University of Colorado at Boulder, and several Australian agencies. The focus of this work will be in the Bay of Bengal, located in the northeast Indian Ocean, along the southern coast of Asia, where scientists will be studying the upper 1,312 feet (400 meters) of the ocean and its connection to the overlying atmosphere. It is hoped that their findings about the interaction between ocean and atmosphere will provide clues to help predict the summer monsoon—giving some advance warning as to the probability of floods or droughts.

## What possible effect does **El Niño** have on **atmospheric carbon dioxide**?

Some scientists have found that during an El Niño event (the periodic warming of the waters off western South America), the oceans—especially the equatorial Pacific Ocean—retain more carbon dioxide than normal. In fact, enough of the gas is held in the waters that there is also a noticeable variation in the amount of carbon dioxide in the atmosphere.

The primary source of atmospheric carbon dioxide is the equatorial oceans. The equatorial Pacific Ocean is a major site for the release of this gas into the atmosphere. Upwelling (pushing of water caused by temperature differences or diverging currents) in this area bring carbon dioxide–rich waters to the surface, where a natural rate of exchange occurs. However, during an El Niño event (the periodic warming of the waters off western South America), the upwelling decreases—reducing the exchange rate of carbon dioxide by 30 to 80 percent. Since this gas is one of the primary greenhouse gases, any natural variation in its amount interests those who study long-term climate trends such as global warming.

## What is the **interaction** between the **atmosphere** and the **Gulf Stream**?



Scientists are using a mathematical model originally developed to study tornadoes to determine what happens between the atmosphere and the

Gulf Stream (a fast-moving North Atlantic current carrying warm water in a narrow band from the eastern coast of Florida along the East Coast of the United States, past Newfoundland, Canada, and eventually arriving at England). This air-sea computer modeling effort is being applied to an area that ranges from a few to hundreds of square miles in size, and is located along the eastern coast of the United States. Although the primary goal is to understand the interactions between the atmosphere and the Gulf Stream, this effort could also improve weather forecasts.

**Research vessels such as the National Oceanic and Atmospheric Administration's *Surveyor* (shown in 1971 against the backdrop of Pago Pago, American Samoa) are critical to the ongoing effort to collect ocean data, allowing scientists to monitor conditions. *NOAA; Lieutenant (JG) Lester B. Smith, NOAA Corps***

During winter, large masses of cold, dry arctic air move south across North America and collide with the warm waters of the Gulf Stream, producing dramatic weather and storms that track along the northeastern coast of the United States. In fact, these collisions can create areas of intense low pressure along the Mid-Atlantic coast just south of New England—and can also lead to the rapid formation of a major cyclone.

Although oceanographers know about the structure and circulation of the Gulf Stream, its role in climate and weather is not well understood. The air-sea modeling effort will attempt to accurately determine (and perhaps predict) what happens when cold air and warmer waters interact. Any computer model is only as good as the data fed into it; so substantial amounts of accurate atmospheric and oceanic data are needed from the study area. Aircraft equipped with sensors and cameras are being used to obtain the data, as are research vessels.

## Is it possible to **predict** an **El Niño** event?

Recently, scientists were able to successfully predict the occurrence of an El Niño event using a combination of observation and computer models. A decade-long international research project, known as the Tropical Ocean Global Atmosphere (TOGA), used a series of buoys across the Pacific to monitor ocean conditions. During the early 1990s, these buoys found an unusual warming beneath the surface of the ocean. This information was then fed into computer models, which interpreted this warming phenomenon as the precursor to an El Niño event—which did indeed happen toward the late 1990s.

## Can the **oceans** be used to **predict long-term weather**?

Yes, it is possible that the oceans *can* eventually be used to predict long-term weather—but the idea needs more study. The variations in ocean conditions might allow researchers to predict the weather for upwards of 10 years—and this predictive ability may evolve within the next 20 years. In fact, in the late-1990s, an international project was established under the United Nation's World Climate Research Program to accomplish this task: The long-term world weather prediction project is called CLIVAR (Climate Variability and Predictability).

Scientists now understand that there are recurrent patterns to weather, which had long been considered chaotic and unpredictable. These patterns are seemingly controlled by the oceans, so an understanding of—and ability to predict—ocean conditions may well translate into an ability to predict the long-term weather. For example, scientists believe that long-term patterns in the North Atlantic Oscillation (NAO)—a fluctuation in pressure over the North Atlantic responsible for up to half of the climate variability in western Europe—are determined by changes in ocean currents. An understanding of ocean current variations may allow prediction of changes in the North Atlantic Oscillation, and thus the long-term weather for that part of the world.

# MAPPING THE OCEAN BOTTOM

## What are **bathymetric maps**?



Bathymetric maps are topographic maps of the ocean floor, but they do not include navigation information. The information is used mainly by

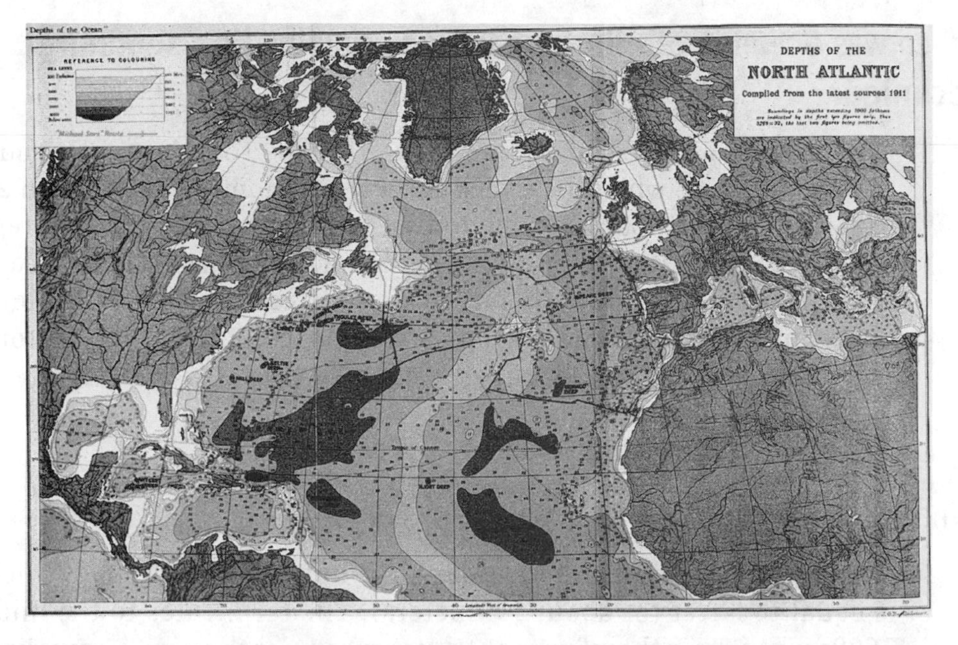


**This chart of the depths of the North Atlantic (compiled by England's Sir John Murray in 1911) accounts for more than 50 years of additional research than did the first attempt at bathymetry—that of American naval officer Matthew Maury in 1855 (*see* page 196.).** ***NOAA Central Library***

researchers—as the maps provide detailed information on the size, shape, and location of significant underwater features. Bathymetry is the measurement of the contours (lines of equal elevation, similar to contours on land) on the ocean floor.

## How did people originally **determine depth** in the **oceans**?

The original way used to determine ocean depth was a technique called line sounding. In this method, a weight was attached to a rope and lowered until it hit bottom; the length of the dropped rope was the depth of the water. One of the problems with this method occurred in deep water: It was hard to tell if the weight had hit bottom. The weight would sometimes sink into soft sediment, giving an erroneous reading. In addition, unless the line was exactly straight up and down (difficult to guarantee in water that is moving), the measured length of the line would be greater than the true depth.

## When was **ocean depth first measured** using the **line-sounding technique**?

Ocean depth was first measured using the line sounding technique around 85 B.C.E., when the Greek thinker Posidonius (also known as Poseidonius) of Rhodes decided to find out the depth of the Mediterranean Sea. He stationed his vessel in the middle of the Mediterranean, and let out a long rope with a large stone attached to one end. More than 1 mile (2 kilometers) of rope had to be let out until the bottom was reached.

## What is a **fathom**?

Sea depths are sometimes measured in a unit called a fathom, which is equal to 6 feet (or about 1.8 meters). The term dates to the mid-1600s when "to fathom" meant to measure a depth using a sounding line, or simply, to make soundings. Today most scientists use meters or feet to express ocean depths, but in some texts, the term fathom still appears.

## What is **echo-sounding**?

A modern way to find ocean depths is called echo-sounding. In this method, sound is transmitted from a ship into the water and bounced off the ocean bottom. The time it takes for the returning sound (known as the echo) to reach the ship can be used to calculate the distance between the ship and the bottom—or the depth. Variations of this echo-sounding technique also can be used to map the topography of the ocean bottom.

## When was **echo-sounding first developed**?

It was not until the 1920s that echo-sounding instrumentation was first developed. This freed scientists from the 2,000-year-old method of line sounding—and finally gave them accurate readings and data showing the true shape of the ocean bottom.

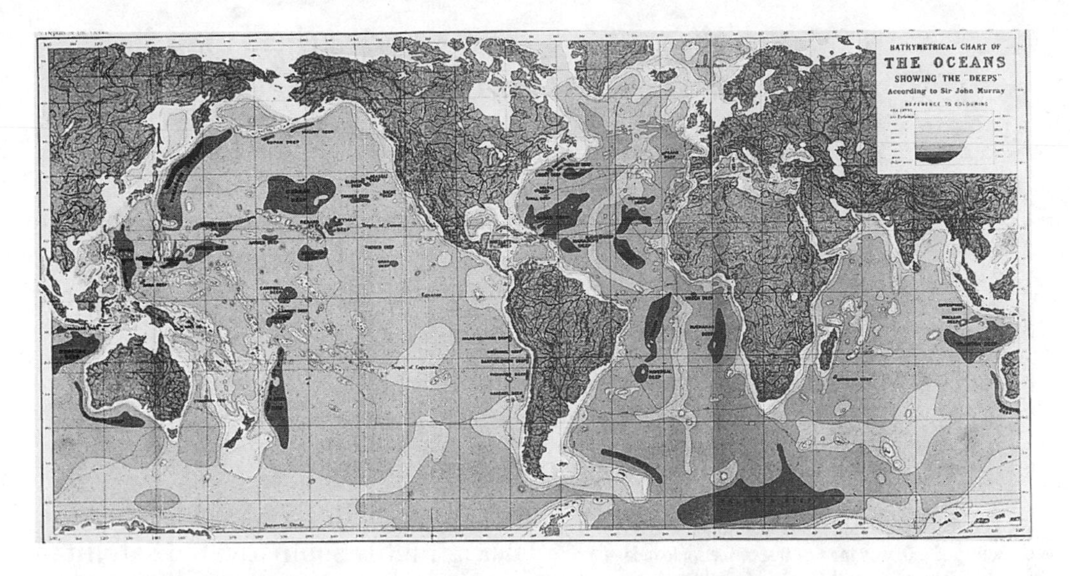


**A 1911 bathymetric map (compiled by Sir John Murray) of the world's oceans; the darkest areas are those of greatest depth.** ***NOAA Central Library***

## **Why** do scientists want to **map the ocean floor**?

Although the original driving force behind mapping the ocean floor was national defense, economics has taken over as the main concern. In 1981, the United States declared the waters and seafloor within 200 miles (322 kilometers) of its shoreline to be an Exclusive Economic Zone (or EEZ)—and other nations soon followed suit. To exploit the vast resources present in these areas, the topography of the ocean floor had to be thoroughly mapped.

## What **information** can be obtained by **mapping the ocean floor**?

New mapping techniques can give a picture of the oceans' bottom as it would appear if all the water were drained out. Using these maps, oceanographers can find areas of the seafloor devoid of strong currents or the danger of underwater avalanches. Such areas would be perfect for laying down communication cables, drilling for oil, or to dispose of solid wastes. By mapping the ocean floor, geologists can locate offshore faults, determining the risk of submarine earthquakes that could potentially affect the shore. And scientists can also use the maps to determine the evolution of the ocean floor—along with the forces that drive it.

The Fessenden oscillator, shown in 1914 aboard the tugboat *Susie D,* was the first test of the sonic radar technology. The crew, headed by Robert Williams, was looking for icebergs, but they also found the ocean bottom. *NOAA Central Library*

## How do researchers **map the ocean floor today**?

Some of the more technologically advanced ways used to map the topography of the ocean floor include satellite altimetry, side-scanning and multi-beam sonar, and underwater photography.

## What is **satellite altimetry**?

Satellite altimetry is a technique using highly sophisticated satellites in Earth-orbit (with the position of the satellite precisely known at all times) to map the sea surface. These satellites cannot measure the seafloor directly, but can record variations in the ocean's surface—which in a broad sense, often translates to the seafloor topography below.

Radar pulses from instruments onboard the satellites are bounced off the ocean waters below to determine the precise distance from the satellite to the water's surface. The combination of the height and location data provides a map of the height of the sea surface, which can vary by as much as 656 feet (200 meters) between locations. These variations in the sea surface height are a result of differences in the Earth's gravity from place to place—most commonly caused by the rugged seafloor topography. For example, a 6,562 foot- (2,000-meter-) high underwater volcano approximately 25 miles (40 kilometers) wide will produce a bulge in the overlying ocean surface about 7 feet (2 meters) high.

However, this technique does have limitations. Seafloor features smaller than approximately 6 miles (10 kilometers) do not produce a detectable variation in the height of the ocean surface—and thus are not able to be mapped using this technique. And sometimes, especially near continental margins, the density of the underlying rock (not the topography) will produce variations in sea height.

## What does **sonar** stand for?

Sonar is an acronym for Sound Navigation and Ranging. It is a technique that uses sound waves to determine the depth of the water underneath a ship, and to detect and determine the position of underwater objects.

## What is **multi-beam sonar**?

Multi-beam sonar is the latest instrument to use sound to determine the depth of the ocean and map the seafloor topography. In this technique, an array of sound sources and listening devices is mounted on the hull of a vessel. The ship travels in a predetermined search pattern, allowing a specific area to be completely mapped. At periodic intervals, a sound burst is emitted from the sources, which are directed toward only a narrow strip of the seafloor; this "illuminated" sound wave strip is oriented perpendicular to the ship's direction. The listening devices are oriented to receive only reflected sound waves coming from narrow seafloor strips oriented parallel to the ship's direction. Thus, the only sound reflections reaching the ship are those that both the emitted sound has reached and the listening devices have sampled. The timing of the sound reflections provides a profile of seafloor depth. The profiles are taken periodically as the ship travels on the surface—building a continuous strip of mapped seafloor.

## What is **side-scan sonar**?

Side-scan sonar is another technique using sound waves to map ocean-floor topography. Two sonar units attached to a sled (one on either side) are towed behind a ship. The sonar units act as both a sound source and listening device, with each one emitting bursts of sound outward to either side. The rough areas of the seafloor scatter the sound waves in all directions, with some returning toward the sleds; these are detected by the listening devices. Using the amplitude of the reflected sound waves and distances from the sled, an image is obtained showing the texture of the seafloor.

## Will scientists really use a **nuclear submarine** to study the **Arctic Ocean**?

Yes, in 1999 scientists were preparing to use the nuclear submarine U.S.S. *Hawkbill* to study the Arctic Ocean. This vessel can travel almost

### Why is underwater photography important?

Underwater photography is important because it provides the most detailed and accurate images of the ocean's bottom. Cameras can be installed on submersibles and remotely operated vehicles (ROVs), or towed along the bottom of a ship. However, the images of the dark ocean bottom are limited to the area that artificial lighting can penetrate. Plus, the ocean is vast—there is too much seafloor for the current number of submersibles and remotely operated vehicles (ROVs) to photograph.

anywhere beneath the Arctic ice pack; it is faster and can cover more area than an icebreaker; and it is quiet and very stable, making it a good platform from which to conduct scientific experiments. It also allows scientists to have access to previously inaccessible areas. The scientists on the U.S.S. *Hawkbill* have two main goals as they travel beneath the Arctic Ocean ice: To map the oceanic ridges and basins on the Arctic Ocean floor and to study the circulation of this body of water.

## How will scientists **map the Arctic Ocean floor**?

The instrument on the nuclear submarine U.S.S. *Hawkbill* used to map the floor of the Arctic Ocean is a sophisticated sonar system called the Seafloor Characterization and Mapping Pods (SCAMP). The data from the SCAMP instrument will provide scientists with three-dimensional images of the Arctic Ocean floor—an area that has been previously unmapped by civilian researchers to this level of sophistication.

This instrument consists of two separate (but complementary) sonar devices: The Sidescan Swath Bathymetric Sonar (SSBS) and the High-Resolution Sub-bottom Profiler (HRSP). These sonar devices are mounted on the underside of the nuclear submarine. The SSBS device measures seafloor depths in an area extending approximately 6 miles (10 kilometers) on either side of the moving submarine. This information can be pieced together to produce large-scale, continuous maps of the Arctic Ocean floor. Backscatter data from this device will highlight outstanding seafloor topography, such as lava flows and scarps (slopes occurring at the boundary between two plates or at a fracture in the Earth's crust). In addition, the HRSP device sends signals that reach under the upper 328 to 656 feet (100 to 200 meters) of the ocean floor—allowing scientists to create images of underlying ocean sediment structure.

Using SCAMP, scientists hope to map the Gakkel Ridge, located between the North American and Eurasian crustal plates—the slowest spreading mid-ocean ridge in the world. There may be less volcanic activity here than at any other mid-ocean ridge, allowing scientists to better understand the ways magma (hot, liquid rock) reaches the surface and how oceanic crust is being created. Another area of interest will be the Chukchi Borderland off Alaska. The presence of glacial scraping and scouring in this area might provide information on the extent and depth of ice cover during the last Ice Age.

## How will scientists study the **circulation of the Arctic Ocean**?

The circulation of water in the Arctic Ocean will be studied by means of sensors mounted on the nuclear submarine U.S.S. *Hawkbill*'s "sail" (the tower on the submarine); by probes that are launched into the water; and by chemical analysis of water samples. The temperature, salinity, and composition data that are collected will help scientists understand the strong circumpolar current that flows around the boundary of the Arctic Ocean—as well as the direction and speeds of the water that circulates throughout the ocean.

## Why is it so important to study **Arctic Ocean circulation**?

The study of the Arctic Ocean circulation is very important not only because the ocean is poorly understood—but because its circulation has global implications. The circumpolar current around the boundary of this ocean takes water from the Atlantic and Pacific oceans and circulates it throughout the Arctic Ocean. Any changes in this current would have an effect on the distribution of the ice pack covering the Arctic. In turn, variations in the ice pack would affect the amount of heat absorbed and reflected in that region of the world—which could affect the overall global climate.

## OCEAN OBSERVATION PROGRAMS

### What is the **Ocean Drilling Program**?

The Ocean Drilling Program (ODP) is an international initiative whose goal is to seek knowledge about the nature of the planet's crust beneath the ocean floor and the history of the ocean basins. This is accomplished mainly by studying cores (samples of the Earth) that are brought up during drilling operations conducted by the program's research vessel, the *Resolution.*

In 1968 an American program began under the Joint Oceanographic Institutions for Deep Earth Sampling (JOIDES), which used the *Glomar Challenger* research vessel to collect data. Since that time, the effort has become international in scope. Besides the United States, full partners in the ODP include Germany, France, Japan, and the United Kingdom; Canada, Australia, Korea, and Chinese Taipei together hold a joint partnership. Other partners include the People's Republic of China and the European Science Foundation, consisting of Turkey, Portugal, Spain, Sweden, Italy, Iceland, Finland, Denmark, the Netherlands, Norway, Belgium, Portugal, and Switzerland.

### How is the **research vessel *Resolution*** used by the **Ocean Drilling Program**?

The JOIDES (Joint Oceanographic Institutions for Deep Earth Sampling) research vessel *Resolution* began her life in 1978 as a conventional oil-drilling ship. But in 1984 she was refitted to accommodate the scientific facilities and equipment needed to conduct research in marine geology.

The ship is 471 feet (144 meters) long and 70 feet (21 meters) wide, with a drilling derrick rising 211 feet (64 meters) above the waterline. While drilling is in process, the ship can maintain her location, even in rough seas, through use of a dynamic-positioning system, which includes 12 computer-controlled thrusters. Drilling can proceed in water depths up to 27,000 feet (8,230 meters) and the ship can handle as much as 30,000 feet (9,144 meters) of drill pipe. The JOIDES *Resolution* has a crew of 62 (including drilling personnel) and carries about 28 scientists, and 20 engineers and technicians. A typical drilling expedition lasts 6 to 10 weeks, and is referred to as a leg. On average, about 6 legs are conducted each year.

There are numerous specialized laboratories onboard to analyze the cores brought up by the drilling rig, such as core handling and sampling, chemistry, X-ray analysis, photography, thin-section preparation, paleomagnetism, physical properties, and paleontology. Also included are refrigerators for core storage and areas for down-hole measurements. A network of computer systems enables scientists to capture and analyze data, and to process reports, tables, and illustrations.

## What **subjects are studied** by the **Ocean Drilling Program**?

The subjects studied by the Ocean Drilling Program (ODP) are numerous and fascinating. For example, ODP researchers are studying the beginnings of the Ice Age, the evolution of continental margins and island arcs, changes in marine sedimentation due to fluctuations in sea level, and the origin and evolution of the oceanic crust (the crust of the Earth underlying the world ocean).

The ODP study of how the Ice Age began is particularly interesting: Before the ODP brought up core samples from the ocean floor, most geologists believed that the Earth entered the Ice Age as a result of glaciation in the Antarctic region. This cooling of the planet's climate took place approximately 2.8 million years ago. But drilling by the ODP in parts of the Norwegian Sea, Baffin Bay, and the West African coastal current system near the Equator, have disproved this theory. All three areas showed relatively sudden changes: The Norwegian Current cooled approximately 2.8 to 2.9 million years ago, Baffin Bay cooled about 2.5 million years ago, and there was a large increase in coastal current upwelling (pushing of water caused by temperature differences or diverging currents) off West Africa from 2.5 to 3.0 million years ago. These changes all happened nearly simultaneously—and therefore could not have been caused by the relatively slow climate cooling from Antarctic glaciation. Instead, the results show that our climate is sensitive to a variety of factors, and more complex than previously thought.

*Resolution* scientists are also studying a cold-water current that is a hundred times the size of the Amazon River—the Deep Western Boundary Current at the southwest Pacific Ocean depths. As part of the global system of ocean circulation that distributes heat around the planet, this current may play a major role in climate change. The scientists followed the current, which carries huge amounts of sediment and sculpts the mud into drifts. It is thought that the drill cores taken by the ODP from these drifts may show an archive of changes in the climate.

## What is the **U.S. Navy's SOSUS**?

SOSUS stands for the Sound Surveillance System—a deep-water, long-range detection system operated by the U.S. Navy, which collects acoustic ocean data. This system was used during the Cold War to track Soviet submarines by their faint acoustic signals. Now SOSUS, with the help of an arrangement of high-gain fixed sensors in the deep-ocean basins, is being used to detect volcanic activity. Acoustic signals from the North Pacific Ocean are monitored and recorded at the Newport, Oregon, facility operated by the National Oceanic and Atmospheric Administration (NOAA). Here, they detect volcanic activity on the Juan de Fuca Ridge, a spreading center (in which two continental plates are splitting apart) in the Pacific Ocean. Scientists can go to the site soon after an event to investigate the volcanic activities.

There are two ways this system gathers data. One is called BEAM, in which 6 hydrophone (instruments that gather sounds underwater) sensors are trained on a certain part of the northeastern Pacific seafloor. In this way, seismic events of magnitudes as low as 1.8 (a very small earthquake) can be detected, as can small volcanic eruptions. The second system called PHONE uses hydrophones from 12 arrays (arrangements of sensors) to detect seismic events as low as magnitude 2.4 in the entire North Pacific basin. Besides detecting small and large seismic events,

Divers prepare to drill into a coral reef—to extract samples (called cores). Cores have been used to help scientists understand changes in the Earth's climates over the past 20,000 years. *NOAA/OAR National Undersea Research Program*

these systems can also catch vocalizations from large marine mammals up to hundreds of miles away—and often pinpoint the location of large, individual animals, such as blue whales.

## What are the **Geophysical Ocean Bottom Observatories**?

The Geophysical Ocean Bottom Observatories (GOBO) are a series of permanent seismograph stations soon to be established on the ocean floor to monitor earthquake activity. After drilling into the floor of the Indian Ocean using the capabilities of the research vessel *Resolution,* Ocean Drilling Program scientists will then install the initial GOBO stations. The Indian Ocean was chosen because of its lack of ocean-bottom seismograph stations.

Although there is already a global network of land-based seismic stations, these only provide earthquake monitoring for the continental regions and some islands. The ocean regions, which represent approximately 70 percent of the Earth's surface, are mostly unmonitored, which creates great gaps in the seismic data and prevents a complete study of the planet's deep interior. The establishment of the GOBO network will help fill these gaps.

## What is the **Hawaii-2 Observatory**?

The Hawaii-2 Observatory (also known as H2O) is the United States' first permanent unmanned deep-ocean floor observatory to monitor and record ocean processes over a period of years. H2O rests on the bottom of the Pacific Ocean almost halfway between Hawaii and California. It sits in 16,400 feet (3 miles, or 5 kilometers) of water on a featureless area of the ocean bottom between the Murray and Molokai Fracture Zones (two long cracks on the ocean floor). The first instruments to go online at the observatory were a seismometer and a hydrophone—allowing scientists to monitor seismic events such as earthquakes and tsunamis.

H2O uses a unique titanium junction box spliced to a retired seafloor telecommunications cable. The cable itself is connected to AT&T's Makaha cable station on the Hawaiian island of Oahu and provides power to the instruments, as well as a direct means of communication. Various types of instruments can be plugged into the junction box using a remotely operated vehicle (ROV) or submersible. When experiments are complete, these instruments can be unplugged and removed, making way for new ones. The oceanographic data received by instruments at H2O will be continuously relayed to land through the cable. This will also permit scientists to directly communicate with the instruments—allowing the equipment to be remotely programmed and easily checked for any problems.



## What is the **Mantle Electromagnetic and Tomography Experiment**?

The Mantle Electromagnetic and Tomography Experiment (MELT) is one of the largest marine geophysical studies ever conducted. Its experiments will discover where hot, liquid rock (magma) is formed and how it moves to the crest of a ridge to form new oceanic crust. In addition, MELT will generate better computer models to show how the magma flows in the Earth's mantle—the layer of the Earth that lies below the crust and above the core and which is comprised mostly of unconsolidated materials.

## How are **nuclear monitoring devices** being used to determine **ocean temperatures**?

Soon, the Comprehensive Nuclear Test Ban Treaty Organization (CNTBTO) will be installing three listening devices in the depths of the Indian Ocean as part of its global network to detect and discourage secret nuclear testing. These devices will be located off Cape Leeuwin in Western Australia, off British Indian Ocean Territory in the northern Indian Ocean, and near the sub-Antarctic island of Crozet.

But nuclear watching is not all the devices will be used for: The Indian Ocean Climate Initiative (IOCI) project will also use these nuclear monitoring devices to measure the speed at which generated sound signals pass through the ocean. The speed of sound is directly related to the temperature of the ocean water—with increasing sound velocity indicating higher water temperatures. The sound signals for these studies will be generated from a site chosen by the IOCI project, probably in the Cocos Islands (a group of 27 small coral islands in the Indian Ocean).

This system will act as a sort of acoustic thermometer, enabling scientists to map ocean water temperatures across thousand of miles. And the resulting data will be extremely valuable for measuring potential global warming and climate changes.

## Will the **sound** from **nuclear monitoring devices** be **detrimental to marine mammals**?

No, the sound generated by the IOCI project has been specifically designed not to disturb marine mammals such as dolphins and whales. It will be a low-level, low-frequency sound emitted at depths of greater than a half mile (1 kilometer).

A similar project using sound to determine water temperature in the Pacific Ocean was delayed for a year, while scientists ran tests to determine the effects this type of sound emission had on marine mammals. The tests were initially conducted using sound generated from a device installed on the Pioneer Seamount off San Francisco; a second sound transmitter was recently installed off Hawaii. Observation of marine mammals, particularly humpback whales, in these areas has turned up no noticeable negative effects on their behavior due to the sound signals.

## What **marine instruments** are used to **measure the oceans**?

There are dozens of marine instruments used to measure the physical and chemical characteristics of the oceans—too many to mention all of them here. Some examples include water quality instruments and sensors that test the chemical quality of seawater; drifting and moored buoys with instruments attached to gather data on the characteristics of water in a certain place; current profilers, such as the acoustic Doppler current profiler that measures the flow of currents in one area; and specifically designed instruments, such as a thermosalinograph, which measures salinity.

## What is the **Sustainable Seas Expeditions** program?



The Sustainable Seas Expeditions program is a 3-phase, 5-year scientific project to thoroughly explore the 12 National Marine Sanctuaries desig-

Moored buoys, such as the deep-ocean buoy (complete with sensors for measuring currents, water temperature, etc.) being launched here, have provided researchers with important ocean data. *NOAA*

nated and protected by the United States government. These sanctuaries are distributed throughout the Atlantic and Pacific oceans and the Gulf of Mexico. They have diverse ecosystems—but only a few have been studied below 100 feet (31 meters).

In the first phase, submersibles will be used to explore each sanctuary down to 2,000 feet (610 meters). The habitats, animals, and plants will be photographed to provide a future resource for scientists and the public. In addition, public interest in the Sustainable Seas Expeditions will be fostered through outreach programs.

The second phase will focus on "charismatic megafauna" such as sharks, turtles, rays, and marine mammals—species that are easily recognizable and fascinating to the general public, as well as scientists. Through the use of advanced technology such as DeepWorker submersibles, Ropecam, and Crittercam (all ways of viewing the sanctuaries), scientists will watch these species in order to help us understand the health of the ecosystems they inhabit. The animals will also be the focus of educational programs—with many of the images and video made accessible to a national audience through live satellite links.

The third phase of the Sustainable Seas Expeditions project will deal with analyzing and interpreting the huge amounts of data obtained. This information will then be used to mount a campaign to raise public awareness of the ocean and the threats it faces.

## How have scientists **studied life** in **extreme conditions**?

One way scientists have studied life in extreme conditions is by traveling to Antarctica during the winter on a two-month-long research trip. As part of the National Science Foundation's Life in Extreme Environments (LExEn) program, scientists journeyed aboard the icebreaker *Nathaniel B. Palmer,* collecting hundreds of samples from the bottom sediment of the ocean and from the frigid waters. They also trudged across the thick ice, collecting ice samples containing microorganisms called protists (made up mainly of algae and protozoa). The largest diversity and density of protists turned out to be in pockets of slush and water in the ice. These pockets were connected to the ocean below by a seep hole—certainly extreme conditions for life to flourish.

The data gathered from these studies, as well as the live cultures that were brought back, will enable scientists to understand these protists. They will determine what temperatures the creatures live in, what they eat, their ecology, how they evolved, and how they are related. This knowledge, combined with studies of organisms living under other extreme conditions (such as hydrothermal vents), will provide a greater understanding of how life can survive under harsh conditions. And perhaps this knowledge will even help to determine if life exists under extreme conditions on other planets.

## Why are **horseshoe crab populations** being **monitored**?

Spawning horseshoe crabs are being counted along Delaware Bay beaches to determine if their populations are declining. Every year, from the end of May until the end of June, the largest population of horseshoe crabs in the world emerges from the Atlantic Ocean to spawning grounds around Delaware Bay. Although there is some evidence that the numbers of these crabs are declining, no official survey has ever been done. Now scientists are surveying along eight beaches in both New Jersey and Delaware to check the population of these creatures.

Horseshoe crabs evolved about 100 million years before the dinosaurs and are related to scorpions and spiders. Not only are they important to commercial fishermen, but also to the field of medicine and the environment. For example, the horseshoe crab's large compound eyes have been studied by scientists in order to learn more about the human eye. A clotting agent called Limulus Amoebocyte Lysatefrom (LAL), extracted from the blood of these crabs, is used to test for bacteria in human blood and is also used in commercially produced intravenous drugs. And environmentally, migratory shorebirds, loggerhead turtles, and some fish depend on the eggs of these crabs as their major source of food.

## What **submarine volcano** might be associated with a **hot spot**?

An active submarine volcano was recently discovered in the Samoa Islands—rising more than 14,100 feet (4,300 meters) above the ocean floor and coming to within 2,000 feet (610 meters) of the surface. This volcano has been named Fa'afafine, a Samoan word that loosely translates as "wolf in sheep's clothing." It is more than 22 miles (35 kilometers) across at its base, a mile (about 2 kilometers) in diameter, and 1,300 feet (400 meters) deep at its central crater. Initially, the location of this volcano was predicted based on an earthquake swarm in the area (an earthquake swarm is a series of minor earthquakes, none of which can be identified as the main shock; they occur in a limited area and usually for a short period of time). Satellite altimetry only showed a small hill-like feature on the ocean floor. However, it took a detailed survey by a research vessel to finally reveal the size and shape of this major underwater volcano.

The volcano's discovery has provided more clues about the formation of hot spot island chains. Some scientists believe that the Samoa Islands are similar to the Hawaiian Islands—another chain of islands that formed far away from a major crustal plate boundary (where most volcanoes are found). It is thought that a hot spot developed in the middle of a plate, sending magma (hot, liquid rock) to the surface. And as the crustal plate moved, the hot spot "stayed" in one place, producing the volcanic island chain.

Fa'afafine is located at the far eastern end of the Samoa Islands—and may be over the present hot spot that created this chain. Researchers plan to return to the volcano using submersibles or remotely operated vehicles (ROVs)—to explore the caldera (crater) and search for hydrothermal hot springs.

## How are **polynyas** being studied in the **Antarctic Ocean**?

A polynya (a word of Russian origin) is an area of open water or thin ice that occurs in the winter near the Antarctic coast in areas usually covered by thick pack ice. A recent mid-winter scientific expedition focused on the Mertz Polynya, which lies near the coast of Adelie Land almost directly south of Tasmania. This area is one of the windiest places on the planet—which may be why this area is relatively ice-free—even amidst an otherwise ice-covered sea.

Scientists gathered data from oceanographic moorings previously laid down in the open water and on pack ice adjoining the polynya. These moorings were placed in depths from 1,626 feet (500 meters) to 9,843 feet (3,000 meters), and contain 34 instruments that measure temperature throughout the water column, salinity, changes in currents, sea-ice thickness, and pressure. Other sources of data included meteorological balloons, probes on helicopters, automatic weather stations, and buoys—all to help discover more about the polynyas.

## Why are **polynyas** important?

The polynyas (regions of open water and thin ice that occur in the winter near the Antarctic coast in areas usually covered by pack ice) may be important for two reasons: They may have an influence on global climate and they may serve as an oasis for life during the harsh South Pole winters.

Understanding the processes occurring within polynyas may be a key to understanding global climate issues. Coastal polynyas are areas of open waters along otherwise ice-bound coasts—often created by strong offshore winds. Any water that is exposed by the wind driving away the ice soon becomes refrozen by thin ice. As the ice forms, salt is left in the underlying water, increasing its density. This cold, dense water, known as Antarctic Bottom Water, sinks deep into the ocean, then spreads along the bottom of all the Earth's oceans. Scientists know this is the major driving force in the circulation of the world's oceans—which, in turn, greatly influences the global climate.

In addition, polynyas, by their very nature, may be areas where krill and other forms of life congregate during the Antarctic winter. Larger animals such as seals, whales, and birds may also be attracted to these oceanic oases—making them important habitats for marine life.

## Is there a proposal to **dam** the **Mediterranean Sea**?

Yes, there is currently a proposal to dam the Mediterranean Sea by using a giant barrage (an artificial bar placed in a watercourse) across the Strait of Gibraltar—the narrow passageway that connects the Mediterranean with the Atlantic, between Spain and Africa. Interestingly enough, scientists are making this suggestion in order to prevent the Northern Hemisphere from being plunged into a new ice age.

This highly speculative theory contends that the damming of the Nile River at Aswan, Egypt, in 1968 deprived the Mediterranean Sea of a large influx of freshwater. In turn, this has led the waters of the Mediterranean to become increasingly saline—and therefore more dense. This dense water is flowing more rapidly than normal out through the Strait of Gibraltar and heading north, in the Atlantic. Some scientists suggest this dense water may be interacting with the Gulf Stream (a warm, fast-moving current originating near the southern tip of Florida and that follows the East Coast of the United States)—pushing warm water from that current into the Labrador Sea, between northeastern Canada and southwestern Greenland.

As a result, the Labrador Sea may become warmer. If this is true, there would be an increase in the rate of evaporation, which would lead to a higher amount of snowfall to build up ice packs in the northern reaches of Canada. These growing ice sheets would reflect more heat, further cooling the Arctic and causing more ice to form. Eventually, we could enter a new ice age in the Northern Hemisphere.

Some scientists think the only way to prevent this from happening is to stop most of the outflow from the Mediterranean. They estimate a barrage to hold back the water's flow would need to have a volume approximately 420 times that of Egypt's Great Pyramid—and would reduce the outflow by about 80 percent. As yet, no one has taken the first step to go through with this controversial project.

## What was the **World Ocean Circulation Experiment**?

The World Ocean Circulation Experiment (WOCE) was an international effort to study the large-scale circulation of the world ocean from 1990 to 1997. It employed dozens of ships, several satellites, and thousands of instruments in order to obtain data about the physical properties and circulation of the global ocean during this limited period. WOCE also supported regional experiments and explored possibilities for longer-term measurements.

The data gathered during the World Ocean Circulation Experiment made possible major improvements in the accuracy of ocean circulation computer models. This is important to our understanding of ocean and atmospheric interactions, and our ability to predict long-term weather and global climate trends. The improved ocean circulation models also enhanced other representations of the ocean and atmospheric circulation. In turn, these models can now better simulate how the ocean and atmosphere together cause long-term weather and global climate changes. It is hoped that these recent models will enable researchers to predict weather months in advance—and the global climate for decades to come.

# MONITORING THE OCEANS FROM ABOVE

## Are **satellites** necessary to **watch the oceans**?

Yes, satellites are necessary to watch the oceans, mainly because they take advantage of the "whole Earth" view—images and data that can only be taken from space. For example, acquiring worldwide data on ocean wave height, sea-surface temperatures, and the abundance of phytoplankton would take the construction and distribution of hundreds of ground data stations to accomplish—but such data are easily measured from space.

## What is the difference between a **polar-orbiting** and **geostationary satellite**?

Of the hundreds of satellites that watch our planet, most are either in a polar or geostationary orbit. The polar-orbiting satellites travel over the Earth from pole to pole; geostationary satellites remain over the same spot on the Earth 24 hours a day.

## What do various **satellite instruments measure from space**?

There are numerous satellite instruments that measure many ocean variables from space. For example, a special instrument on an Earth-orbiting satellite called the Acoustic Doppler Current Profiler can measure the speed of water at various levels in the ocean—all at once. The information gathered from this instrument is used to provide real-time transmission of current and water levels in major harbors around the United States. It has also been used to watch the waters of the Chesapeake Bay during the "Great Chesapeake Bay Swim," making this charity swim that much safer for participants.

Another instrument is the Advanced Very High Resolution Radiometer—a 5-channel scanning device that measures electromagnetic radiation. This information allows scientists to determine cloud cover and surface temperature, mainly of the oceans.

## When did scientists finally measure worldwide sea-surface temperatures?

It was not until the 1980s that a series of satellites was sent into space to record sea-surface temperatures of the entire world ocean.

## What is the **Mission to Planet Earth**?

The National Aeronautics and Space Administration's (NASA) Mission to Planet Earth is a project to establish a global baseline of information about the Earth using not only satellites, but ground-based data. This information will be used to understand how our Earth works, including whether or not our climate is changing—and if it is, if the change is caused by natural and/or human activities.

There may be some interesting technical difficulties as scientists collect such data. The amount of data to be gathered is immense: Some estimate that the entire system watching our world will produce data at a rate equivalent to 3 sets of the *Encyclopedia Britannica* per minute. Scientists need to figure out how to deal with so much data, including sorting, storing, and analyzing the measurements in a reasonable amount of time—and with a minimal amount of effort. There even needs to be a method for determining what information should be thrown away.

## What is NASA's **Earth Sciences Enterprise**?

The National Aeronautics and Atmospheric Administration's (NASA) Earth Sciences Enterprise (ESE) is a long-term research and technology program to look at the Earth's land, oceans, atmosphere, ice, and life—all as a totally integrated, global environmental system. It is built on the foundation of the Mission to Planet Earth—to take in our global environment as a whole.

Participants in the ESE will strive to discover patterns in climate that will allow us to predict and respond to environmental events—such as floods and severe winters—well in advance of their occurrence. Nations, specific regions, and individuals can then use this information to prepare for such events—likely saving countless lives and resources. In addition, the results of this research may provide an objective starting point for the development of sound global environmental policies.

The ESE has three main components: A series of Earth-observing satellites; an advanced data system to hold and process collected information; and teams of scientists who will study the data. The major areas of study include clouds, water and energy cycles, oceans, chemistry of the atmosphere, land surface, water and ecosystem processes, glaciers and polar ice sheets, and the solid Earth.

The first phase of the ESE, which included experiments on the Space Shuttle and various airborne and ground-based studies, has already begun. The next phase began with the launch of the first Earth Observing System (EOS) satellite, Terra (formerly AM-1) and Landsat 7. Scientists hope that the end result of ESE will be to improve human interaction with the Earth's environment. To develop that understanding, the ESE will rely on the EOS Data and Information System (EOSDIS)—designed to archive, manage, and distribute Earth science data worldwide.

## What is the **Earth Observing System**?

The Earth Observing System (EOS) is the centerpiece of Earth Systems Enterprise. It is a program that includes multiple spacecraft and interdisciplinary science investigations. Its purpose is to provide a 15-year data set of key parameters and advances in the scientific knowledge needed to understand global climate change. EOS will feature a series of polar-orbiting and low-inclination satellites for global observations of the land surface, biosphere, solid Earth, atmosphere, and oceans.

The initial Earth Observing System (EOS) plan was to launch a platform with 5 different instruments that would continuously monitor certain facets of the worldwide solid Earth, oceans, and atmosphere. But the high cost of EOS put this plan in trouble and funding was cut. Now, smaller and cheaper satellites are being built to take the place of the single EOS platform.

## What is the **Earth Probes Program**?

The Earth Probes Program is also part of NASA's Earth Science Enterprise (ESE). It addresses unique, highly-focused Earth science research—and is designed to be flexible, adapting to new technologies or sudden events. It complements the Earth Observing System (EOS) by allowing scientists to gather data on certain processes that can only be understood while the craft (satellites) are in special orbit—or by means of an experiment that has a unique requirement.

## What are some of the **satellites** that will **watch the Earth**?

Some satellites and their instruments that have (and will) watch the Earth—particularly the oceans—include TOPEX/Poseidon, QuikSCAT, SeaWiFS, and certain instruments on Terra.

## What is **Terra**?

Terra (formerly called AM-1, named to indicate its morning equatorial crossing time) was a spacecraft launched as part of the Earth Observing System (EOS) in August 1999. Its instruments are designed to obtain data on the major characteristics of global climate change: The physical and radiative properties of clouds; air-land and air-sea exchanges of energy, carbon, and water; measurements of important trace gases; and volcanology.

## What is the **TOPEX/Poseidon** satellite?

The TOPEX/Poseidon satellite was jointly developed by NASA (the National Aeronautics and Space Administration) and the French Centre National d'Etudes Spatiales (CNES). It bounces radar signals off the surface of the oceans to obtain an exact measurement of the distance between the surface and the satellite. TOPEX/Poseidon was launched in the summer of 1992; a follow-on mission, called Jason-1, is scheduled for a year 2000 launch.

The purpose of the TOPEX/Poseidon satellite is threefold: To monitor ocean circulation, reveal the link between the oceans and the atmosphere, and to improve predictions of the global climate. The satellite's instrument is so sensitive, it can detect elevation changes on the order of a few centimeters—across thousands of miles of ocean surface. Because the exact location of the satellite is always known, scientists use this information to generate a topographic map of the global ocean surface, showing "hills and valleys" that cannot otherwise be detected.



## What has **TOPEX/Poseidon** revealed about **El Niño**?

Data from the TOPEX/Poseidon satellite showed that the El Niño event during 1997 and 1998 may have caused an increase in the average global sea level. This level rose approximately 0.8 inches (2.0 centimeters) before eventually returning to normal.

Data from the TOPEX/Poseidon satellite revealed that a rise in sea level was not confined to the "home" of the El Niño event—the tropical Pacific Ocean off the west coast of South America. Sea levels also rose in the Indian and the southern Pacific Oceans as a result of the periodic warming of the seawater. Using this information, scientists were able to calculate the average global sea level—discovering the 0.8-inch- (2.0 centimeter-) increase.

This is the first time scientists have been able to show that an El Niño event caused a change in average global sea level. This was important because understanding such short-term changes will perhaps lead to better knowledge and detection of longer-term climate changes caused by such patterns.

## What did **TOPEX/Poseidon** reveal about **Indian Ocean circulation**?

Data from the TOPEX/Poseidon satellite were used to confirm the presence of large ocean eddies in the Indian Ocean northwest of Australia. These eddies are larger than the island of Tasmania—and approximately a half mile (about 1 kilometer) deep. Their presence was first discovered by Australian and American oceanographers taking part in the World Ocean Circulation Experiment (WOCE), a project to build a decade-long snapshot of the world's oceans. The data from the TOPEX/Poseidon satellite were used to confirm their measurements.

These eddies form every two months as the South Equatorial Current—flowing west toward Africa—builds in strength. The current becomes unstable and the turbulent eddies peel off, developing a life of their own. Scientists also know that the presence of these large, natural features is another key clue needed to understand the climate of this region.

## What is **QuikSCAT**?

QuikSCAT, short for Quick Scatterometer, is a NASA (National Aeronautics and Space Administration) satellite that will provide oceanographers, meteorologists, and climatologists daily detailed images of the swirling winds above the world's oceans. Launched in June 1999, this satellite weighs 1,910 pounds (870 kilograms) and contains SeaWinds, a 450-pound (200-kilogram) state-of-the-art radar instrument called a scatterometer. QuikSCAT will be placed in a circular, near-polar orbit, making a complete rotation around the planet every 101 minutes. It will orbit at a height of 500 miles (800 kilometers) and have a ground speed of 14,750 miles per hour, or 4 miles per second (6.6 kilometers per second).

## How does a **scatterometer** work?

The state-of-the-art radar instrument called a scatterometer transmits high-frequency microwave pulses toward the ocean's surface from orbit, then measures the pulses that bounce back to the satellite. From these "back-scattered" pulses, the scatterometer can sense the ripples caused by winds near the ocean's surface; it then uses this information to compute the speed and direction of the winds. Such an instrument provides high-resolution, continuous, and accurate measurements of wind speeds and directions in all weather conditions—and greatly exceeds the amount of data that can be collected by ships or buoys.

## How will the **data from QuikSCAT be used** on the ground?

The information beamed down to ground stations 15 times a day from QuikSCAT will provide scientists with a continuous image of the direction and speed of the world's ocean winds. This information will be invaluable in weather forecasting, storm detection and tracking, and discerning changes in global climate.

For example, the National Oceanic and Atmospheric Administration (NOAA) will use the data from QuikSCAT to improve its weather forecasting and storm warnings. Meteorologists will also use the information to determine the location, strength, and track of severe ocean storms—popularly known as cyclones, hurricanes, or typhoons.

Scientists will also use the satellite as an El Niño (and La Niña) watcher. The changes in the winds over the equatorial Pacific Ocean are a critical factor in the development of El Niño and La Niña events, the periodic warming and subsequent cooling of those waters. QuikSCAT will be able to see any variations in the winds that may be precursors to these climate patterns. The combination of QuikSCAT's wind data and ocean height data will also give scientists an almost real-time image of wind patterns and their effects on ocean currents and waves.

## What is **SeaWiFS**?

SeaWiFS stands for Sea-viewing Wide Field Sensor, a project under NASA's Mission to Planet Earth that will provide data on global ocean bio-optical properties—changes in the color of the oceans caused by organisms. There are subtle changes in ocean colors that signify that various types and numbers of marine phytoplankton (tiny marine plants) are present. This type of data has both scientific and practical applications. For example, ocean color data can be especially useful for the study of marine plant production at various times of the year. Because an orbiting sensor can see every square mile of cloud-free ocean every 48 hours, ocean color data becomes valuable for determining the abundance of these plants on a global scale. And because the plants release carbon into the oceans and atmosphere, the information on their location and population can be used to better understand the ocean's role in the global carbon cycle.

## How else do **scientists observe the ocean**?

Scientists have other ways to watch the oceans, including balloons with attached instruments (similar to meteorological weather balloons), airplanes outfitted with special instruments, and even buoys.

## How are radio sondes used to watch the oceans?

Radio sondes—instruments used to collect weather data—can be used to determine changes in temperature, pressure, and humidity from high above the oceans. The instruments are usually suspended from helium-filled balloons and they relay data to the ground using radio transmissions.

## What is the **National Data Buoy Center**?

The National Data Buoy Center maintains a network of ocean buoys that automatically log and transmit real-time meteorological and oceanographic data to the National Ocean Service (NOS) in Maryland.

During the 1960s, data from buoys were collected by a variety of oceanographic agencies. But in March 1966, the Ocean Engineering Panel of the Interagency Committee on Oceanography recommended that the United States Coast Guard (USCG) investigate ways to consolidate the national data buoy system. As a result of that investigation, the National Council for Marine Research Resources and Engineering Development endorsed the formation of the National Data Buoy Development Program (NDBDP) in 1967. The NDBDP was created and was placed under the control of the U.S. Coast Guard. In 1970, the National Oceanic and Atmospheric Administration (NOAA) was formed and the NOAA Data Buoy Office (NDBO) was created under the National Ocean Service at Stennis Space Center in Mississippi. In 1982, the NDBO was renamed the National Data Buoy Center (NDBC) and was placed under the National Weather Service (NWS), which is part of the National Oceanic and Atmospheric Administration (NOAA).

## Is **Global Positioning Satellite** technology being used to **watch oceans**?

Yes, techniques are now being perfected using Global Positioning Satellite (GPS) technology to eventually watch the oceans. One such experiment is being conducted by the Marine Observatory at Scripps Institute of Oceanography. Scientists know that signals propagating from GPS satellites to ground-based GPS receivers are delayed by atmospheric water vapor. Methods have been developed to estimate this delay and to measure the surface temperature and air pressure readings at the GPS receiver. This information enables scientists to determine the water vapor at a given point. Currently, this method has only been developed on land, but scientists hope to eventually use this method over the more complicated marine surface.

## What has the ***Galileo* spacecraft** discovered about **Callisto**, a moon of Jupiter?

Data from the *Galileo* spacecraft indicate that Callisto, the second-largest moon of the planet Jupiter, may have a liquid ocean under its icy crust. Initial studies of Europa, the planet's largest moon, showed that electric currents flowing near the moon's surface caused changes in its magnetic field; a salty liquid ocean under the ice is often mentioned as the conductor for these currents. Data from *Galileo* revealed Callisto's magnetic field was also variable—in ways similar to Europa—leading some scientists to theorize the presence of a salty liquid ocean on this moon, too.

# THE PUBLIC & OCEANOGRAPHY

## READING ABOUT THE OCEAN

### What **fiction books** feature the **ocean**?

There are hundreds of works of fiction about the ocean or in which the ocean is an important aspect of the plot. The following list includes some of the more popular or enduring novels involving the sea. For even more selections, check your local library, neighborhood bookstore, or favorite online bookstore.

*The Hunt for Red October* by Tom Clancy
*Jaws* by Peter Benchley
*Moby Dick* by Herman Melville
*The Odyssey* by Homer
*The Old Man and the Sea* by John Steinbeck
*Raise the Titanic!* by Clive Cussler (and other books by the author about the sea)
*Robinson Crusoe* by Daniel Defoe
*South Sea Tales* by Robert Louis Stevenson
*Swiss Family Robinson* by Johann Wyss
*Tales of the South Pacific* by James Michener
*Treasure Island* by Robert Louis Stevenson
*2000 Leagues Under the Sea* by Jules Verne
*Two Years Before the Mast* by Richard Henry Dana

## Which **magazines** feature articles about the **ocean**?

There are many popular magazines that are either devoted to or occasionally publish articles and news about the ocean and ocean exploration. Check your local library, newsstand, or favorite bookstore for the following periodicals: *Audubon Magazine, Discover, International Wildlife* (published by the National Wildlife Federation), *Natural History* (published by the American Museum of Natural History), *Popular Science, Science News, Scientific American,* and *Wildlife Conservation Magazine* (published by the Wildlife Conservation Society).

## Which **children's magazines** feature articles about the **ocean**?

Many magazines for children and young adults publish ocean-oriented articles and news. Check your local library, newsstand, or favorite bookstore for the following periodicals: *Boys' Life* (published by the Boy Scouts of America), *Dolphin Log* (published by the Cousteau Society), *Muse, National Geographic World* (published by the National Geographic Society), *Science World* (published by Scholastic), *Scientific American Explorations,* and *3-2-1 Contact* (published by CTW, the Children's Television Workshop).

## Where can information on **oceans** be found at the **local library**?

If you go to your local library, there are several places where you can find information or books on the ocean. In the Dewey Decimal System, which libraries use to arrange books, the sciences are found in the 500s: Earth science books will be shelved in the 550s (here you'll find information on the physical features of the oceans), botany in the 580s (for information on the plants in the ocean), and zoology in the 590s (for information on penguins or sea turtles). For geographical and historical information on the oceans, start with the 900s; in these shelves you'll find books about famous ocean voyagers and voyages. If you need help, ask your librarian; he or she will be happy to assist you in locating a shelf for browsing or in pointing you to the exact book you're looking for. Additionally, many public libraries have made their catalogs available electronically, meaning you can use a computer terminal (at the library, and sometimes from a remote location such as your own home) to look up books by subject, title, or author. It's a great way to find new reading material—and keep up to date with what's being published about the underwater world.

## Which **nonfiction books** can you recommend to **adults** who want to read more about the **ocean**?

There are hundreds of nonfiction works on the ocean. The following lists some of the more recent popular books. For more information, consult your local library, neighborhood bookstore, or favorite online bookstore.

*Against the Tide: The Battle for America's Beaches* by Cornelia Dean; Columbia University Press (1999).

*Coral Seas* by Roger Steene (Photographer); Firefly Books (1998).

*Deep Atlantic: Life, Death, and Exploration in the Abyss* by Richard Ellis; Knopf (1996).

*The Enchanted Braid: Coming to Terms With Nature on the Coral Reef* by Osha Gray Davidson; John Wiley & Sons (1998).

*Lament for an Ocean: The Collapse of the Atlantic Cod Fishery: A True Crime Story* by Michael Harris; McClelland & Stewart (1998).

*Life on the Edge: Amazing Creatures Thriving in Extreme Environments* by Michael Gross; Plenum Press (1998).

*The Perfect Storm* by Sebastian Junger; W.W. Norton and Company (1997).

*The Sea Around Us* by Rachel L. Carson, Ann H. Zwinger (Introduction), Jeffrey Levinton (Photographer); Oxford University Press (1991).

*Sea Change: A Message of the Oceans* by Sylvia Earle, Joelle Delbourgo (Editor); Fawcett Books (1996).

*Secrets of the Ocean Realm* by Michele Hall, Howard Hall, Peter Benchley (Foreword); Beyond Words Publishing & Carroll & Graf (1997).

*Song for the Blue Ocean: Encounters Along the World's Coasts and Beneath the Seas* by Carl Safina; Henry Holt & Company, Inc. (1998).

## What are some of the **children's books** about the **ocean**?

There are many books about the ocean that are geared toward children and young adults. The following list includes some of the more recent or highly acclaimed titles. For more information, consult your local library, neighborhood bookstore, or favorite online bookstore.

*Adventures of the Shark Lady: Eugenie Clark Around the World,* by Ann McGovern; Scholastic (1998).

*Animals of the Oceans* (Animals by Habitat) by Stephen Savage; Raintree/Steck Vaughn (1997).

*Beneath the Oceans* (Worldwise) by Penny Clarke, Carolyn Scrace (Illustrator); Franklin Watts (1997).

*Beneath the Waves: Exploring the Hidden World of the Kelp Forest* by Norbert Wu; Chronicle Books (1997).

*By the Seashore* by Maurice Pledger; Advance Marketing Services (1998).

*The Caribbean Sea* (Life in the Sea) by Leighton Taylor, Norbert Wu (Illustrator); Blackbirch Marketing (1998).

*Coral Reef* (Watch It Grow) by Kate Scarborough, Michael Woods (Illustrator); Time Life (1997).

*The Coral Reef Coloring Book* by Katherine S. Orr; Stemmer House Publishing (1988).

*Crafts for Kids Who Are Wild About Oceans* by Kathy Ross, Sharon Lane Holm (Illustrator); Millbrook Press (1998).

*Dive to the Deep Ocean: Voyages of Exploration and Discovery* by Deborah Kovacs; Raintree/Steck Vaughn (2000).

*Diving Into Darkness: A Submersible Explores the Sea* by Rebecca L. Johnson; Lerner Publications Company (1989).

*Exploring the Oceans: Science Activities for Kids* by Anthony D. Fredericks, et. al.; Fulcrum Publishing (1998).

*Fountains of Life: The Story of Deep-Sea Vents* (First Book) by Elizabeth Tayntor Gowell; Franklin Watts (1998).

*Great Barrier Reef* (Wonders of the World) by Martin J. Gutnik, Natalie Browne-Gutnik; Raintree/Steck Vaughn (1994).

*The Hawaiian Coral Reef Coloring Book* (Naturencyclopedia Series) by Katherine S. Orr; Stemmer House Publishing (1992).

*How to Be an Ocean Scientist in Your Own Home* by Seymour Simon, David A. Carter (Illustrator); Lippincott-Raven Publishers (1988).

*Incredible Facts about the Ocean: How We Use It, How We Abuse It* by W. Wright Robinson; Dillon Press (1990).

*Learning About Shells* by Sy Barlowe, Richard Bonson; Dover Publications (1997).

*Ocean* (Eyewitness Books) by Miranda MacQuitty, Frank Greenaway (Photographer); Knopf (1995).

*Ocean* by Andrea Posner; Ladybird (1996).

*Oceans* by Seymour Simon; Mulberry Books (1997).

*Oceans* (First Starts) by Alex Pang (Illustrator), Joy A. Palmer; Raintree/Steck-Vaughn (1994).

*Oceans* (Interfact) by Lucy Baker, Francis Mosley (Illustrator); World Book Inc (1997).

*Oceans* (Make It Work Geography) by Andrew Haslam, Barbara Taylor; World Book Inc. (1997).

*Oceans & Rivers* (Changing World) by Frances Dipper; Thunder Bay Press (1996).

*Our Mysterious Ocean* by Peter D. Riley, et. al.; Readers Digest (1998).

*Seas and Oceans* (Usborne Understanding Geography) by Felicity Brooks, et. al.; E D C Publications (1994).

*Seaweed Book: How to Find and Have Fun With Seaweed* by Rose Treat, et. al.; Star Bright Books (1995).

*The Shell Book* by Barbara Hirsch Lember; Houghton Mifflin (1997).

*This Is the Sea That Feeds Us* by Robert F. Baldwin, Don Dyen (Illustrator); Dawn Publications (1998).

*Under the Sea in 3-D!/With 3-D Glasses* by Rick Sammon, Susan Sammon; Elliott & Clark Publishing (1995).

*Usborne Book of Ocean Facts* by B. Gibbs, Anita Ganeri; E D C Publications (1991).

*What's Under the Sea?* (Usborne Starting Point Science) by Sophy Tahta, Stuart Trotter (Illustrator); E D C Publications (1994).

# THE OCEAN IN TELEVISION AND MOVIES

## What **television programs** have been about the ocean?

Popular TV series that featured the ocean included *Sea Hunt, Flipper, Voyage to the Bottom of the Sea, The Man from Atlantis, The Undersea World of Jacques Cousteau,* and, most recently, *SeaQuest,* although most of these were fictionalized (dramas). As of this writing, there are no television series about the ocean, but The Discovery Channel, The Learning Channel, and, to a lesser extent, The History Channel (Modern Marvels) feature special programs about the sea and underwater exploration. Additionally, Public Television's *Nova,* a popular science program, occasionally covers the subject. Check local listings or visit their websites at www.discovery.com, www.learningchannel.com, www.historychannel.com, and www.pbs.org/wgbh/nova to learn about past and upcoming programs.

## What **popular movies** feature the **oceans**?

Since Hollywood began making movies, filmmakers have often turned to the ocean for subject matter. The following list includes films (mostly adventure dramas and science fiction) about underwater exploration, fishing, shipwrecks, submarines, and ocean life. (All films listed here were given a PG-13 rating or lower by the Motion Picture Association of America, preceded the ratings system, or were made for television.) For even more options, check the category indexes of *VideoHound's Golden Movie Retriever* (published annually by Visible Ink Press), under the headings Deep Blue, Go Fish, Sail Away, Sea Disasters, Shipwrecked, and Submarines.

The popular television show *Flipper* ran from 1964 to 1967. Here actors Tommy Norden, from left, Brian Kelly, and Luke Halpin are pictured with one of the seven dolphins that played the title role. *AP Photo/NBC*

*Above Us the Waves* (1956)

*The Abyss* (1989)

*The Day Will Dawn* (1942)

*The Deep Blue Sea* (1999)

*Enemy Below* (1957)

*Flipper* (the original, which preceded the popular TV series, was made in 1963; it was updated and remade in 1996)

*Flipper's New Adventure* (1964)

*Flipper's Odyssey* (1966)

*For Your Eyes Only* (1981); a James Bond adventure

*Free Willy* (1993); there were two sequels, *Free Willy 2: The Adventure Home* and *Free Willy 3*

*The Hunt for Red October* (1990); based on Tom Clancy's blockbuster novel

*It Came from Beneath the Sea* (1955)

*Jason and the Argonauts* (1963)

*Jaws* (1975); there were three sequels, *Jaws 2, Jaws 3,* and *Jaws: The Revenge*

*Mission of the Shark* (1991)

*Moby Dick* (1956)

*Namu, the Killer Whale* (1966); based on a true story

*A Night to Remember* (1958); about the sinking of the *Titanic*

*The Old Man and the Sea* (1958)

Swimmers flee the ocean in panic in this scene from the blockbuster movie *Jaws*—the story of a New England seacoast town dealing with an unwelcome summer visitor—a menacing shark. *Kobal*

*Poseidon Adventure* (1972)

*The Spy Who Loved Me* (1977); a James Bond movie

*Titanic*; the 1953 film stars Clifton Webb and Barbara Stanwyck; the box-office hit of 1998 stars Leonardo di Caprio and Kate Winslett

*20,000 Leagues Under the Sea* (1954)

*Voyage to the Bottom of the Sea* (1961)

*Waterworld* (1995)

## What **non-fiction videos** are available about the **ocean**?

While movies are engaging and some of them strive for historical accuracy, they are, after all, works of fiction or, at best, fictionalized accounts of actual events. But you can also rent or buy videos about the real-life ocean explorers and adventures. Here are several good ones to start with.

*Audubon Video: If Dolphins Could Talk* (1990)



*Cousteau: The Great White Shark, Lonely Lord of the Sea* (1992)

The action movie *The Abyss* told a riveting tale of underwater exploration. In this scene, three divers investigate the depths while a research submarine (or ROV, remotely operated vehicle) lights the area. *Kobal*

*Eyewitness Fish* (1996)

*Eyewitness Ocean* (1996)

*Great Minds of Science: Oceanography* (1997), with Dr. Sylvia Earle, et al.

*In the Company of Whales* (1992)

*In the Wild: Dolphins with Robin Williams* (1999)

*Jacques Cousteau's Voyage to the Edge of the World: An Arctic Adventure* (1991)

*National Geographic: Amazing Planet: Shark-a-Thon* (1997)

*National Geographic: The Great Whales* (1978), directed by Nick Noxon

*National Geographic: Killer Whales: Wolves of the Sea* (1993), with David Attenborough

*National Geographic: Ocean Drifters* (1999)

*Raging Planet: Hurricane* (1998)

*Raging Planet: Tidal Wave* (1997)

*Secrets of the Ocean Realm,* (1998; 5 volumes) directed by Howard Hall, Michele Hall

*Visions of Nature: Undersea World/Ocean Breeze* (1998)

# GETTING INVOLVED

## What **oceanic expeditions** are open to **public participation**?

There are many oceanic expeditions open to the public. The short list that follows describes some of the more popular or well-established groups that welcome the interested adventurer.

Expedition Research
13110 NE 177th Place, Suite 144
Woodinville, Washington 98072
Tel.: (360) 668-2670
Fax: (360) 668-9370
http://www.expeditionresearch.org/english/
These worldwide expeditions are open to interested adventurers and professionals who wish to help scientists with their experiments. Participants can assist on archaeological digs or sign on for one of the oceanography projects (some of which include ocean sailing and polar research). And there is a range of other projects in between. The Expedition Research links parties with common interests and goals—and eventually will bring the excitement of these expeditions into classrooms and living rooms worldwide.

The Exploration Company
Whale Watching Expedition
Eleele Shopping Center
Kaumuali'i Highway
Eleele, Kauai
This whale-watching expedition aboard the *Na Pali Explorer* takes participants on a 2-1/2-hour tour to see the whales along Kauai's majestic South Shore. The watch also goes past coastal towns and beaches; in addition to whales, participants also frequently sight flying fish, dolphins, and green sea turtles.

**The bow of the *Titanic,* at rest on the ocean floor of the North Atlantic (about 400 miles, or 640 kilometers, southeast of Newfoundland, Canada). In September 1998, the first tourists, aboard a tiny submersible, glimpsed the ocean-liner's wreckage.** ***AP Photo/Ralph White***

Ocean Voyages
1709 Bridgeway
Sausalito, California 94965
Tel.: (415) 332-4681
Fax: (415) 332-7460
sailing@oceanvoyages.com
Founded by Mary T. Crowley in 1979, this group enables people all over the world, at all levels of skill, to participate in adventure sailing programs; the group also offers a wide range of adventure travel and educational programs. Programs are offered each year in more than 14 countries—in places like the Aegean, Corsica, Sardinia, the Grenadines, the Galapagos, Chile, Venezuela, the California Channel Islands, Hawaii, the Pacific Northwest, the San Juan Islands, British Columbia, Alaska, French Polynesia, New Zealand, Australia, Fiji, and Vanuatu. Ocean Voyages' worldwide network of professional vessels exceeds 400. Special naturalist and scuba diving programs are also offered in certain areas.

**The life of the aquanaut: Divers conduct experiments outside *Hydrolab,* on the seafloor near St. Croix (U.S. Virgin Islands). The research vessel was "home" to more than 700 aquanauts between 1965 and 1985.**
***NOAA/OAR National Undersea Research Program***

Zegrahm DeepSea Voyages
1414 Dexter Avenue North, Suite 2001
Seattle, Washington 98109
Tel.: (206) 285-3743 or 888-772-2366
Fax (206) 285-7390
http://www.deepseavoyages.com/home.htm
This group offers several programs for the deep-sea adventurer. Recent adventures have included a working scientific expedition to view the wreckage of the legendary R.M.S. *Titanic*; exploring undersea volcanoes and hydrothermal vent sites off the Azores archipelago; visiting the wreck of the H.M.S. *Breadalbane,* under the ice of the Arctic Circle; and traveling on expeditions off the coast of Vancouver Island, British Columbia, to observe the rare 25-foot (7.6-meter) Sixgill sharks and the Giant Pacific octopus.

## How can I **participate** in the **Aquanaut program**?

The Aquanaut program offers a 7-day Aquanaut Adventure through Zegrahm Expeditions and Space Voyages (the same group mentioned above)—allowing the adventure traveler to have a hands-on experience underwater. This program features high-tech underwater equipment and systems, and is designed for people of all skill levels—from those with little snorkeling experience to certified divers. Their base is at the Man in the Sea research facility and lagoon in Key Largo, Florida. Each day, participants can pilot a mini-submarine, work in an undersea laboratory, or sleep overnight at the Jules' Undersea Lodge, under 22-feet (7 meters) of water. (The Man in the Sea program was co-founded by Scott Carpenter, the *Mercury 7* astronaut and SeaLab commander.) Space is limited to only 18 participants at a time.

**Proof positive that the field of oceanography is open to women as well as men: The all-female crew of *Tektite II* does "re-breather" training for a mission in 1970. *NOAA/OAR National Undersea Research Program***

## What is **Project Oceanography**?

Project Oceanography is a distance-learning educational television program for middle school science students. It's offered through the University of South Florida's Department of Marine Science in St. Petersburg. The weekly, half-hour live television broadcasts feature a range of oceanography topics and hands-on experiments taught by practicing scientists. The programs are broadcast to registered schools in more than 20 states—with the primary goal to enhance science education in middle schools, especially in oceanography. (You can visit their Internet site at http://www.marine.usf.edu/pjocean)

The university also runs an oceanography camp, a three-week program for girls poised to enter high school. It offers a setting that helps encourage young women to consider opportunities in science—and to understand the natural world around them. (You can visit their Internet site at http://www.marine.usf.edu/girlscamp)

## How do Florida's **specialty license plates** help **protect dolphins**?

In the state of Florida, specialty license plates are now available with a picture of a leaping dolphin and the words "Protect Wild Dolphins." These plates are available to residents for an additional twenty dollars per year. The money collected is used to fund ongoing wild dolphin research projects and educational programs. Some funds will also be given to nonprofit agencies and organizations involved in the rescue, rehabilitation, and release of injured or sick dolphins. The program is administered by the Harbor Branch Oceanographic Institution.

## Are there problems with **humans feeding** and interacting with **marine mammals**?

Yes, there are a variety of problems caused by humans feeding and otherwise interacting with marine mammals such as whales, sea lions, seals, and wild dolphins. For one thing, these animals are wild, and can attack and bite people. But the most serious consequences are experienced by the marine mammals themselves—not by humans.

When people feed these animals, or repeatedly get close to them, their natural fear of humans can be lost. This can leave them vulnerable to injury or death due to collisions with boats. Feeding animals, such as dolphins, from boats also leads them to associate these craft with food. Aside from causing the animal to incur possible injury, such off-boat feeding may train the animals only to beg for food, prompting them to neglect their young. Getting too close to seal pups may prevent the return of the mother from her foraging. Walking near or swimming with marine mammals may harass and stress them. Many people don't realize there are federal laws (especially those under the Marine Mammal Protection Act) that seek to protect these animals by prohibiting this type of activity—and that make this kind of activity illegal and punishable. The best way to interact with marine mammals is the same as for interacting with other wildlife—from a distance; observing but not disturbing.

## Are any **ocean or marine societies** open to the public?



Yes, there are several ocean societies that welcome new members. The following lists a few.

Center for Marine Conservation
725 DeSales Street, N.W., Suite 600
Washington, D.C. 20036
Tel.: (202) 429-5609
Fax: (202) 872-0619
http://www.cmc-ocean.org/

The Cousteau Society
870 Greenbrier Circle,
Suite 402
Chesapeake, Virginia 23320
Tel.: 1-800-441-4395
http://www.cousteau
society.org/

The Oceania Project
P.O. Box 646
Byron Bay NSW 2481
Australia
Tel.: 61 2 6687 5677
http://www.nor.com.au/users/oceania/

Oceanic Society
Fort Mason Center, Building E
San Francisco, California 94123
Tel.: 1-800-326-7491
Fax: (415) 474-3395
http://www.oceanic-society.org/

Save the Manatee Club
500 N. Maitland Avenue
Maitland, Florida 32751
Tel.: 1-800-423-5646 in U.S.; (401) 539-0990 outside the U.S.
Fax: (407) 539-0871
http://www.savethemanatee.org/

A visitor to the Monterey Bay Aquarium views some of the marine life—sea nettles and jellyfish—on display. *AP Photo/Paul Sakuma*

## What are some of the more popular **aquariums** in the **United States**?

There are many aquariums throughout the country that are open to visitors and offer excellent learning opportunities. The following list, which is organized alphabetically by state, includes some of the larger or more well-known aquariums. As hours and exhibits change, it's always advisable to call before planning a visit. For more listings, consult a city travel guide or local visitors' bureau.

Cabrillo Marine Aquarium
3720 Stephen White Drive
San Pedro, **California** 90731
Tel.: (310) 548-7562.
Specializes in the marine life of southern California.

Monterey Bay Aquarium
886 Cannery Row
Monterey, **California** 93940
Tel.: (408) 648-4888
Features the rich marine life found in Monterey Bay, including sea otters and a kelp forest.

Sea World of California
1720 South Shores Road
San Diego, **California** 92109
Tel.: (619) 226-3901
Marine zoological park with shows.

Stephen Birch Aquarium Museum at the Scripps Institution of Oceanography
2300 Expedition Way
La Jolla, **California** 92093
Tel.: (619) 534-3474

Features marine animals from California and Mexico as well as the Indo-Pacific.

Clearwater Marine Aquarium
249 Windward Passage
Clearwater Beach, **Florida** 33767
Tel.: (813) 441-1790
Features local marine life; exhibits include sea turtles, river otters, dolphins, as well as mangrove and seagrass habitats.

The Maritime Aquarium
10 North Water Street
South Norwalk, **Connecticut** 06854
Tel.: (203) 852-0700
Features an aquarium, museum, and IMAX Theater. Many educational programs, including trips on the R/V *Oceanic*.

Mystic Aquarium
55 Coogan Boulevard
Mystic, **Connecticut** 06355
Tel.: (860) 572-5955
More than 6,000 sea creatures, including seals, dolphins, penguins, and whales, are shown in 49 exhibits.

Museum of Natural History
The Smithsonian Institution
10th Street at Constitution Avenue
Washington, **D.C.** 20560
Tel.: (202) 357-2700
Caribbean reef and turtle grass ecosystem exhibits.

The National Aquarium
14th and Constitution Avenues, NW
Washington, **D.C.** 20230
Tel.: (202) 482-2825
Fax: (202) 482-4946
Over 80 exhibits, including a 6,000-gallon shark tank.

Sea World of Florida
7007 Sea World Drive
Orlando, **Florida** 32821

Chicago's Shedd Aquarium, which includes the largest indoor aquarium in the world, is a destination for all ages. *CORBIS*

Tel.: (407) 351-3600
Marine zoological park with shows.

Waikiki Aquarium
2777 Kalakaua Avenue
Honolulu, **Hawaii** 96815
(808) 923-9741
Features about 2,000 animals representing 350 species; includes coral reef exhibits. Visitors can also see endangered species such as the Hawaiian monk seal, threatened species like the Hawaiian green sea turtle, and rare fish such as the masked angelfish from the remote Northwest Hawaiian Islands.

The Shedd Aquarium
1200 South Lakeshore Drive
Chicago, **Illinois** 60605
Tel.: (312) 939-2438
More than 6,000 aquatic animals from every region of the world. A re-creation of a Pacific Northwest environment features beluga whales, sea otters, Pacific white-sided dolphins, and harbor seals.

Visitors at the Oregon Coast Aquarium view Keiko, the killer whale (Orcinus orca) that was used in the movie *Free Willy*. *CORBIS/Kevin Schafer*

The Aquarium of the Americas
1 Canal Street
New Orleans, **Louisiana** 70130
Tel.: (504) 861-2537
Exhibits include one of the largest collections of sharks.

The National Aquarium
Pier 3, 501 East Pratt Street
Baltimore, **Maryland** 21202
Tel.: (410) 576-3800
Features an Atlantic coral reef, tropical rain forest, sharks, and a seal pool.

The New England Aquarium
Central Wharf
Boston, **Massachusetts** 02110-3399
Tel.: (617) 973-5200
More than 2,000 aquatic animals from starfish to piranha, including turtles, sharks, penguins, and sea lions.

Thomas H. Kean New Jersey State Aquarium
1 Riverside Drive
Camden, **New Jersey** 08103-1060
Tel.: (609) 365-3300/8332
Emphasis on the environment of New Jersey, with more than 40 different species of fish, harbor and gray seals.

The Aquarium for Wildlife Conservation
West 8th Street, Coney Island
Brooklyn, **New York** 11224
Tel.: (718) 265-FISH
Home to thousands of fish and a multitude of other marine creatures, including beluga whales, California sea otters, sea turtles, and bottlenose dolphins.

Sea World of Ohio
1100 Sea World Drive
Aurora, **Ohio** 44202
Tel.: (216) 562-8101
Marine zoological park with shows.

Oregon Coast Aquarium
2820 SE Ferry Slip Road
Newport, **Oregon** 97365
Tel.: (503) 867-3474
More than 190 animal species that thrive in Oregon's unique marine habitats; it was also home to Keiko, the orca who "starred" in the movie, *Free Willy.*

Sea World of Texas
10500 Sea World Drive
San Antonio, **Texas** 78251
Tel.: (210) 523-3606
Marine zoological park with shows.

Virginia Marine Science Museum
717 General Booth Boulevard
Virginia Beach, **Virginia** 23451
Tel.: (757) 425-FISH
Features sea turtles, seals, an aviary, shark tank, and a 3-D IMAX theater.

Seattle Aquarium
Pier 59
Seattle, **Washington** 98101
Tel.: (206) 386-4320
Hundreds of species of invertebrates, fish, birds, and marine mammals from Puget Sound and around the world, including the giant Pacific octopus.

## What are some of the more popular **aquariums** in **other countries**?

There are many aquariums around the world—more than can be listed in the pages of this book! Consult travel guides or your travel agent to find out if there is an aquarium in the foreign city you plan to visit. These sources will provide you with directions on how to get there as well as information on hours and fees, which are subject to frequent, and often seasonal, change. (It's always advisable to call ahead.) The following short list of aquariums includes a sampling of "ocean" experiences that await world travelers.

The Great Barrier Reef Aquarium
2-68 Flinders Street
P.O. Box 1379
Townsville, Queensland **Australia**
Tel: in Australia 07 4750 0800; outside Australia +61 7 4750 0800
Fax: in Australia 07 4772 5281; outside Australia +61 7 4772 5281
A collection of the creatures along the Great Barrier Reef—and more.

Sydney Aquarium
Aquarium Pier
Darling Harbour NSW 2000
Sydney, **Australia**
Tel.: +61 2 9262 2300
Fax: +61 2 9290 3553
Exhibits include the Open Ocean Oceanarium, with underwater tunnels featuring stingrays, fish schools, and sharks; a Seal Sanctuary with Australian fur seals, harbor seals, and Sub-Antarctic fur seas; a Great Barrier Reef display; and the Sydney Harbour Oceanarium, with underwater tunnels featuring fish, eels, turtles, and sharks.

Vancouver Aquarium
Stanley Park
Vancouver, British Columbia, **Canada**
+(604) 268-9900
This aquarium is home to more than 8,000 marine creatures.

Fenit Sea World
County Kerry, **Ireland**
Tel.: +353-66-36544
Highlights species that inhabit the underwater world where Tralee Bay meets the Atlantic, including tiny delicate prawns, ferocious conger eels, cod, and sharks.

Port of Nagoya Public Aquarium
Nagoya, **Japan**
One of the largest aquariums in Japan. Features approximately 36,000 living organisms representing 540 species, which live in the five oceans from Japan to the Antarctic.

Deep-Sea World
Fife, **Scotland**
Tel.: 44 1383 411411
Fax. 44 1383 410514
National Aquarium of Scotland features British Coastal Life as well as other Tropical and Sub-Tropical exhibits. Has an Underwater Safari with a 1-million-gallon (4-million-liter) exhibit including the world's longest underwater tunnel, and displays of more than 3,000 fish and crustaceans, including large Sand Tiger Sharks.

## What are the **National Marine Sanctuaries**?

The National Marine Sanctuaries (NMS) were designated by the U.S. Congress as protected marine areas because of their unique ecological, research, educational, and conservation qualities. Since the program was established in 1972, 12 areas have been designated National Marine Sanctuaries. Although they have been set aside for conservation, these areas can be visited by recreational divers.

The 12 currently designated National Marine Sanctuaries are the Channel Islands, Cordell Bank, Gulf of the Farallones, and Monterey Bay (all in California); the Florida Keys; Gray's Reef (Georgia); Hawaiian Islands Humpback Whale (around the islands of Lanai, Maui, and Molokai); Flower Garden (in the Gulf of Mexico, off Louisiana); Stellwagen (Massachusetts); Great Lakes (Michigan); Monitor (off Cape Hatteras, North Carolina); Olympic Coast (Washington); and Fagatele Bay (American Samoa).

In the summer of 1999, it looked like a 13th National Marine Sanctuary was about to be designated: the Great Lakes NMS near Thunder Bay, Michigan. This proposed sanctuary encompasses 808 square miles (2,092 square kilometers) of Thunder Bay and surrounding waters in Lake Huron, which contain nationally significant underwater cultural resources—including more than 160 shipwrecks. Congressional approval of the newest NMS was expected in late 1999 or early 2000.

## What is the **Channel Islands National Marine Sanctuary**?

The Channel Islands NMS is a 1,252-square-nautical-mile portion of the Santa Barbara Channel located off the coast of southern California. It encompasses the Pacific waters that surround Anacapa, San Miguel, Santa Cruz, Santa Rosa, and Santa Barbara islands, extending from mean high tide to 6 nautical miles offshore. The primary purpose of this sanctuary, designated in 1980, is to protect the natural beauty and resources found in the area.

## What is the **Cordell Bank National Marine Sanctuary**?

The Cordell Bank NMS is located in the Pacific Ocean approximately 60 miles (97 kilometers) northwest of San Francisco, California; the boundaries of this sanctuary were established in 1989 and encompass 526 square miles (1,362 square kilometers). Cordell Bank is an offshore seamount that rises out of the ocean depths to within 115 feet (35 meters) of the surface. The southeastward-flowing California Current causes upwellings of nutrient-rich, deep-ocean waters in this area, leading to the growth of all types of marine organisms. Many marine mammals and seabirds use this area as a feeding ground.

## What is the **Gulf of the Farallones National Marine Sanctuary**?

The Gulf of the Farallones NMS was designated a sanctuary in 1981, and is located just north of San Francisco, California. It consists of 1,235 square miles (3,200 square kilometers) of near-shore and offshore waters, and has diverse habitats including wetlands, intertidal, and deep-sea. The Gulf of the Farallones NMS is used for surfing, whale-watching, sailing, and fishing. One of the busiest shipping lanes in the world also passes through it, leading toward San Francisco Bay.

## What is the **Monterey Bay National Marine Sanctuary**?

The Monterey Bay NMS is the largest of the National Marine Sanctuaries, covering an area of 5,328 square miles (13,800 square kilometers) along the central California coast; it was designated a sanctuary in 1992. The habitats included in the Monterey Bay NMS are numerous and diverse, ranging from rocky shores and sandy beaches, to kelp forests and one of the deepest submarine canyons on the West Coast. There is also a great diversity of life here, including tiny plants, giant kelp, sea otters, and whales. Because of its large size and diversity, this sanctuary is used for marine education and research.

## What is the **Florida Keys National Marine Sanctuary**?

The Florida Keys NMS encompasses the entire length of the Florida Keys, a chain of islands extending from the southern tip of Florida along an arc toward the southwest. This entire area is home to a number of diverse marine plants and animals living in habitats such as coral reefs, seagrass meadows, and mangroves.

## What is the **Gray's Reef National Marine Sanctuary**?

Gray's Reef National Marine Sanctuary is one of the largest near-shore live-bottom reefs in the southeastern United States. It is located 17.5 nautical miles (32 kilometers) off Sapelo Island, Georgia, and has an area of 17 square nautical miles (58 square kilometers). This reef consists of a series of discontinuous rock ledges ranging from 6 to 10 feet (2

**What is the Flower Garden National Marine Sanctuary?**

The Flower Garden NMS can be likened to an oasis in the desert—the pair of underwater "gardens" that comprise this sanctuary are actually banks rising up from the ocean bottom—the result of salt domes beneath the seafloor. Located approximately 100 miles (161 kilometers) south of the Texas-Louisiana border in the Gulf of Mexico, this 56-square-mile (145-square-kilometer) area was designated a sanctuary in 1993, and is considered a premier diving destination. Flower Garden NMS contains the northernmost coral reefs in the United States, and is home to numerous shallow-water Caribbean reef fish and invertebrates. Also found in this area are manta rays, loggerhead turtles, and hammerhead sharks.

to 3 meters) in height, with flat-bottomed, sandy troughs in between. The resulting topography contains overhangs, caves, and burrows—a haven for all sorts of temperate and tropical marine flora (vegetation) and fauna (animal life).

The rock surfaces at the sanctuary support algae and invertebrates such as sea fans, sea stars, hard coral, sponges, barnacles, crabs, snails, shrimp, and lobsters; the reef is also a magnet for fish such as mackerel, grouper, and snapper. Gray's Reef provides year-round food for loggerhead sea turtles, and is part of the only known winter calving ground for the endangered northern right whale.

## What is the **Hawaiian Islands Humpback Whale National Marine Sanctuary**?

The Hawaiian Islands Humpback Whale NMS was established on November 4, 1992, and extends from the high-water mark to the 600-foot (183-meter) isobath (line of equal depth) around the islands of

Lanai, Maui, and Molokai; the Pailolo Channel, Penguin Banks, and a small area off Kauai's Kilauea Point are also included. The warm, shallow waters of the sanctuary are one of the world's most important habitats for the endangered humpback whale. It has been estimated that approximately two-thirds of the humpback whales in the North Pacific—about 2,000 to 3,000 whales—travel to the waters off the Hawaiian Islands to breed, or give birth and nurse.

The Hawaiian Islands Humpback Whale NMS seeks to protect the humpback whales and their habitat, while educating the public about these marine mammals and their unique relationship to the islands' waters. Human use of these waters is tightly controlled, and nearby areas are being researched for eventual inclusion in the sanctuary.

## What is the **Stellwagen Bank National Marine Sanctuary**?

The Stellwagen Bank NMS is located off the coast of Massachusetts, and covers approximately 842 square miles (2,180 square kilometers) of open waters and the seafloor below. This sanctuary encompasses all of Stellwagen Bank and Tillies Bank, and southern portions of Jeffreys Ledge.

The topography of this area leads to the upwelling of nutrient-rich bottom water; plankton flourish here, attracting a tremendous variety of animals. Stellwagen Bank has a large and diverse fish population, which attracts seabirds; more than 30 species of seabirds can be seen here, ranging from herring gulls to the endangered Roseate tern. This area is also one of the most important in the North Atlantic for whales, serving as a feeding and nursery ground for minkes, northern rights, pilots, orcas, fins, and humpbacks. Many dolphins, including the bottlenose, striped, common, white-sided, and white-beaked, are seasonal visitors. Infrequent visitors include leatherback turtles, gray seals, and harbor seals.

## What is the **Monitor National Marine Sanctuary**?

The Monitor NMS is doubly unique. This was the first National Marine Sanctuary, established in 1975, *and* is the location of a famous shipwreck. The purpose of this sanctuary is to protect the wreck of the Civil War ironclad U.S.S. *Monitor,* known for its battle with the Confederate ironclad C.S.S. *Virginia* (formerly the U.S.S. *Merrimack*) in Hampton

Roads, Virginia, on March 9, 1862. On December 31, 1862, the *Monitor* sank in a storm off Cape Hatteras, North Carolina. Its wreckage is the subject of intense underwater archaeological investigation—with the sanctuary seeking to preserve this site for future generations.



### What is the **Olympic Coast National Marine Sanctuary**?

The Olympic Coast NMS covers 3,310 square miles (8,570 square kilometers) of water off the coast of Washington's Olympic Peninsula, encompassing a good part of the continental shelf. This area reaches from the mouth of the Copalis River to Cape Flattery along the coast, and extends approximately 38 miles (61 kilometers) out to sea. The sanctuary was designated in 1994, and boasts kelp forests, islands, seastacks, rocky and sandy shores, as well as the open ocean. The animals living here include tufted puffins, northern sea otters, dolphins, and gray whales. The Olympic Coast NMS protects both a critical area for birds in the Pacific flyway, and a habitat that is home to a diversity of marine mammals.

### What is the **Fagatele Bay National Marine Sanctuary**?

Fagatele Bay NMS is located on the island of Tutuila, American Samoa, and is the most remote and the smallest of all the sanctuaries. This area, designated in 1986, encompasses a fringing coral reef ecosystem located within an eroded volcanic crater, and is home to numerous native species of flora (vegetation) and fauna (animal life). Among the animals living here are sharks, whales, giant clams, and turtles.

## OCEANOGRAPHIC INSTITUTIONS AND ORGANIZATIONS

### What is the ***Aquarius***?

The *Aquarius* is the world's most advanced underwater laboratory; it is located at Conch Reef in the Florida Keys' National Marine Sanctuary. The

*Aquarius*, situated on the ocean floor off the Florida Keys (in the National Marine Sanctuary located there), is the world's only underwater habitat. This August 1998 photo shows Dr. Sylvia Earle, a National Geographic Explorer-in-Residence, peaking through one of the habitat's windows. *AP Photo/Victor R. Caivano*

lab is owned by the National Oceanic and Atmospheric Administration (NOAA) and is operated through the National Undersea Research Center of the University of North Carolina at Wilmington. The *Aquarius* is the world's only underwater laboratory from which diving scientists can live and work beneath the sea. Research missions last up to 10 days at a time.

The *Aquarius* operates about 60 feet (18 meters) below the surface, at the base of a coral reef wall off Key Largo, Florida. The underwater laboratory weighs 81 tons and measures about 43 by 20 by 16 feet (13 by 6 by 5 meters)—which is divided between space where scientists can conduct experiments and living quarters that provide many of the comforts of home.

The deep-sea laboratory was created so that underwater researchers could conduct saturation diving: This special kind of diving allows scientists to adapt to the lab's underwater environment, permitting them to work on the reef for up to nine hours a day without fear of getting the bends—decompression sickness caused by nitrogen bubbles that form after emerging from compression too quickly. If researchers dove from the ocean surface to the depth of the underwater laboratory, they could only work for about an hour because of the bends.

## What does the ***Aquarius*** do?

The environmental research program at the *Aquarius* aims to better understand and preserve the health of coral reefs and other near-shore ecosystems. Studies of the surrounding coral reef ecosystem are conducted at the laboratory: For example, researchers are examining how corals cleanse themselves of sediment and they are observing the effects of global change and pollution on the reef. Armed with this knowledge, scientists can develop recommendations for ecologically sound coastline development—that protect our important near-shore ecosystems. *Aquarius*'s recent findings and current projects include:

Effects of ultraviolet radiation: Ultraviolet radiation, on the increase worldwide because of decreases in the ozone layer, damages the coral reef environment. Most corals produce chemicals that act as a sunblock—but if they do not produce enough, and if the water is clear and calm, ultraviolet radiation is able to penetrate the water and damage the corals.

Effects of pollution: *Aquarius* scientists are measuring the natural cycles of nutrient chemistry. Using monitoring wells, they are now trying to determine the variations in this cycle—especially how sewage-contaminated groundwater that seeps into the reef affects the organisms that build the reef and those that live around it. Scientists have also documented water quality variations when Gulf Stream water makes its way into the shallow reefs.

Pharmacological benefit: Using reef organisms such as sponges and corals, scientists on the *Aquarius* are trying to determine if the chemicals produced by these organisms have any pharmacological potential—that is, medicinal benefits to humans. Most of these chemicals are released to protect the organisms from predators—and scientists believe they may be of help to humans one day.

Study of fossils: The key to a reef's past is often its fossils—and reefs are unusually well-preserved. Scientists are examining fossil reefs and comparing them with today's reefs, trying to determine whether changes in the underwater ecosystem are cause for concern or simply part of the ocean's natural cycle.

## Which **U.S. institutions, laboratories, universities, and government agencies** are involved in **oceanography**?

There are many American institutions, laboratories, and universities that are involved in the field of oceanography. Those on the forefront include the following (listed in alphabetical order):

Atlantic Oceanographic and Meteorological Laboratory

Coastal and Hydraulics Laboratory

Florida Institute of Technology

Florida State University

Humboldt State University

Johns Hopkins University: Ocean Remote Sensing Group

Joint Oceanographic Institutions

Lamont-Doherty Geological Observatory

Massachusetts Institute of Technology

NASA Goddard Space Flight Center

NASA/JPL Air-Sea Interaction and Climate Team

National Center for Atmospheric Research: Oceanography Section

National Center for Supercomputing Applications

National Data Buoy Center

National Ice Center

National Oceanic and Atmospheric Administration (NOAA)

National Oceanic and Atmospheric Administration (NOAA) Geophysical Fluid Dynamics Laboratory

National Science Foundation, Ocean Sciences Division

Naval Meteorology and Oceanography Command

Naval Postgraduate School

Naval Research Laboratory

Nova Southeastern University

Office of Naval Research

Old Dominion University

Oregon State University

Pacific Marine Environmental Laboratory

Princeton University

San Diego State University

Scripps Institution of Oceanography

Skidaway Institute of Oceanography

State University of New York at Stony Brook

Texas A&M University

U.S. Geological Survey: USGS Marine Sciences

U.S. Naval Academy

University of California, Santa Cruz: Ocean Sciences Department

University of California, Santa Barbara

University of Chicago

University of Colorado

University of Delaware

University of Hawaii

University of Maine

University of Miami

University of North Carolina, Wilmington (the National Undersea Research Center)

University of Rhode Island

University of South Carolina

University of South Florida

University of Southern California

University of Southern Mississippi

University of Washington

Virginia Institute of Marine Science

Woods Hole Oceanographic Institution

## Which **Canadian institutions, laboratories, and universities** are involved in **oceanography**?

A selection of the Canadian institutions, laboratories, and universities engaged in the field of oceanography are listed below (in alphabetical order).

Canadian Hydraulics Centre

Dalhousie University

Institute of Ocean Sciences

Maurice Lamontagne Institute

McGill University

National Institute of Scientific Research

University of British Columbia

University of New Brunswich, Ocean Mapping Group

University of Quebec at Rimouski

University of Victoria

## What **other countries** have **institutions, laboratories, and universities** involved in **oceanography**?

The following lists, in alphabetical order by country, some of the oceanographic institutions, laboratories, and universities around the world.

Argentina

University of Buenos Aires

Argentina Institute of Oceanography (IADO), Bahia Blanca

Australia

University of Adelaide

CSIRO Marine Laboratories

Curtin University

Flinders University

James Cook University

Monash University

University of New South Wales

University of Sydney

University of Tasmania, Institute of Antarctic & South Ocean

University of Western Australia

Belgium

University of Leige: Mediterranean Oceanic Data Base

Brazil

University of Sao Paulo

China

Chinese Academy of Sciences

University of Hong Kong

Croatia

Institute of Oceanography and Fisheries

Denmark

Niels Bohr Institute

Ecuador

Escuela Superior Politecnica del Litoral

Estonia

Estonian Marine Institute

Finland

University of Helsinki

Finnish Institute of Marine Research

France

Institute Pierre Simon Laplace: Campus Jussieu: Paris

Centre d'Oceanologie de Marseille

The French Institute of Research and Exploitation of the Sea

National Center for Scientific Research (CNRS)

Oceanologic Observatory of Villefranche-sur-Mer

Germany

Institute fur Meereskunde Kiel

The Alfred Wegener Institute

University of Hamburg

Baltic Sea Research Institute

Federal Maritime and Hydrographic Agency

Iceland

Marine Research Institute

India

National Institute of Oceanography

Ireland

National University of Ireland, Galway

University College, Cork

Italy

University of Trieste

Institute of Marine Geology

Joint Research Center, Ispra

Japan

Hokkaido University

Japan Marine Science and Technology Center (JAMSTEC)

Kyushu University

National Institute of Polar Research

Shizuoka University

University of Tokyo

Mexico

Center for Scientific Research and Higher Education, Ensenada

The Netherlands

Institute for Marine and Atmospheric Research, Utrecht

Netherlands Institute for Sea Research (NIOZ)

New Zealand

University of Otago

University of Waikato

National Institute of Water and Atmospheric Research

Norway

University of Bergen

University of Oslo



Norwegian University of Science and Technology

Poland

Institute of Oceanology: Sopot

Portugal

Technical University of Lisbon

Russia

Russian Academy of Sciences

Shirshov Institute of Oceanology

Slovenia

University of Ljubljana

South Africa

University of Cape Town

University of Natal

South Korea

Seoul National University

Spain

University of the Balearic Islands

University of Cadiz

Universidad de las Palmas de Gran Canaria

Spanish Oceanography Institute

Sweden

Goteborg University

Swedish Meteorological and Hydrological Institute

Stockholm Marine Research Center

Umea Marine Sciences Center

Taiwan

National Taiwan University

Turkey

Middle East Technical University

United Kingdom

Oxford University

British Antarctic Survey

Proudman Oceanographic Laboratory

Reading University

Southampton Oceanography Centre

University of East Anglia

University of Edinburgh

University of Liverpool

University of Plymouth

University of Southampton

University of Wales, Bangor

University of Birmingham

## What is the **world's largest collection** of **public oceanographic data**?

The United States National Oceanographic Data Center (NODC) has the world's largest collection of publicly available oceanographic data. The NODC facility is located at the National Oceanic and Atmospheric Association's (NOAA) building complex in Silver Spring, Maryland. NODC also operates the World Data Center A, Oceanography.

## Which **scientific organizations** are involved in **oceanography**?

There are many scientific organizations involved in oceanography. The following lists only some—and although many of these organizations are not 100-percent oceanographic, many of them have special sections in ocean science. For more information on each organization, check out its website.

American Geophysical Union
http://earth.agu.org/

American Meteorological Society
http://www.ametsoc.org/AMS/

American Society of Limnology and Oceanography
http://www.aslo.org/

Australian Meteorological and Oceanographic Society
http://www.amos.org.au/

Bureau Gravimetrique International
http://bgi.cnes.fr:8110/

Canadian Meteorological and Oceanographic Society
http://www.meds-sdmm.dfo-mpo.gc.ca/cmos/default.html

European Geophysical Society
http://www.copernicus.org/EGS/EGS.html

European Science Foundation
http://www.esf.org/

Geological Society of America
http://www.geosociety.org/index.htm

ICSU Scientific Committee on Antarctic Research (SCAR)
http://www.scar.org/

ICSU Scientific Committee on Oceanic Research (SCOR)
http://www.jhu.edu/~scor/

Intergovernmental Oceanographic Commission
http://ioc.unesco.org/iocweb/

International Association of Geodesy
http://www.gfy.ku.dk/~iag/

International Association of Geomagnetism and Aeronomy
http://www.ngdc.noaa.gov/IAGA/iagahome.html

International Association of Hydrological Sciences
http://www.wlu.ca/~wwwiahs/

International Association of Meteorology and Atmospheric Sciences
http://iamas.org/

International Association of Seismology and Physics of the Earth's Interior
http://www.seismo.com/iaspei/

International Association of Volcanology and Chemistry of the Earth's Interior
http://geont1.lanl.gov/HEIKEN/one/iavcei_home_page.htm

International Council for the Exploration of the Sea
http://www.ices.dk/

International Ocean Institute
http://is.dal.ca/~ioihfx/

International Tsunami Information Center
http://www.shoa.cl/oceano/itic/frontpage.html

National Academy of Sciences, Earth Sciences (USA)
http://www.nationalacademies.org/subjectindex/ear.html

National Academy of Sciences, Environment (USA)
http://www.nationalacademies.org/subjectindex/env.html

North Pacific Marine Science Organization (PICES)
http://pices.ios.bc.ca/

The Oceanography Society
http://www.tos.org/

Royal Meteorological Society, United Kingdom
http://itu.rdg.ac.uk/rms/rms.html

The Royal Society, United Kingdom
http://www.royalsoc.ac.uk/

SCAR Task Group on Ice Sheet Mass Balance and Sea Level
http://www.antcrc.utas.edu.au/scar/ismass.html

Tsunami Society
http://www.ccalmr.ogi.edu/STH/society.html

World Meteorological Organization (WMO)
http://www.wmo.ch/

# INTERNET RESOURCES

*Note: Internet site addresses (URLs) change often. The information listed here was checked prior to publication. We regret any inconvenience for changes since that time.*

## Where can **international oceanographic data** be found?



There are numerous sources for international oceanographic data—too many to mention in these pages. But the following is a good representative sampling.

British Oceanographic Data Centre (BODC)
http://www.nbi.ac.uk/bodc/bodcmain.html

Global Sea-Level Observing System (GLOSS)
http://www.nbi.ac.uk/psmsl/gloss.info.html

IAPSO Standard Seawater Service
http://www.oceanscientific.com/

Interactive Marine Observations (USA)
http://www.nws.fsu.edu/buoy/

Japan Oceanographic Data Center
http://www.jodc.jhd.go.jp/

Marine Information Service (MARIS)--Netherlands
http://www.maris.nl/start.htm

NASA/JPL Physical Oceanography Distributed Active Archive Center
http://podaac-www.jpl.nasa.gov/

NOAA Global Wind and Wave Model Output
http://polar.wwb.noaa.gov/waves/waves.html

NOAA National Oceanographic Data Center (NODC)
http://www.nodc.noaa.gov/

NOAA/ NGDC Marine Geophysics On-Line System (GEODAS)
http://www.ngdc.noaa.gov/mgg/gdas/go_sys.Html

## Which Internet sites offer information on **careers in oceanography**?

There are lots of places to find information on careers on the Internet. The following lists a few of the more relevant sites. (For even more websites, key the words "career and oceanography" into a search engine.)

Careers in Marine Science: Advice to Students and Parents From the Scientists of Mote Marine Laboratory
http://www.marinelab.sarasota.fl.us/careers.phtml

Careers in Oceanography: The U.S. Office of Naval Research
http://www.onr.navy.mil/onr/careers/

Marine Science Careers: A Sea Grant Guide to Ocean Opportunities
http://www.marine.stanford.edu/HMSweb/Career_booklet.html

So You Want to Become a Marine Biologist: Scripps Institution of Oceanography
http://www.siograddept.ucsd.edu/Web/To_Be_A_Marine_Biologist.html

## Is there an **online directory** of **aquariums** around the **world**?

Yes, a website that contains a directory of the world's aquariums can be found at http://infofarm.cc.affrc.go.jp/~mtoyokaw/aquarium/world.html

## Which **websites** have information on **endangered** or **threatened marine species**?

Among the many websites offering information on endangered or threatened marine species, are:

The Congressional Research Service
http://www.cnie.org/nle/biodv-18.html

NOAA's National Marine Fisheries Service
http://www.nmfs.gov/

U.S. Fish and Wildlife Endangered Species Home Page
http://www.fws.gov/r9endspp/endspp.html

## Which **Internet sites** have information on **coral reefs**?

The following websites are devoted to coral reefs:

Action Atlas: Coral Reefs
http://www.motherjones.com/coral_reef/

## Where can information on marine reptiles be found on the Internet?

There are a few sites you can visit, but one is particularly interesting: It tracks the movements of a loggerhead sea turtle. When this sea turtle recently showed up on Roosevelt Beach, Delaware, it was cold, stunned, and hypothermic. The animal, a juvenile turtle named Perdida (which is Spanish, for "lost"), was nursed back to health at the National Aquarium in Baltimore, Maryland. After rehabilitation, researchers returned her to the wild off Assateague Island, just off the Maryland coast. But she left with a reminder of her visit: Perdida was fitted with a satellite transmitter that will allow the public and researchers to monitor her migration, which they do over the Internet. By studying the movements of this particular turtle, researchers with the Sea Turtle Survival League and the U.S. Army Corps of Engineers will learn more about how sea turtles feed, forage, and behave along the East Coast.

Within the Sea Turtle Survival League's Sea Turtle Migration-Tracking Education Program, Perdida is one of more than 20 sea turtles being actively tracked, helping people around the world, especially children, learn more about sea turtle populations, their habits, migration—and the importance of protecting our coastal waters. You can link to the Rehabilitated Loggerhead Satellite Tracking Project through the Internet address: http://cccturtle.org/satperd.htm

Coral Health and Monitoring Program
http://coral.aoml.noaa.gov/

Hawaii Coral Reef Network
http://www.coralreefs.hawaii.edu/ReefNetwork/default.htm

Reef Relief
http://www.blacktop.com/coralforest/

## What **online sources** are there for learning more about **marine mammals**?

There are many websites that provide information on marine mammals. The following are good places to begin your online reading and research.

Beluga Whales
http://oceanlink.island.net/Beluga2.html

British Columbia's Killer Whales
http://oceanlink.island.net/kwhale.html

Dolphins and Porpoises
http://oceanlink.island.net/dolphin.html

Gray Whales
http://oceanlink.island.net/graywhale.html

Hawaiian Islands Humpback
http://www.nos.noaa.gov/nmsp/hinms/

Hawaiian Monk Seals
http://leahi.kcc.hawaii.edu/~et/wlcurric/seals.html

Hawaii Whale Research Foundation
http://www.hwrf.org/hwrf/index.html

An Introduction to Whales
http://www.edu-source.com/marine/whales.html

Sea Otters
http://oceanlink.island.net/seaotter.html

WhaleNet
http://whale.wheelock.edu/

Whale-Watching-Web
http://www.physics.helsinki.fi/whale/

Other sites about marine reptiles include Oceanlink's (at http://ocean link.island.net/seaturtle.html) and Sea World's site (http://www.seaworld.org/Sea_Turtle/stclass.html).

## Are there any **Internet sites** about **seahorses**?

Yes, two of the better ones are Project Seahorse at http://www.seahorse.mcgill.ca/intro.html and *Nova*'s site at http://www.pbs.org/wgbh/nova/seahorse/

## Which websites offer information on **oceans and climate**?

There are many ocean and climate sites on the Internet. The following lists some of the more popular ones.

Climate (Physical Processes)
http://www.aoml.noaa.gov/general/nclim.html

International Year of the Ocean
http://www.un.org/Depts/los/IYO/Climate_and_Oceans.htm

National Ocean Service
http://www.nos.noaa.gov/

National Oceanographic Library, UK
http://www.soc.soton.ac.uk/LIB/java.html

Ocean Observing System for Climate
http://www.ocean.tamu.edu/OOSDP/FinalRept/t_of_c.html

U.S. Joint Global Ocean Flux Study
http://www1.whoi.edu/jgofs.html

U.S. Naval Pacific Meteorology and Oceanography Center (FL)
http://www.nlmof.navy.mil/

U.S. Navy's Fleet Numerical Meteorology and Oceanography Center
http://152.80.56.202/index.html

World Meteorological Organization
http://www.wmo.ch/

## Which sites offer information on **El Niño**?

The following websites publish information on El Niño, the periodic ocean-weather event:

El Niño Resources
http://www.coaps.fsu.edu/lib/elninolinks/

El Niño Scenario
http://www.crseo.ucsb.edu/geos/el_nino.html

NASA Facts: El Niño
http://eospso.gsfc.nasa.gov/NASA_FACTS/el_nino/Elnino.html

Observations of the 1997 El Niño/Southern Oscillation
http://darwin.bio.uci.edu/~sustain/ENSO.html

What is an El Niño?
http://www.pmel.noaa.gov/toga-tao/el-nino-story.html

As the phenomenon of El Niño continues to garner worldwide interest, it's worth checking for new sites by keying the term (El Niño or El Nino) into the text box of one of your favorite search engines.

## Are there any **Internet sites** that deal with **coastal zones**?

Yes, there are many Internet sites devoted to publishing information on coastal zones. They include the following:

Atlantic Coastal Zone Database Directory
http://is.dal.ca/~mbutler/aczisc.htm

Dorset CoastLink
http://csweb.bournemouth.ac.uk/consci/coastlink/

Estuaries: Gateway to the Sea
http://www.enn.com/features/1998/10/100298/estuaries.asp

Florida Bay and Adjacent Marine Systems
http://www.aoml.noaa.gov/flbay/

HazNet
http://www.haznet.org/

Massachusetts Coastal Zone Management
http://www.magnet.state.ma.us/czm/

Oregon Coastal Management Program
http://www.lcd.state.or.us/coast/index.htm

Tillamook Bay National Estuary Project
http://www.orst.edu/dept/tbaynep/nephome.html

U.S. Environmental Protection Agency National Estuary Program
http://www.epa.gov/nep/nep.html

Wetlands International
http://www.wetlands.agro.nl/



## How can I find out about **beach pollution in the United States** on the **Internet**?

For the most recent information, go to http://www.epa.gov/ost/beaches/beachwatch%20EPA

## Are there any **Internet resources** about **plankton**?

The following websites publish information on plankton:

Emiliania Huxley Home Page
http://www.soc.soton.ac.uk/SUDO/tt/eh/

Institute of Ocean Sciences, Plankton Productivity
http://www.ios.bc.ca/ios/plankton/default.htm

Online Plankton Key
http://biology.rwc.uc.edu/HomePage/BWS/planktonkey/phytozoo.html

Plankton Net
http://www.uoguelph.ca/zoology/ocean/

Zooplankton Production Laboratory
http://www.ios.bc.ca/ios/plankton/ios_tour/zoop_lab/zoopllab.htm

## Which websites offer information on **octopuses and their relatives**?

Visit the following websites to learn more on this subject:

About Octopi...
http://www.marinelab.sarasota.fl.us/OCTOPI.HTM

The Cephalopod Page: Octopi, Squids, Cuttlefish
http://is.dal.ca/~ceph/TCP/index.html

Squid
http://seawifs.gsfc.nasa.gov/squid.html

## Which websites offer information on **sharks**?

The following are a few of the many Internet sites about sharks:

Shark Realities
http://www.discovery.com/area/nature/sharks/sharks1.html

Sharks: Information and Conservation
http://www.brunel.ac.uk/admin/alumni/sharks/home.html

## Is there an **Internet site** for the **National Marine Sanctuaries program**?

Yes, there is a website maintained by the National Oceanic and Atmospheric Administration (NOAA) that describes the 12 National Marine Sanctuaries and the proposed 13th sanctuary (Great Lakes NMS), and contains up-to-date news about them. The address is: http://www.sanctuaries.nos.noaa.gov/

Additionally, there are websites for each National Marine Sanctuary; their content varies, but all are informative and worth a look.

Channel Islands National Marine Sanctuary
http://www.cinms.nos.noaa.gov/home.htm

Cordell Bank National Marine Sanctuary
http://www.nos.noaa.gov/ocrm/nmsp/nmscordellbank.html

Fagatele Bay National Marine Sanctuary
http://www.nos.noaa.gov/nmsp/FBNMS/

Florida Keys National Marine Sanctuary
http://www.nos.noaa.gov/nmsp/fknms/

Flower Garden National Marine Sanctuary
http://www.nos.noaa.gov/ocrm/nmsp/nmsflowergardenbanks.html

Gray's Reef National Marine Sanctuary
http://www.graysreef.nos.noaa.gov/

Great Lakes National Marine Sanctuary (proposed)
http://www.glerl.noaa.gov/glsr/thunderbay/

Gulf of the Farallones National Marine Sanctuary
http://www.nos.noaa.gov/nmsp/gfnms/welcome.html

Hawaiian Islands Humpback Whale National Marine Sanctuary
http://www.t-link.net/~whale/

Monitor National Marine Sanctuary
http://www.nos.noaa.gov/nmsp/monitor/

Monterey Bay National Marine National Marine
http://bonita.mbnms.nos.noaa.gov/

Olympic Coast National Marine Sanctuary
http://www.nos.noaa.gov/ocrm/nmsp/nmsolympiccoast.html

Stellwagen Bank National Marine Sanctuary
http://vineyard.er.usgs.gov/

### Is there an **Internet site** about **invasive species in the United States**?

Yes, to find out more about invasive species in the United States, link to the website for the Nonindigenous Aquatic Species (NAS) information resource of the United States Geological Survey, located at the Florida Caribbean Science Center (http://nas.nfrcg.gov/).

## CAREERS IN OCEANOGRAPHY

### What is **oceanography**?

Oceanography is a multi-disciplinary science aimed at improving our understanding of the marine environment—including its waters, life,

**Oceanographic work can be varied: This plane table crew, at work on a 1928 survey on Alaska's Kenai Peninsula, was headed by NOAA Rear Admiral Paul Smith of the Coast & Geodesic Survey.** ***NOAA/Family of Rear Admiral Paul A. Smith, C&GS***

and sediments. Many sciences can be applied to oceanography, including physics, geology, chemistry, biology, meteorology, geography, and geodesy. The science is also involved in the development of related technologies—to improve data gathering and research.

## What is an **oceanographer**?

An oceanographer is a scientist who studies the chemical, physical, geologic, meteorological, or biological features of the ocean. Some scientists also blend several of these fields together in a multi-disciplinary approach.

## Where do **oceanographers work**?

Most oceanographers, marine scientists, engineers, and support personnel work in government agencies (where they conduct research), at colleges and universities (teaching and conducting research), and in private industry (as part of research and development initiatives). Some work as independent consultants and writers.

Many oceanographers work aboard ships, taking measurements and gathering samples at sea. Others gather data during short ocean voyages, and then spend much of their time in the laboratory, interpreting the data and conducting experiments. Still others spend most of their time in the laboratory, where they receive remote-sensing data from satellites; they analyze the data, using it to create computerized simulations (models) of ocean phenomena.

## What do **physical oceanographers** do?

Physical oceanographers study the ocean's physics—especially the interaction of the various forms of energy: light, heat, sound, and wind. These scientists are interested in the interaction between the ocean and atmosphere, and the relationship among the sea, weather, and climate.

Some physical oceanographers study and describe the causes behind the motions of the ocean waters, water masses, and currents, as well as the driving forces that produce such patterns in the sea. These studies include the currents, tides, winds, and certain periodic events, such as El Niño. Other physical oceanographers examine the physical characteristics of the ocean, including its temperature, salinity, and pressure. Still others are interested in the ocean's fluid dynamics (water movement) and how this relates to transport of sediment, changes in temperature, and the processes involved in coastal erosion.

## What do **marine physicists** do?

Marine physicists develop ways for researchers to interact with the ocean environment in order to learn more about it. They often design and build specialized technologies such as remotely operated vehicles

(ROVs), sophisticated instruments to examine the seafloor—and even remote-sensing instruments to explore the oceans. Some also work with marine engineers, developing equipment to control the erosion of beaches or to draw energy from ocean waves.

## What do **chemical oceanographers** do?

Chemical oceanographers study the chemistry of the oceans. Their work is multifaceted: It includes determining the chemistry of ocean water, developing theories on how chemical compounds affect the overall ocean, and studying how organic and inorganic compounds interact. Some chemical oceanographers study the interaction among the Sun's energy, atmospheric compounds, dissolved and suspended oceanic organic and inorganic material, sealife, and the seafloor. Others investigate the impact of natural substances (such as seafloor seeping petroleum) and humanmade substances (such as deposits of wastes or pollution) on the ocean's chemistry. Analytical, physical, bio-, nuclear, and industrial chemistry are applied in ocean-mapping studies and in determining the ocean's overall major cycles.

Chemical oceanography may be vital to the future of the Earth: If we can better understand the chemistry of the oceans, we may be able to identify new food sources, find new types of energy, better monitor our environment, improve health, and find substitutes for dwindling resources on Earth.

## What do **marine geologists and geophysicists** do?

Marine geologists study the topographic features (mountains, canyons, and valleys), rocks, sediments, and physical composition of the ocean floor. These scientists relate their observations to phenomena seen under the oceans. For example, they may examine the movement of suspended sediments by currents in the water (such as turbidity currents); determine how volcanoes and island arcs originated; examine and understand the creation of new ocean floor at mid-ocean spreading centers; or piece together the movement of the seafloor plates over the past millions of years.

### What do marine biologists do?

Marine biologists study the diverse forms of life in the sea. They are concerned with the cycling of nutrients in the marine food chain, animal behavior, how fish communicate, how energy is produced for the organisms in the seawater and on the seafloor, and how life can exist on the harsh ocean bottom. Marine biology is also concerned with human impact on the ocean environment.

Like marine geologists, marine geophysicists also deal with the ocean floor, but with more physics thrown in. These scientists search for answers to questions such as why the Earth's magnetic field reversed itself at least three times in the last million years. They also examine the ocean floors—its earthquakes and the spots where heat escapes—to determine what processes are taking place in the planet's interior.

## What does a **biological oceanographer** do?

There is a very fine line between a marine biologist and a biological oceanographer. The biological oceanographer is mainly interested in the complex interactions of groups of marine organisms—with their environment and one another. They also study such ideas as how warm and cold currents affect the availability of food for fish and other marine life.

## What do **atmospheric scientists** do?

Though this occupation may not sound like it is involved in oceanography, it is: The ocean impacts atmosphere and therefore influences world climate. Atmospheric scientists examine large-scale weather conditions and determine how the oceans are involved; they may also study sea surface temperatures to determine climate conditions; and they study the buildup of atmospheric pollutants and how they affect weather over the oceans—and sometimes the long-term global climate.

A technician communicates with the shore base about life support systems aboard the underwater research laboratory and habitat *Aquarius. NOAA/OAR National Undersea Research Program*

## What do **marine engineers** do?

Marine engineers are usually involved in designing a structure that must withstand conditions at sea or along the coasts. For example, towers used to drill oil offshore on the continental shelf must be designed to withstand ocean currents, saltwater corrosion, marine life interaction, and other factors. Marine engineers also design equipment to make oceanographic measurements—usually in cooperation with other marine scientists. For example, physical oceanographers have to find the best way to capture water samples from different parts of the ocean water column. The marine engineer works with the physical oceanographer to determine what is the best design to accomplish that task.

## What do **ocean technicians** do?

Ocean technicians are important to the study of the sea. They are often responsible for equipment calibration and preparation, taking measurements and samples at sea, repairing and maintaining equipment used for taking measurements and sampling, and processing the resulting data.

## What **courses** should a person interested in oceanography **study in high school**?

A high school student who is interested in pursuing oceanography should study mathematics and the sciences—including physics, chemistry, and biology.

## How does a person **pursue a career in oceanography**?

If a career in oceanography is your goal, you should follow one of three paths. The first option is to attend a four-year college or university that offers a program in oceanography or a related marine discipline, obtaining a bachelor's of science degree. The second is to obtain a bachelor's of science (B.S.) degree in one of the basic sciences such as geology, chemistry, biology, physics, or engineering. From there it is often possible to go directly to work using your basic scientific knowledge to solve ocean-oriented problems; or you can continue studying and obtain your master's or doctorate degree in oceanography (the time required to complete graduate programs varies by institution and by student). Finally, obtaining an associate of arts degree in technology or science from a junior college or technical school can also provide you with the background you need to enter the field.

## Which **colleges and universities** offer **degrees in oceanography**?

Many colleges and universities offer courses in oceanography and other marine sciences. A publication titled *Curricula in the Atmospheric, Oceanic and Related Sciences* is one of many that lists the schools and the courses they offer. It is available in libraries or may be purchased from the American Meteorological Society, 45 Beacon Street, Boston, Massachusetts, 02108.

## What **college courses** should I take to become an **oceanographer**?

Undergraduate students should take courses in one of the subfields of oceanography (physics, chemistry, biology, geology, geophysics) or in meteorology, mathematics, or engineering. These courses of study can be completed at any educational institution that is strong in the sciences, math, or engineering.

In the United States, there are more than 80 schools that offer undergraduate degrees in marine science. Many of these degree programs are restricted in scope, with the majority focused on estuarine and intertidal biology; others offer a broader approach to oceanography. And there are more than 50 institutions in the United States that offer doctoral degrees in various oceanographic disciplines. A book describing university curricula in oceanography and related fields can be obtained by writing The Marine Technology Society, 1828 L Street NW, Suite 906, Washington, DC 20036-5104. In addition, you can check out the latest edition of the annual *Peterson's 4-Year Colleges,* available at your library or local bookstore.

## Where are **oceanography courses** offered in the **United States**?

The following are some of the educational institutions that offer courses in the field of oceanography and related areas. They are arranged alphabetically by state.

Alabama

Dauphin Island Sea Lab
P.O. Box 369-370
Dauphin Island, Alabama 36528
Tel.: 205-861-2141

California

Scripps Institution of Oceanography
University of California, San Diego
La Jolla, California 92093-0210
Tel.: (619) 534-2830
Fax: (619) 534-5306

Delaware

College of Marine Studies
University of Delaware
Robinson Hall
Newark, Delaware 19716-3501
Tel.: 302-831-2841
Fax: 302-831-4389

Florida

University of Miami
Rosenstiel School of Marine and Atmospheric Science
4600 Rickenbacker Causeway
Miami, Florida 33149-1098
Tel.: 305-361-4000
Fax: 305-361-4711

Hawaii

School of Ocean and Earth Science and Technology
The Hawaii Undersea Research Laboratory
University of Hawaii
1000 Pope Road
Honolulu, Hawaii 96822
Tel.: 808-956-6036 (HURL's administrative specialist)
Fax: 808-956-9772

Massachusetts

Cooperative Marine Education and Research
Blaisdell House
University of Massachusetts
Amherst, Massachusetts 01003-0820
Tel.: 413-545-2842
Fax: 413-545-2304

Woods Hole Oceanographic Institution
Woods Hole, Massachusetts 02543
Tel.: 508-457-2000

Mississippi

Gulf Coast Research Lab
(administered by the University of Southern Mississippi)
P.O. Box 7000
Ocean Springs, Mississippi 39564-7000
Tel.: 601-872-4201

New Hampshire

Shoals Marine Lab
G-14Y Stimson Hall
Cornell University
Ithaca, New York 14853-7101
Tel.: 607-254-4636

North Carolina

Duke University Nicholas School of the Environment Marine Laboratory
135 Duke Marine Lab Road
Beaufort, North Carolina 28516-9721
Tel.: 919-504-7503
Fax: 919-504-7648

Oregon

Hatfield Marine Science Center
Oregon State University
2030 Marine Science Drive
Newport, Oregon 97365
Tel.: 503-867-0212

Oregon Institute of Marine Biology
University of Oregon
Eugene, Oregon 97403
Tel.: 503-888-2581

Rhode Island

Graduate School of Oceanography
University of Rhode Island
Narragansett, Rhode Island 02882
Tel.: 401-874-6222

Texas

Marine Science Institute
The University of Texas at Austin
750 Channelview Drive
Port Aransas, Texas 78373
Tel.: 361-749-6711
Fax: 361-749-6777

Virginia

Virginia Institute of Marine Science
School of Marine Science
College of William and Mary
Box 1346
Gloucester Point, Virginia 23062
Tel.: 804-684-7000

Washington

Friday Harbor Laboratories
University of Washington
620 University Road
Friday Harbor, Washington 98250
Tel.: 206-543-1484

## Where are **oceanography courses** offered in **Canada**?

The following are some of the Canadian educational institutions that offer courses in oceanography and related fields. The list is in alphabetical order by province.

British Columbia

Department of Earth and Ocean Sciences
The University of British Columbia
6339 Stores Road
Vancouver, British Columbia V6T 1Z4
Tel.: 604-822-2449
Fax: 604-822-6088

School of Earth and Ocean Sciences
University of Victoria
P.O. Box 3055
Victoria, BC, Canada V8W 3P6
Tel.: (604)721-6120
Fax: (604) 721-6100

New Brunswick

Ocean Mapping Group
Saint John Campus
University of New Brunswick
P.O. Box 5050
Saint John, NB, Canada E2L 4L5
Tel.: 506-453-3577
Fax: 506-453-4943

Newfoundland

Fisheries and Marine Institute of Memorial University of Newfoundland
P.O. Box 4920, St. John's, Newfoundland—A1C 5R3
Tel.: (709) 778-0200
Fax: (709) 778-0346

Nova Scotia

Department of Oceanography
Dalhousie University, Halifax, Nova Scotia B3H 4J1
Tel.: (902) 494-3557
Fax: (902) 494-3877

## Do any **colleges or universities** offer courses in **oceanography as it relates to astrobiology**?

The University of Washington will become the first institution to launch a doctoral program specifically designed to train scientists to search for life on other planetary bodies, such as Mars or Jupiter's icy moon, Europa. The curriculum for the highly interdisciplinary degree includes astronomy, genetics, microbiology, biochemistry, geological science—and even oceanography. The School of Oceanography at the university will provide the dedicated laboratory space for the study of organisms that live in extreme conditions. In fact, some of the professors in the department have already closely studied organisms living in high-temperature and high-pressure conditions in ocean environments. Another participant, the Pacific Northwest National Laboratory in Richland, Washington, will offer students the chance to study microbial life in subterranean basalt formations in the eastern part of the state. Zymo Genetics, Inc. of Seattle, is offering internships so students can study enzymes from unusual bacteria.

The key to discovering life in forbidding space environments is understanding how life exists on Earth under harsh conditions, such as undersea vents, pools of brine within polar sea ice, volcanic basalt formations, or hot springs similar to those found in Yellowstone National Park. Scientists believe that the strange and unusual creatures living in these extreme environments may have been precursors to advanced life on Earth. The presence of organisms on other planets could give clues to the evolution of life there.

## What is the **Joint Oceanographic Institutions**?

The Joint Oceanographic Institutions (JOI) is a consortium of the largest academic oceanographic institutions in the United States. It was established as a private, non-profit corporation in 1976. Currently, JOI manages the international Ocean Drilling Program (ODP); the U.S. Science Support Program (USSSP, associated with ODP); the SeaNet planning office; and the Secretariat for the Nansen Arctic Drilling Program. The address for the JOI is: Joint Oceanographic Institutions, 1755 Massachusetts Avenue, NW, Suite 800, Washington, DC 20036-2102; telephone (202) 232-3900; fax (202) 232-8203.

The JOI's member institutions are:

Scripps Institution of Oceanography of the University of California

Rosenstiel School of Marine and Atmospheric Science, University of Miami (Florida)

School of Ocean and Earth Science Technology of the University of Hawaii

Woods Hole Oceanographic Institution (Woods Hole, Massachusetts)

Rutgers, The State University of New Jersey

Lamont-Doherty Geological Observatory of Columbia University (New York, New York)

College of Oceanic and Atmospheric Sciences, Oregon State University

Graduate School of Oceanography, University of Rhode Island

College of Geosciences and Maritime Studies, Texas A&M University

Institute for Geophysics, The University of Texas at Austin

College of Ocean Fishery Sciences, University of Washington

# Index

A **bold** page number refers to a photo or other illustration.

## D



## E

## F



## G

## H



## I

## J



## K



## L



## M

## N

## P

## W



## Z

# HANDY-ology: The science of learning

## These four Handy Answer Books™ help you understand the secrets of the Earth, wind, science and more

### *The Handy Geography Book*

Take a tour of the world the whole family will enjoy. Discover the difference between England, Great Britain and the United Kingdom. Find out who owns the oceans. Name the seven seas. *Handy Geography* answers 1,000 common and not-so-common questions about the natural features of the world and the ever-changing mark humans make on our planet. The nontechnical explanations will appeal to adults and students alike. Entry topics cover a lot of ground — from trivia (such as "the highest" or "the deepest") to how terrain affects the location of countries and the people who inhabit them.

1998 • Matthew Todd Rosenberg • Paperback • 462 pp. with 16-page color insert • ISBN 1-57859-062-0 • $19.95

### *The Handy Science Answer Book*

Can a bird fly upside down? Is white gold really gold? Why do golf balls have dimples? *Handy Science* holds the answers to those questions and the nearly 1,400 others. The answers were compiled from the ready-reference files of the Science and Technology Department of The Carnegie Library of Pittsburgh. Everyone will enjoy hours of discovery on topics such as cars, outer space, the inner workings of the human body, computers and much more.

1996 • Paperback • 598 pp. • ISBN 0-7876-1013-5 • $16.95

### *The Handy Weather Answer Book*

Find out if mobile homes really attract tornadoes, the difference between sleet and freezing rain and answers to nearly 1,000 other questions. *Handy Weather* covers such confounding and pertinent topics as hurricanes, thunder and lightning, droughts and flash floods, earthquakes and volcanoes. There is also coverage on weather-related phenomena like El Niño, the greenhouse effect, aurora borealis and St. Elmo's fire.

1996 • Walter A. Lyons, Ph.D. • Paperback • 398 pp. • ISBN 0-7876-1034-8 • $16.95

### *The Handy Physics Answer Book*

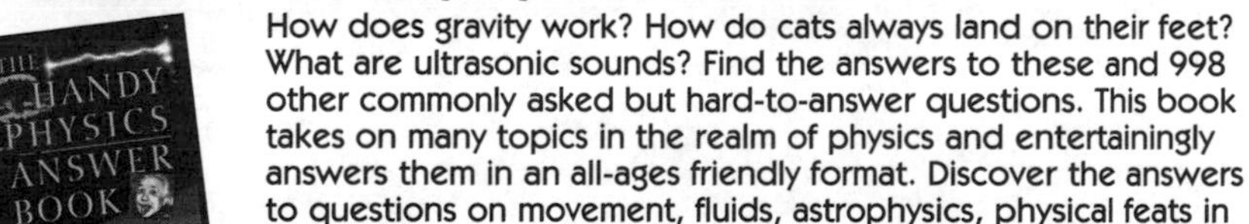

How does gravity work? How do cats always land on their feet? What are ultrasonic sounds? Find the answers to these and 998 other commonly asked but hard-to-answer questions. This book takes on many topics in the realm of physics and entertainingly answers them in an all-ages friendly format. Discover the answers to questions on movement, fluids, astrophysics, physical feats in engineering, famous physicists and many more.

1998 • Paperback • 400 pp. • ISBN 1-57859-058-2 • $19.95

Available at fine bookstores everywhere
In the U.S. call 1-800-776-6265